# THE CITY THAT ATE ITSELF

**Mining and Society Series**
Eric Nystrom, Arizona State University
Series Editor

"If it can't be grown, it must be mined," as the bumper sticker reminds us. Attempting to understand the material basis of our modern culture requires an understanding of those materials in their raw state and the human effort needed to wrest them from the earth and transform them into goods. Mining thus stands at the center of important historical and contemporary questions about labor, environment, race, culture, and technology, which makes it a fruitful perspective from which to pursue meaningful inquiry at scales from local to global.

Books published in the series examine the effects of mining on society in the broadest sense. The series covers all forms of mining in all places and times, building from existing press strengths in mining in the American West to encompass comparative, transnational, and international topics. By not limiting its geographic scope to a single region or product, the series helps scholars forge connections between mining practices and individual sites, moving toward broader analyses of the global mining industry in its full historical and global contexts.

*Seeing Underground:*
*Maps, Models, and Mining Engineering in America*
Eric C. Nystrom

*Historical Archaeology in the Cortez Mining District:*
*Under the Nevada Giant*
Erich Obermayr and Robert W. McQueen

*The City That Ate Itself:*
*Butte, Montana and Its Expanding Berkeley Pit*
Brian James Leech

# THE CITY THAT ATE ITSELF

## Butte, Montana and Its Expanding Berkeley Pit

Brian James Leech

UNIVERSITY OF NEVADA PRESS *Reno & Las Vegas*

University of Nevada Press | Reno, Nevada 89557 USA
www.unpress.nevada.edu

Jacket photograph courtesy of the World Museum of Mining.
© World Museum of Mining.
Cover design by Louise OFarrell

Library of Congress Cataloging-in-Publication Data
Names: Leech, Brian James, author.
Title: The city that ate itself : Butte, Montana and its expanding Berkeley Pit / by Brian James Leech.
Description: Reno : University of Nevada Press, [2017] | Series: Mining and society | Includes bibliographical references and index. |
Identifiers: LCCN 2017036829 (print) | LCCN 2018013554 (ebook) | ISBN 9781948908290 (pbk. : alk. paper) | ISBN 9780874175981 (ebook) | ISBN 9781943859429 (cloth : alk. paper)
Subjects: LCSH: Anaconda Company. Montana Mining Division--History. | Mineral industries--Montana--Butte--History. | Copper mines and mining--Montana--Butte--History.
Classification: LCC HD9539.C7 (ebook) | LCC HD9539.C7 A53 2017 (print) | DDC 338.2/7430978668--dc23
LC record available at https://lccn.loc.gov/2017036829

The paper used in this book meets the requirements of American National Standard for Information Sciences—Permanence of Paper for Printed Library Materials, ANSI/NISO Z39.48-1992 (R2002).

First Printing

Manufactured in the United States of America

# CONTENTS

# ACKNOWLEDGMENTS

It would not be a far stretch to say that this book has its origins in my visits to Butte as a kid. Our soccer team would travel to play the Butte team, which meant encounters with the unavoidable Berkeley Pit. I often asked what had been there before the pit, only to receive vague comments about people living near it. These visits also meant some exposure to Butte's stigma. Parents who accompanied the team made snide comments about Butte, claiming that everything was poisoned and insisting we bring our own water to town, as Butte's was supposedly toxic. My parents, Alan and Karen Leech, shared what they knew about Butte's past and present. They therefore deserve the first set of thanks, because they never let me accept Butte's negative stereotypes and they always encouraged my curiosity.

I drove by the pit even more often when I was an undergraduate at the University of Montana in Missoula. Traveling between Missoula and my hometown of Bozeman meant passing by the Berkeley Pit. At the university, I started taking history courses, which helped me understand that I could study the places around me. I clearly remember sitting in Harry Fritz's "Montana History" course as a freshman and hearing him say that you had to study Butte if you wanted to be taken seriously as a Montana historian. One historian he was thinking of, David M. Emmons, is more responsible for me attending history graduate school than anyone else at the university. He treated undergraduates in his research classes as if they were already professionals. His zeal for Butte's social history was clearly contagious. His colleague, Dan Flores, introduced me to environmental history, while Sean O'Brien, then at the Davidson Honors College, convinced me that I could write. By the end of college, I had determined that I wanted to go to graduate school, and that I needed to study Butte.

The long process of writing this book began during graduate school at the University of Wisconsin, Madison. The history department faculty there, especially Bill Cronon, Colleen Dunlavy, Camille Guerin-Gonzales, Susan Lee Johnson, and my ever-patient advisor, Thomas Archdeacon, served as model scholars and great mentors. Fellow grad students also pushed me to the finish line of my PhD, including Andrew Case, Linda Brindeau, Todd Dresser, Nicole Kvale, Christine Lamberson, Muggy Lee, Maureen Mahoney, Nic Mink, David Rodriguez, and James Thompson.

Other people who deserve thanks are those who housed and helped me during my many research trips. Ray and Kay Campeau as well as Kelly and Roy Crosby let me stay at their places in Butte, even though I was there for weeks or months at a time. Others put me up elsewhere across the U.S. West, including Ryan Berger, Aaron Bouschor, David Dunbar, and Taryn Hall. Both Jim Michelotti and Pauline deBarathy helped me to identify and contact interviewees. Without their help, this book clearly would not have been possible. Staff at the Butte-Silver Bow Public Archives, including Director Ellen Crain and staff members Lee Whitney and Harriet Schultz, went out of their way to assist me—even moving my materials to a separate research room when the archives were being renovated. Staff at the Montana Historical Society, particularly Brian Shovers, and at the American Heritage Center, especially Ginny Kilander, proved similarly helpful. Some of those research trips were financially supported by the following organizations: the University of Wisconsin, Madison, the Rock Island Arsenal Historical Society, Phi Kappa Phi, the Mining History Association, and the Charles Redd Center for Western Studies.

After moving away from Wisconsin, I began to turn my research into a book. While at Augustana College in Rock Island, Illinois, I received welcome encouragement from my fellow faculty members. Thanks in particular to Tom Brown, Lendol Calder, David Ellis, Jane Simonsen, Chris Strunk, Molly Todd, Stephen Warren, and, especially, Matthew Fockler, who created a couple of maps for the book. I've received excellent research tips from other mining historians. Matthew Basso, Kent Curtis, Janet Finn, Pat Kearney, Laurie Mercier, and Mary Murphy deserve thanks for taking time to visit with me when I was early in this project, while Chris Huggard, Terrence Humble, Timothy LeCain, and Fred Quivik gave even more time and energy. As I began crafting book chapters, presentations at and assistance from members of the Mining History Association were particularly meaningful.

Thanks to the journals that previously published material later adapted into this book (and the journals' editors, Fred Quivik and Molly Holtz). Portions appeared as "Boom, Bust, and the Berkeley Pit: How Insiders and Outsiders Viewed the Mining Landscape of Butte, Montana," *IA: The Journal of the Society for Industrial Archaeology* 37, nos. 1 and 2 (2011): 153–70; and "Protest, Power, and the Pit: Fighting Open-Pit Mining in Butte, Montana," *Montana, The Magazine of Western History* 62, no. 2 (Summer 2012): 24–43.

It has been a pleasure to work with the University of Nevada Press, particularly with Director Justin Race, who ensured an easy process for this first-time author. Thanks also to former acquisitions editor Matthew Becker, who convinced me to publish in the Mining & Society series. Series editor Eric Nystrom has been both a friend and a mentor throughout the writing of this book.

Last, but clearly not least, are my family members. My wife, Siri, gave birth to two amazing kids while I was writing this book: Espen and Elin. Siri has kept us all sane and so it is to her that I dedicate this book.

# THE CITY THAT ATE ITSELF

# INTRODUCTION

EARLY IN 1974 VINCENT CIABATTARI, a retired machinist for the Anaconda Mining Company, sat in the disintegrating neighborhood of East Butte. He spoke with an out-of-town reporter who wanted to hear about his experience living in one of the old ethnic neighborhoods being devoured by the Berkeley Pit mine. East Butte faced the fate of the working-class districts of Meaderville, McQueen, and Finn Town before it. Expansion of the Anaconda Company's Berkeley Pit brought blasting noise, dumps of overburden, and dust-kicking trucks right next to Ciabattari's small house. The reporter described the East Butte neighborhood as "a moribund suburban wasteland."[1]

Of the 260 families who had called East Butte home, Ciabattari was one of seventy who had not yet sold their houses and moved down into the valley. Although the Anaconda Company had given the residents what many considered to be a fair deal for their homes, Ciabattari wondered why so many had accepted the company's offers. Why had they left their close-knit group of friends and family, losing both their Catholic Church and their school? He remarked to reporter Don Moser, "The way this community was, you'd think the people would have got together and fought. People said they'd fight, but then they jumped."[2]

Ciabattari's disillusionment with open-pit mining at that moment emerged out of the method's often unintended effects—effects that themselves were the result of moving industrial work above ground, out into the open, and directly into the community. As a former company employee, Ciabattari recognized how different, and often alienating, work in the open was in comparison to the underground labor that had long defined Butte, but he was even more concerned about the natural tendency of an open-pit mine to act as an expanding factory. The Berkeley Pit had brought industrial work farther into his neighborhood, producing new industrial hazards and forcing the relocation of his neighbors.

From the 1950s to the 1970s, the Anaconda Company switched the majority of its Butte operations from underground to aboveground copper mining. This book tells the social and environmental history of people like Vincent Ciabattari who lived through this change. Those who lived outside Butte had long been disturbed by the city's status as an industrial center in the middle of the rural West. In his book *English Creek*, regional author Ivan Doig suggests, "Parts of Montana like ours were apprehensive, actually a little scared, of Butte. There seemed to be something spooky about a place that lived by eating its own guts."[3] That the city seemed to be consuming itself through underground and, later, aboveground mining only added to this fascination and worry. Residents, on the other hand, had much more mixed feelings. Some were disgusted as the Berkeley Pit mine ate away at the community's eastern edge. Others recognized that, even if they were upset by some of its effects, open-pit mining kept the city economically alive. Not only did open-pit mining provide a way to cope with increasingly lower-grade deposits, but, in comparison to older underground practices, it was a quick and safe way to gather the minerals necessary for electric wires, automobiles, and other modern conveniences.

By covering a community that experienced both a long history of underground mining followed by a fiercely contested history of open-pit mining, this book recognizes the progress and the price of mass mining by looking closely at the ways that it changed space. Because an open-pit mine has to expand outward in order for workers to process ore, this new workplace's effects similarly spread outward, dramatically changing the lives of workers, residents, and visitors, even long after the mine concluded operations. In other words, an open-pit mine took industrial work, typically hidden behind factory walls, and forced people to continually confront its results. By tracing this process in Butte, the book will show that open-pit mining had surprisingly dramatic, long-lasting, and often detrimental consequences for both miners and their communities.

Historians have only recently paid attention to open-pit mining, which has long taken a back seat to the more romanticized front seat that underground mining holds in the popular (and scholarly) imagination.[4] More recently, however, historian Timothy LeCain's *Mass Destruction* has argued that open-pit copper mining technology meant much worse effects for the environment than underground mining. Instead of bringing together various pieces of raw material to produce goods as rapidly and efficiently as possible, as a mass-production facility would, hard-rock, open-pit mines became factories devoted to "mass destruction." In LeCain's words, open-pit mines "took apart the natural world in order to produce raw materials as rapidly and efficiently as possible." Results of this process included not only valuable minerals like copper, but also often-poisonous refuse, deposited into a mining landscape's many "dead zones."[5]

As more a social history than a history of technology, this book approaches the environmental history of mining in a way that is different from LeCain's work. Back in 1998, historians Jeffrey Stine and Joel Tarr identified the intersection of mining and the environment as a significant gap.[6] A particular problem was the lack of histories that did more than just recount negative environmental effects. This situation has quickly changed. In 2003 historian Christopher Huggard labeled mining a "nascent subfield of environmental history."[7] By 2010 Katherine Morrissey was able to compile an impressive list of new resources on mining and the environment.[8] One of her key examples is Thomas Andrews's award-winning *Killing for Coal*. Its thoughtful combination of labor and environmental history was groundbreaking, helping to inspire this book, as did other studies in the new and growing body of research on the history of class and environment. In particular, this book has drawn on historian Gunther Peck's suggestions about what a geographically and environmentally sensitive history of the working class should look like. His geographies of labor would combine the study of neighborhoods, work, and landscape with attention to environmental and social action.[9]

Following Peck's advice allows this book to make four major contributions to the history of the American West. First, it is the only full scholarly study to examine the social effects of the shift from underground to open-pit mining—a shift that occurred across the West's hard-rock industry. While surface coal mining has multiple chroniclers, few have touched the social history of hard-rock open pits.[10] Once exposed above ground, mining itself changed dramatically. As Butte shows, bringing mining up allowed for bigger machines and hence more-rapid production. Visitors in the 1950s and 1960s could stand in awe on an observation deck, experiencing the increasing power of this labor. Surface mining also took workers away from the underground's incredible dangers: ground falls, fires, and accidents were far more deadly in the darkness below.

The rise of the open pit meant that labor was done in the open, however, exposing workers to greater oversight from both management and union leaders. This decrease in miners' autonomy led to jurisdictional disputes among unions, and challenged the independent, masculine culture of the profession. Increased managerial control and worker specialization, along with a devaluation of skill, are similar results to those posited by other histories of mass production, but such changes in Butte often affected a new group of workers. These effects were typically unintended, as shown by the company's lack of preparation for and annoyance with jurisdictional disputes.[11]

Instead of seeing the switch to mass mining through a Marxist lens, it is therefore perhaps more useful to show how pits reorganized the space of work—what historian Thomas Andrews calls a "workscape."[12] As scholars involved in visual culture studies have shown, there are often dramatic shifts in power when something becomes visible. The same can be said for the mining industry, where the increasing visibility of men's work puts employees under increased scrutiny.[13] Although they were eventually forced to accept such changes, Butte's workers were often surprisingly resistant, even during the supposed height of Cold War complacency.[14] Mine mechanization's social effects, including enormous job losses, became more frightening to mine workers who could see massive equipment in a pit than it previously was when these same kinds of changes were occurring underground and out of sight.

The switch to open-pit mining was perhaps even more jarring for people who lived near the Berkeley Pit than it was for workers. Open-pit mines demand a lot of space—some of it for the mine, but much of it for waste storage, equipment, and transport. The problem for Butte was that an open pit's demands continue to grow. Going deeper means expanding outward, because machines have to be able to dig and then get out of the mine. Because of the proximity of residential and commercial areas to the growing factory, the threat of open-pit mining caused many people to move. Many western towns have disappeared into iron, copper, and other types of open-pit mines—either moved by a mining company a couple of miles down the road, like Kimberly, Nevada, or cleared entirely away, like Bingham, Utah, leaving the population to find new homes elsewhere. Indeed, open-pit mining continues to provide instances of what social scientists label development-induced displacement and resettlement, a topic of great concern to regions around the world currently experiencing the siting of massive industrial projects in landscapes full of people.[15] What scholars are only now more thoroughly considering is that North American history is also replete with examples of this phenomenon. Megaprojects—dams are the most well-known of these—erased communites from the map and radically altered people's daily lives.[16] Recent histories of the Jeffrey Mine in Asbestos, Quebec, and the Santa Rita Mine in Santa Rita, New Mexico, suggest that North American mining projects were among the worst at forcing continual relocations and land-use conflicts.[17]

This book's second major contribution to the history of the American West is therefore to examine in more detail how the region's companies instituted these kinds of projects, looking at how workers and residents managed the attendant hazards, risks, and relocation. The city of Butte never disappeared completely, but the expanding workplace of the Berkeley Pit did devour small ethnic communities on the eastern outskirts. In Butte the problem of "urban removal," as one resident termed it, affected daily life.[18] The company had previously moved its smelters outside the city, an act that had shielded Butte residents from the worst of mining pollution, but open-pit mining introduced a new set of hazards.[19] Noisy blasts, constant dust, and large vibrations became part of daily life. Residents complained about their new sensory experiences, as well as about the possible property damage that seemed to come along with it. Possible risks from these hazards seemed much more dispersed in the pit, making residents' ability to prove such damage difficult. As the mine continued to grow, residents used financial inducements from the Anaconda Company to relocate. The moving process scattered longtime neighbors, which weakened their communities' ability to maintain ethnic traditions. The all-encompassing nature of open-pit mining gave truth to one journalist's claim that "the Pit is life in Butte."[20]

The movement of industrial work directly into Butte forced the Anaconda Company to adjust its overall approach to the community. The company's successful efforts to relocate people provide a window into the exercise of corporate power in postwar resource communities. Anaconda switched to a more modern, gentle form of public relations management when compared to the bluntness of its earlier company-town approach. Even Anaconda's extraordinary legal power of eminent domain was seldom used, as the threat was more powerful than direct force. Instead, the company operated through economic threats to the community's well-being. Engineers and managers also increasingly tried to limit residents' exposure to the sensory experience of open-pit mining so as to reduce the company's financial liability. Anaconda's management never intended to target working-class people, but the structural problems facing working-class people on the eastern outskirts do look much like those facing other communities involved in environmental justice issues.[21]

The Berkeley Pit became so big that it altered, if not dominated, many insiders' and outsiders' perceptions, both of mining and of the city itself. This book's third contribution is therefore to expand on previous research about mining landscapes by exploring the power of the open pit as an icon.[22] As the shift to open-pit mining became prevalent across the American West and the world, people's ideas of it and its effects on the land changed too. The Berkeley Pit first functioned as an icon of industrial power and technological sublimity, then it became a symbol of corporate villainy. To journalists and newcomers in the 1970s and 1980s, the Berkeley Pit made Butte a pariah, representative of everything that was wrong with the extractive Old West, including environmental devastation and inevitable deindustrialization. Residents alternately saw it as a savior and an enemy, or, as its intrusive nature became normalized, an expanding organism outside of human control. City planners became convinced of the inevitability of the pit's continued growth, no matter what the company that controlled it said. The open pit exposed the often-ugly reality of mass industrial work, making it visible to the public.

Amidst declining copper prices and a raft of corporate problems, the company shut down the Berkeley Pit in 1982, but the pit still looms over Butte. Its presence as a large toxic lake, now filled with metallic, acidic water, continues to affect residents' search for a postindustrial future. The story of the Berkeley Pit's decline and closure in the final chapters of this book provides a fourth and final contribution to the history of the American West. Butte's early years have been extensively studied, but its later history has been barely touched.[23] Tackling this later history allows the book to cover the period of bust, which has long been overlooked by historians more excited with natural resource booms. Indeed, because open-pit mining itself introduced dramatically lower labor requirements, it is possible to say that the entire story of open-pit mining in Butte is one of deindustrialization, as Jeffrey Manuel claimed for Minnesota's iron country after the introduction of taconite technology.[24]

Many previous studies of natural resource communities have portrayed the postindustrial transition as a quick move from mining or logging into tourist-friendly, skiing utopias, but the pit's story shows that the process of economic, environmental, and social revival has often been more challenging, although it can be just as successful.[25] As Finis Dunaway has shown, icons like the Berkeley Pit can galvanize the public toward environmental and social action, but people's understanding of such icons can often be incomplete, hiding underlying concerns and possible solutions.[26] In Butte, even as residents have figured out ways to readapt after mining's decline, revitalization has sometimes been stifled by the mining landscape. The ability of an open-pit mine to force people to continually confront industry therefore lasts long after companies and workers have left.

To explore the lived experience of open-pit mining, this book uses company records, union files, city records, and personal collections held in a variety of archives. Oral history interviews, however, form the core of many chapters. Since much of residents' daily lives happened without anyone making contemporary recordings, oral histories allow for a much fuller exploration of neighborhood life, mining labor, and common experiences with hazards and relocation. When it comes to everyday life and routines, people typically recall with a kind of detail and precision that is less likely when it comes to single events. Interviews also provide this history with something impossible to gain otherwise—a better understanding of how people felt about their daily activities. Both performing interviews and using oral histories takes practice. Interviewers can influence interviewees, nostalgia can color recollections, and people remember similar events differently. Oral historians therefore need to carefully consider the interviews' context, appreciate their subjective and conversational nature, and cross-check for empirical facts. Oral memories are active creations, not just objective recordings of the past, so the final chapter also considers these interviews as recent interpretations of Butte's past.[27]

This book tells Butte's story in four parts. The first part establishes Butte as an underground mining community, explaining how underground mining shaped both workers' lives and the city's geography. Parts II and III cover the major era of open-pit mining in Butte, which lasted from 1955 to 1975. Part II discusses the first stages of open-pit mining work, showing that surface mining initially led to both labor and community protest. It then focuses on how the shift to the surface exposed employees to greater surveillance and a social reorganization of the workplace. Part III looks at the impact of open-pit mining on Butte's neighborhoods, discussing how residents and company officials managed industrial hazards and the process of relocation. Part IV concludes the book, looking at Butte's 1970s economic troubles and its movement, starting in the 1980s, into a post–Anaconda Company, post–Berkeley Pit era. It argues that people's increasingly negative perceptions of open-pit mining shaped the region's environmentalism, outsiders' opinions, and Butte residents' attempts to move the city toward environmental, economic, and social rebirth.

The pit reshaped Vincent Ciabattari's personal story as well. In a search for good jobs, the Ciabattari family had relocated to Butte from Lucca, a beautiful, walled city in Tuscany, in the early twentieth century. At that time, the eastern edge of Butte had seemingly held everything a family could need, from commercial services to social pleasures, to, most importantly, decent-paying jobs in labor-intensive mines. Sixty years later, a more expansive, invasive form of mining had claimed the neighborhood. A couple of years after Ciabattari's interview, East Butte no longer required a label on the city map or a section in the city directory. It was gone. Even Vincent, the holdout, had finally moved. Before this book turns to the protests, labor, hazards, and relocation that so concerned Vincent's generation, it begins its tale where Ciabattari's Montana experience began—with Butte's thriving ethnic neighborhoods and underground mines.

## Notes

1. Don Moser, "The Monster That's Devouring Butte, Montana," *New Times* 2, no. 4 (Feb. 22, 1974), 48–50, quote on 50.

2. Ciabattari quoted in Moser, "Monster That's Devouring," 50.

3. Ivan Doig, *English Creek* (New York: Penguin, 1984), 236.

4. On the romanticized understanding of underground mining in comparison to open-pit mining, see Ronald M. James, *The Roar and the Silence: A History of Virginia City and the Comstock Lode* (Reno: University of Nevada Press, 1998), 255.

5. Timothy LeCain, *Mass Destruction: The Men and Giant Mines that Wired America and Scarred the Planet* (New Brunswick, NJ: Rutgers University Press, 2009), quotes on 132, 172.

6. Jeffrey K. Stine and Joel A. Tarr, "At the Intersection of Histories: Technology and the Environment, Updated with a Postscript by the Authors," *H-Environment Global Historiography Series* (Dec. 2001). The article was originally published in *Technology and Culture* 39, no. 4 (1998): 601–40.

7. Christopher J. Huggard, "Squeezing out the Profits: Mining and the Environment in the U.S. West, 1945–2000," in *The American West in 2000: Essays in Honor of Gerald D. Nash*, eds. Richard W. Etulain and Ferenc Morton Szasz (Albuquerque: University of New Mexico Press, 2003), 105–26, quote on 123.

8. Katherine G. Morrissey, "Rich Crevices of Inquiry: Mining and Environmental History," in *A Companion to American Environmental History*, ed. Douglas Cazaux Sackman (Malden, MA: Wiley-Blackwell, 2010), 394–409.

9. Thomas G. Andrews, *Killing for Coal: America's Deadliest Labor War* (Cambridge, MA: Harvard University Press, 2008); Gunther Peck, "The Nature of Labor: Fault Lines and Common Ground in Environmental and Labor History," *Environmental History* 11 (Apr. 2006): 212–38. Key research on the history of class and environment includes Richard White, "'Are You an Environmentalist or Do You Work for a Living?': Work and Nature," in *Uncommon Ground: Rethinking the Human Place in Nature*, ed. William Cronon (New York: W. W. Norton, 1996); Andrew Hurley, *Environmental*

*Inequalities: Class, Race, and Institutional Pollution in Gary, Indiana, 1945–1980* (Chapel Hill: University of North Carolina Press, 1995); Alan Taylor, "Unnatural Inequalities: Social and Environmental Histories," *Environmental History* 4 (Oct. 1996): 6–19.

10. Among many coal examples, see Chad Montrie, *To Save the Land and People: A History of Opposition to Surface Coal Mining in Appalachia* (Chapel Hill: University of North Carolina Press, 2003).

11. On mining mechanization, see Wallace Clement, *Hardrock Mining: Industrial Relations and Technological Change at INCO* (Toronto: McClelland and Stewart, 1981), 251–345; Keith Dix, *What's a Coal Miner to Do? The Mechanization of Coal Mining* (Pittsburgh, PA: University of Pittsburgh Press, 1988); Logan Hovis and Jeremy Mouat, "Miners, Engineers, and the Transformation of Work in the Western Mining Industry, 1880–1930," *Technology and Culture* 37, no. 3 (1996): 429–56. For critiques of mass production and its effects on labor, see David A. Hounshell, *From the American System to Mass Production: The Development of Manufacturing Technology in the United States* (Baltimore, MD: Johns Hopkins University Press, 1984), 316–27; Harry Braverman, *Labor and Monopoly Capital: The Degradation of Work in the Twentieth Century* (New York, Monthly Review Press, 1974); Melvyn Dubofsky, "Technological Change and American Worker Movements," *Technology, the Economy, and Society*, ed. Joel Colton and Stuart Bruchey (New York: Columbia University Press, 1987), 162–85; Richard Edwards, *Contested Terrain: The Transformation of The Workplace in the Twentieth Century* (New York, Basic, 1979). A number of historians have challenged the declensionist portrayal by writers like Braverman of the effects of mass production on workers. See Daniel Nelson, "Mass Production and the U.S. Tire Industry," *The Journal of Economic History*, 47, no. 2 (June 1987): 329–39; Michael Nuwer, "From Batch to Flow: Production Technology and Work-Force Skills in the Steel Industry, 1880–1920," *Technology and Culture* 29, no. 4 (Oct. 1988): 808–38.

12. Andrews, *Killing for Coal*, 125.

13. On the ability to gaze at something to make the subject knowable and hence manipulable, see Michel Foucault, "Panopticism," in *Discipline and Punish: The Birth of the Prison*, trans. Alan Sheridan (New York: Vintage, 1979), 195–228. Applying visual culture studies about power to the history of mining is Eric C. Nystrom, *Seeing Underground: Maps, Models, and Mining Engineering in America* (Reno: University of Nevada Press, 2014).

14. On Cold War complacency, see David F. Noble, *Forces of Production: A Social History of Industrial Automation* (New York, Alfred A. Knopf, 1984); Stephen J. Whitfield, *The Culture of the Cold War* (Baltimore, MD: 1996); Bob H. Reinhardt, "Drowned Towns in the Cold War West: Small Communities and Federal Water Projects," *Western Historical Quarterly* (Summer 2011): 149–72.

15. Representative of a large literature of development-induced displacement is Theodore E. Downing, "Avoiding New Poverty: Mining-Induced Displacement and Resettlement" (London: International Institute for Environment and Development, 2002); Robert Muggah, "A Tale of Two Solitudes: Comparing Conflict and Development-induced Internal Displacement and Involuntary Resettlement," *International Migration* 41, no. 5 (2003): 5–31; Tulsi Charan Bisht, "Development-Induced

Displacement and Women: The Case of the Tehri Dam, India," *Asia Pacific Journal of Anthropology* 10, no. 4 (2009): 301–17.

16. The term *megaproject* is from Joy Parr, *Sensing Changes: Technologies, Environments, and the Everyday, 1953–2003* (Vancouver: UBC Press, 2010). For other scholarship about vanished communities and/or relocation in the North American West, see Douglas Seefeldt, "Creating Kearney: Forging a Historical Identity for a Central Arizona Mining Community," *Journal of Arizona History* 46, no. 1 (2005): 1–32; J. Douglas Porteous and Sandra E. Smith, *Domicide: The Global Destruction of Home* (Toronto: McGill Queens, 2001); Reinhardt, "Drowned Towns."

17. Jessica Van Horssen, *A Town Called Asbestos: Environmental Contamination, Health, and Resilience in a Resource Community* (Vancouver: University of British Columbia Press, 2016); Christopher J. Huggard and Terrence M. Humble, *Santa Rita Del Cobre: A Copper Mining Community in New Mexico* (Boulder: University Press of Colorado, 2012).

18. Butte Forward Minutes, May 15, 1974, Box 3A, Fldr. "Meeting Minutes," Butte Forward Collection (hereafter BFC), Butte–Silver Bow Public Archives, Butte, MT (hereafter BSBPA). A large literature discusses hazards, or possible sources of danger, and risks, which involves an attempt to measure the chance that a hazard will cause harm.

19. A primary reason why this book focuses on mining work and not on ore processing is that processing was largely completed outside Butte during the post–World War II era. See more on Anaconda's Montana smelter town in Laurie Mercier, *Anaconda: Labor, Community, and Culture in Montana's Smelter City* (Urbana: University of Illinois Press, 2001).

20. Bruce Johansen, "Butte: Sitting Atop a Fortune," *Seattle Times*, May 6, 1973.

21. Arn Keeling and John Sandlos, "Environmental Justice Goes Underground? Historical Notes from Canada's Northern Mining Frontier," *Environmental Justice* 2, no. 3 (2009): 117–25.

22. The most in-depth considerations of the mining landscape include David Robertson, *Hard as the Rock Itself: Place and Identity in the American Mining Town* (Boulder: University Press of Colorado, 2006); Liza Piper, *The Industrial Transformation of Subarctic Canada* (Vancouver and Toronto: UBC Press, 2009), 113–64; Richard Francaviglia, *Hard Places: Reading the Landscape of America's Historic Mining Districts* (Iowa City: University of Iowa Press, 1991); Thomas Dublin, *When the Mines Closed: Stories of Struggles in Hard Times* (New York: Cornell University Press, 1998).

23. There are many excellent studies of Butte in earlier eras, such as Michael P. Malone, *The Battle for Butte, MT: Mining and Politics on the Northern Frontier, 1846–1906* (Seattle: University of Washington Press, 1981); David M. Emmons, *The Butte Irish: Class and Ethnicity in an American Mining Town, 1875–1925* (Urbana and Chicago: University of Illinois Press, 1989); Mary Murphy, *Mining Cultures: Men, Women, and Leisure in Butte, 1914–41* (Urbana: University of Illinois Press, 1997); Matthew L. Basso, *Meet Joe Copper: Masculinity & Race on Montana's World War II Home Front* (Chicago: University of Chicago Press, 2013). The few studies that cover some of Butte's postwar period include Janet Finn, *Tracing the Veins: Of Copper, Culture, and Community from*

*Butte to Chuquicamata* (Berkeley: University of California Press, 1998); Bode J. Morin, *The Legacy of American Copper Smelting: Industrial Heritage versus Environmental Policy* (Knoxville: University of Tennessee Press, 2013); Janet L. Finn, *Mining Childhood: Growing Up in Butte, Montana, 1900–1960* (Helena: Montana Historical Society Press, 2012).

24. Jeffrey T. Manuel, *Taconite Dreams: The Struggle to Sustain Mining on Minnesota's Iron Range, 1915–2000* (Minneapolis: University of Minnesota Press, 2015).

25. Other than Manuel, the most complete and complex studies of mining busts and deindustrialization include Eric L. Clements, *After the Boom in Tombstone and Jerome, Arizona: Decline in Western Resource Towns* (Reno: University of Nevada Press, 2003); Donald Reid, *The Miners of Decazeville: A Geneaology of Deindustrialization* (Cambridge, MA: Harvard University Press, 1985); Larry Lankton, *Hollowed Ground: Copper Mining and Community Building on Lake Superior, 1840s–1990s* (Detroit, MI: Wayne State University Press, 2010). On mining towns becoming ski towns, see Kristen Smart Rogers, "'We Didn't Think He Was Gonna Build It': Skiing Hits a Mining Town," *Utah Historical Quarterly* 69, no. 4 (2001): 310–25; Hal K. Rothman, *Devil's Bargains: Tourism in the Twentieth-Century American West* (Lawrence: University Press of Kansas, 1998), 202–26.

26. Finis Dunaway, *Seeing Green: The Use and Abuse of American Environmental Images* (Chicago: University of Chicago Press, 2015).

27. On oral history in practice and theory, see Lynn Abrams, *Oral History Theory* (New York: Routledge, 2010), especially 78–105; Linda Shopes, "Oral History and the Study of Communities: Problems, Paradoxes, and Possibilities," *Journal of American History* 89, no. 2 (Sept. 2002): 588–98; Alessandro Portelli, "On the Peculiarities of Oral History," *History Workshop Journal* 12 (Spring 1981): 96–107; Ashley Ward, "Reclaiming Place through Remembrance: Using Oral Histories in Geographic Research," *Historical Geography* 40 (2012): 133–45; Daniel James, *Doña María's Story: Life History, Memory, and Political Identity* (Durham, NC: Duke University Press, 2001), 119–56; Orlando Figes, "Private Life in Stalin's Russia: Narratives, Memory, and Oral History," *History Workshop Journal* 65 (2008): 117–37.

# PART I

# MINING IS LIFE, 1864–1954

CHAPTER 1

# UNDERGROUND

THE STORY OF BUTTE'S era of underground mining is usually told as a lusty, rip-roaring tale. Historians typically begin the story with a gold rush, like they do for so many other cities in the West. People had known about gold for decades, but it was only in 1862 that the first diggings opened up in what is now southwestern Montana. Migrants had been traveling across the West looking for gold for more than a decade at this point; the California Gold Rush, which began in 1849, had encouraged a rush mentality, accelerating western settlement. This westward movement was not the mythic frontier later described by historian Frederick Jackson Turner, with people moving successively westward to start their own farms. Instead, mining rushes meant a messy series of migrations that were "widely separated geographically and chronologically rather than a single entity," as western historian Rodman Paul later put it.[1]

Nor were mining rushes peaceful. By 1862 southwestern Montana had met the two preconditions to a gold rush. First, violence led to legal agreements between the natives living in each area and the expansionist United States. Second, transportation networks had been built, providing an easier path to and around the area. The rush affected native peoples the most, many of whom had lived in western Montana for centuries. In fact, more than 12,000 people lived in western Montana in 1862, very few of whom were European Americans.[2]

New people from around the world found their way to the site of each new boom, forming complex communities.[3] Enough people had arrived in southwestern Montana—slightly more than 20,000—that in 1864 Abraham Lincoln signed a bill creating Montana Territory.

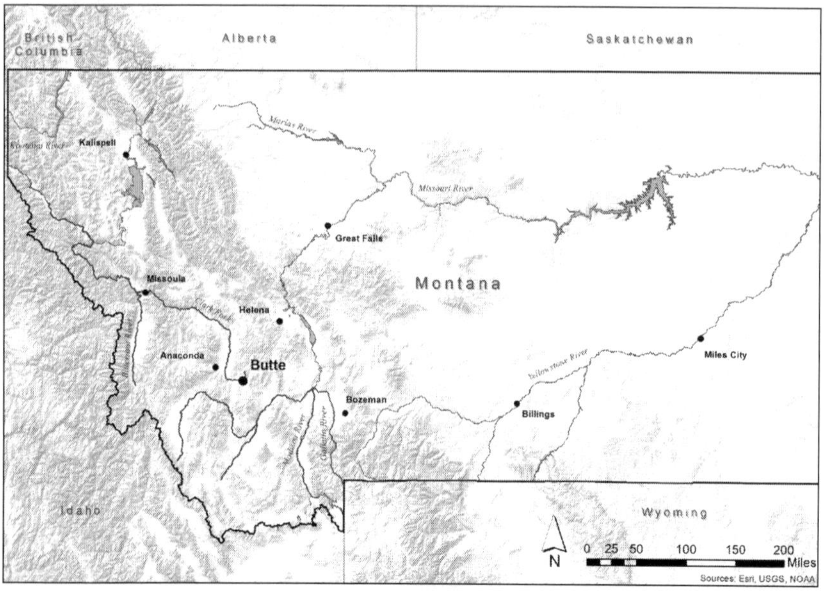


MAP OF MONTANA. Map created by geographer Matthew Fockler, 2017.

At first, many migrants engaged in placer mining, which involved a simple act—washing gold out of gravels, streambeds, and banks.[4]

Mining began in the territory's Summit Valley in 1864. Surrounded by ridges and peaks, this mile-high mountain valley sits just to the west of the continental divide. On the northeastern edge of the valley, prospectors came across a small pit along the Missoula Gulch. The trader who originally discovered the pit, Caleb E. Irvine, found worn elk horns nearby. He thought that perhaps members of a nearby tribe had been mining it, using the horns as picks. In 1864 prospectors William Allison and G. O. Humphries dug even deeper into the pit, setting off more mining in the area. A boomtown named Silver Bow rapidly formed a few miles west of the site. From 1865 to 1867, placer miners would arrive in the spring and summer to mine along Silver Bow Creek. They began a small community of three hundred to five hundred people at the creek's headwaters. The community gained the name Butte because of a tall nearby landmass. As in most gold rushes, the boom was short-lived. By 1874 the area had lost much of its population. The easy-to-grab gold had played out and the rest was bonded with other minerals, like silver, copper, manganese, and zinc, and hence difficult to separate. The economic panic of 1873 and a gold rush to the Black Hills provided further incentive to leave.[5]

After experiencing its first boom and bust in just ten years, Butte faced the prospect of becoming a ghost town, but before that could happen silver and then copper rescued the area. The history of mining in Butte thus stands out partly because of its longevity; most mining camps fail to last past their first bust. Butte's longevity came partly because of its mineral wealth, but also because of the skilled miners who later arrived and made great sacrifices in a hazardous underground, and the intelligent mine managers who capitalized on new discoveries, navigated fluctuating markets, and continually updated mining methods. The result was a dramatically altered landscape and a local culture generated from underground work. Even as mine dangers bound workers together in camaraderie and industrial unions, the mining workplace gave miners a sense of independence. The mines similarly encouraged resilience in community members: they had a sense that most problems, whether industrial hazards, strikes, economic troubles, or outsiders' attacks, were commonplace occurrences through which the community would persevere. This culture was challenged only in the mid-twentieth century, during an attempt to shift Butte's operations toward mass mining.

## Copper Becomes King

Butte's underground era began with a silver boom. The processing of Butte's new silver finds in the 1870s required a lot of capital investment and technological know-how. New corporate investors and skilled miners hence arrived to deal with the district's complex quartz ores. Revitalization allowed residents to incorporate Butte as a city.[6] The city's wide, southward-facing hill, on which most of Butte sat, would soon make the area famous, but because of copper, not silver. The switch to copper happened partly because of two important figures who arrived in the 1870s.

The first was a second-generation Scots-Irishman, William A. Clark, who entered Butte in 1872. He quickly snatched up mining claims from people who could not figure out how to deal with the hill's ores. As a city grew around him, Clark obtained an interest in the first mill to process those ores, learned some metallurgy, gained financial backing from eastern interests, founded a mercantile, and, eventually, bought a newspaper to help promote his interests. Favoring skilled Cornish miners and other Protestants, Clark quickly amassed an empire.[7]

The man who was to become his rival, Marcus Daly, arrived four years after Clark. The Irish-born Daly came as a representative of the Walker Brothers, merchant bankers in Utah. He had already enjoyed a short, but fruitful, career as a practical silver mining man in Nevada and Utah. After suggesting that the Walkers purchase the Alice Mine, he moved to Butte to pursue other properties. Daly had interest in the Anaconda Mine, which had been named for a Civil War metaphor about General George M. McClellan's plan to surround Confederates and, like a giant anaconda, to squeeze them with his forces. Daly raised capital by appealing to mining magnate George Hearst, who, along with his California partners James Haggin and Lloyd Tevis, was forming a major mining conglomerate, with properties in Nevada, South Dakota, and Utah. As the silver boom died, Daly sneakily bought up surrounding claims because he had located a massive, fifty-foot-wide copper vein in his new purchase: the Anaconda Mine.[8]

Founded on this initial discovery, the Anaconda Copper Mining Company quickly became the largest corporation in Butte. Both Anaconda and other copper producers raised metal production to great heights, both as a way to gain market share and as a way to offset the major investments needed for lode mining. Low-priced copper thus became available for new uses, like electricity. By the 1890s copper wire had become vital to transmitting electricity, which itself had become a major industry, benefitting Butte's expanding mines.[9] With more Butte miners turning to copper, production from the area's mines soon overtook that of upper Michigan, which previously had been the biggest copper producer in the United States.[10] Daly relied on his fellow Irishmen and other Catholics to work the rich properties. Butte soon became the most Irish city in the United States, with more than a quarter of its population either first- or second-generation Irish.[11]

Having Catholics, Protestants, and money-hungry industrialists suddenly shoved together in one city meant tension. After Montana became a state in 1889, Clark and Daly fought over the location of the new capital. Daly favored the town of Anaconda, a company town that bore the corporate name that he had established, twenty-six miles west of Butte. Using money from Anaconda's new investors, the European Rothschild banking family, Daly had built a large smelter to process his ore there. Clark favored a silver boomtown, Helena, for the capital. Clark's forces slung accusations about Anaconda's overpowering greed and its brutish immigrant workers, which helped carry Helena to the win. (The enormous amount of money that Clark put into bribing thousands of people likely helped as well, although Daly also used that trick.) In the coming years the two copper barons also battled over a U.S. Senate seat, which Clark desired as a final step to becoming more than just a copper baron, but a famous millionaire.[12]

In the War of the Copper Kings, as it became known, another competitor entered the fray: Augustus F. Heinze. Trying to take advantage of the intense fracturing of Butte's ore bodies, Heinze made use of the Apex Law, federal legislation from the 1860s that allowed the owner of a claim to follow a vein of ore as far below ground as it goes, as long as it apexes, or breaks the surface, on that person's property. Cultivating a populist persona, Heinze wrapped Anaconda up in the courtroom, while his miners, having tunneled into an Anaconda Mine, actually turned to underground fisticuffs with Anaconda's at one point.[13] Anaconda eventually won most of the legal battles, thanks especially to engineer David W. Brunton and geologists Horace Winchell and Reno Sales, who were gaining acclaim for developing a systematic method for geological mine mapping. Working to consolidate all of Butte's mines, Anaconda eventually wore down Heinze, buying him out for $12 million in 1906. Having gained his Senate seat and making piles of money from an Arizona copper mine, Clark lost interest in Butte, spending most of his time and money outside the state. After besting its top two competitors, the Anaconda Company was victor.[14]

Victory's spoils in Butte were not enough for Anaconda. The company briefly became subsumed under the Amalgamated Copper Company, when William Rockefeller and Henry Huttleston Rogers of Standard Oil tried to corner the market on copper. By the time Amalgamated dissolved in 1915, Anaconda had become an industrial powerhouse. It both bought forest around Montana and trespassed onto the public domain so that it could provide logs to hold up its underground mine tunnels. Company management devoted increasing amounts of energy to lobbying state politicians, providing legislators in Helena with drinks and steaks. Beyond buying influence, Anaconda also practiced union infiltration to ensure its workforce stayed in line; it also used cruder coercion, sometimes including gunplay, as it did when company detectives fired on striking miners in 1920, killing two and wounding sixteen.[15]

Anaconda ran into a burgeoning labor movement on many other occasions as well. Called the Gibraltar of Labor, Butte featured many labor organizations, from the conservative Butte Miners' Union (BMU), which founded the Western Federation of Miners (WFM), to the more radical International Workers of the World (IWW), who, in the 1910s, were trying to gain a foothold in the area. Butte even elected a socialist mayor in 1914. That year, however, the conservative miners' union, which had worked hard to protect steady jobs for Irish men in particular, broke apart due to tensions between old and new immigrants, and conservative and radical unionists. An angry mob blew up the BMU hall. Declaring martial law, the governor sent in the Montana National Guard. Anaconda took advantage of the situation, instituting a rustling-card system, which were permits that allowed men to look for work at each mine (hence giving the company more control over who worked where). The company then declared that the union's closed shop was over because the WFM lacked legitimacy.[16]

Union organizers continued to stream into Butte because they could see discontent that needed an outlet, but they faced a strong company challenge. The company had political backing because copper was vital, and was needed for bullets and other war matériel. Many immigrants, particularly the Irish, were upset about U.S. entry into the Great War on the side of Britain, which made others in the region suspicious. Socialists, including many Finns, were angry with the company, particularly since it had fired five hundred miners for radical beliefs a few years earlier. The Speculator Mine disaster in June 1917, in which more than 160 men lost their lives in a mine fire, accelerated the tumult. It was not only the worst metal mining disaster in history, but it also triggered a major strike. The supposed initiator of the accident had a German name, and so many Montanans turned against the immigrant workforce and its Irish-German influence. Into this situation entered IWW organizer Frank Little, who was murdered in August 1917. The existence of radical political views among workers continued to seem anti-American during the national Red Scare just after the war ended. Butte remained under martial law until 1921, while the labor movement took years to recover from the scare, both in Butte and across the nation.[17]

## The Underground

Unions were important to workers who desired higher wages, fewer hours, and better conditions. Such protection was necessary because, as one miner put it, Butte was "noted by all boomers as the toughest of mining districts because of its combination of dangerous heavy ground, heat and humidity, acid mine waters, and hard-nosed bosses."[18] A man who wanted to get involved in the physically demanding process of mining, first had to get his rustling card, make sure to get a lunch packed, and then dress in work clothes: bib overalls, a hardhat, a headlamp, a battery pack to power the lamp, and heavy boots, made, by the 1940s, of rubber with a hard steel toe. At the mine shaft he would find his partner and receive instructions from the shift boss, who would outline worksites and tasks. The partners would then get into a nine-man cage. The hoisting engineer would receive a signal, then vertically drop the cage down a big steel cable to the appropriate level of the mine. It was a noisy and fast way to travel, dropping at up to twenty-five miles an hour. Eventually the cages would be retired and replaced with skips, which were the steel receptacles with which ore was hoisted out of the mine.[19]

Having arrived at the correct station, an enlarged room next to the shaft, the two partners would exit the cage. Then they would walk either along the drifts, which were horizontal tunnels along ore veins, or crosscuts, which went in between drifts, until they found their worksite. At the worksite the dust would have cleared from blasting during the previous shift, so their first job was to bar down, which meant using metal bars to clear the ceiling and sides of the tunnel from loose rock slabs; these slabs were known in Butte as Duggans, after the local mortician. Butte's veins sharply inclined at about 70 degrees. The top part of the vein was the hanging wall and the bottom portion the foot wall. Miners typically mined between the hanging wall and the footwall, progressing upward, creating what miners called a raise. Every now and then, miners would instead dig downwards from the drift, in a process known as sinking a winze.[20]

In the next step, the broken-up rock would need to be dug, or mucked, out of the blast area. Because most mining went into the ore vein from the drift below, gravity helped broken material to fall either down a chute compartment that miners had built or to the tunnel floor itself. Workers would then use hand or mechanical shovels to push the rest of the rock to the floor. By the 1930s and 1940s, a mucking machine had replaced hand labor to pick all of this material off the floor and put it into tram cars. Electric, battery-powered trains had replaced donkeys, who had originally taken tram cars to the station for hoisting. Along the tram tracks, a number of men would remove rock and mud, in an attempt to ensure safe passage. Back at the station, next to the shaft, other men received the newly blasted rock. Their job was to work the grizzlies, which involved breaking large slabs of rock with a sledge so that they could fit through ten-inch openings in I-beams, known as grizzlies, making the material small enough to hoist back up in the skip.[21]

Miners back at the blast site next tried to ensure that their workspace would remain both safe and open for mining. As the mines continued farther and farther underground, falling rocks and caving became common. Miners often used timber to support dangerous ground, using "millions and millions of board feet of good Montana timber" from forests owned by Anaconda.[22] Underground, the miners would typically construct heavy cubes made out of lumber, known as a square set. These cubes would support the ground, enabling miners to perform their work with increased safety. The twenty-foot-wide open chamber that they created as they mined was known as a stope; stopes consisted of multiple levels, about ten feet apart, that the miners would climb up and down via ladders. Other mining methods were used in Butte because the weakness of the ground and the shape of the ore veins differed widely from place to place. By midcentury it became increasingly common for Anaconda's miners to instead use a cut-and-fill method: as ore removal progressed, workers filled each mined-out void with waste material, to make sure the ground would remain relatively well supported.[23] No matter what precautions they took, the walls could still come crashing in. According to one 1950s Anaconda manual, "More serious and fatal injuries are caused by falling ground and rocks than from any other hazard underground."[24]

Danger was also ever-present when handling the drill and explosives, which represented the next stage in the mining process. Miners would come equipped to drill with their own tips, though if they needed any other materials they would get them from a nipper, who shuttled tools back and forth from the surface. To drill, the miner and his partner would set up a pneumatic jackhammer, which was often so large that it had to be mounted to a standing leg; they would then buzz a pattern of holes in the rock. Typically, the miner would place explosives into each hole and tap dirt on top of it. The famous Butte goodbye, "tap 'er light," refers to the miners' goal not to blow themselves up when they covered each hole. Next, miners hooked the holes up with blasting caps and fuses; then they would get ready to go back to the surface. After they yelled "Fire in the hole!," they would explode the dynamite and head back up in the cage. Their eight-hour shift having passed, the two miners would finally stop at the dry, a change house, to put on dry clothes, then they would head home.[25]

## Hazards and Happiness

In the almost three thousand miles of tunnels that made up the underground world, workers created what some scholars have called an envirotechnical system and others a workscape; whatever it is called, the point is essentially the same: human thought, tools, and labor interacted with nature to form a mine.[26] The work and technology of timbering serves as a good example of a workscape: Nature above, in the form of trees, was brought below ground. Workers then used both this lumber and their technological know-how to build supports for crumbling ceilings, fashioning a relatively livable space. Butte's subterranean workers found this workscape to be a wholly new environment, understood through all the senses. Miners proved their

Two Butte Miners with Rock Drill Underground. Photo courtesy of Butte–Silver Bow Public Archives, C. Owen Smithers Collection, 10.037.03.

increasingly intimate knowledge of the underground through one common practice: the use of sounding, which meant tapping around the work area, then listening for movement, to see if serious rock falls were likely.[27]

When asked about their workscape, many miners first bring up the underground climate, which the interaction of natural processes and human modifications created. No matter what the temperature on the surface, some levels of the mines were naturally hot: below four thousand feet, Butte's igneous granite could maintain temperatures in excess of 100 degrees. Other levels were dreadfully cold. The tunnels were quite humid because they ran below the water table. As one miner explained, "Everybody had hard toe shoes on, and the boot came up to mid-calf so that the copper water wouldn't eat your foot."[28] Due to acidic copper water, the humidity, and the decaying timber that supported tunnels in almost all Butte mines, the underground workscape had its own distinct smell.[29]

The most deadly component of the air, however, was not something miners could sense: it was the fine particles of silica that came from rapid drilling and blasting. The resulting buildup of particles in the lungs caused a disease called silicosis, or miner's consumption. The deadly disease inflamed and scarred the lungs, which could not rid themselves of the tiny particles. The U.S. Bureau of Mines found evidence of silicosis in 42.4 percent of the Butte miners who asked to be examined between 1916 and 1918.[30] Anaconda introduced water into the drilling process to keep down the dust, which lessened the danger, but the condition still killed many miners.[31] The deadly conditions in early Butte's underground were rivaled only by mines in the Transvaal of South Africa.[32]

To deal with what ventilation expert Floyd Bossard called these "unseen hazards," Anaconda tried to change the underground climate. Big fans blew air from the surface down below.[33] In 1965 Anaconda bragged that the "3,000,000 cubic feet of fresh air per minute" moving through the mines "requires as much horsepower as would be needed to supply the electrical energy requirements for a city of 25,000 people."[34]

Besides the problem of "not enough air," Anaconda also worked on the issue of "too much water."[35] To enable deeper mining, pumps pushed water out of tunnels. The pump station contained eight powerful pumps that drove underground water to the surface through stainless steel pipe columns. Each 1,500-horsepower motor pump could push a thousand gallons per minute.[36]

The pumps were just one part of this underground city, which the company described as a "vast and complex system of shafts, drifts, and crosscuts," all of which were "supplied and serviced by railway tracks and underground trains; ventilation systems and giant air-conditioning plants, hoisting facilities, communication lines, safety and fire-protection devices, electrical and water lines, pumps, and air lines which bring compressed air from plants on surface to provide the power for drills and other equipment."[37]

Beyond technology, Anaconda officials began additional safety precautions in the 1910s. Despite company opposition, the state passed a workmen's compensation bill in 1914, which encouraged the company to reduce injuries. Along with training first-aid and rescue teams, the company hosted first-aid competitions, giving out cash prizes. It publicized winners, safety regulations, pictures of bad mine practices, and photos of each month's "safest crew" in a company publication called *The Anode*, which ran from 1915 to 1942.[38] The company encouraged miners to take salt tablets, which lessened cramps when men went into hot mines, and it eliminated open flames underground in 1934, transitioning entirely from carbide to electric headlamps, thus cutting down significantly on fire risk.[39] Reflecting on his midcentury work, Anaconda's chief safety engineer proudly noted that Butte had gone from the district with the most recorded underground mine fires to a place that saw few deadly accidents.[40] In 1940 another safety engineer claimed that increased safety and conveniences in a newly inhabitable workscape had turned the job into "department store mining."[41]

Yet the underground still offered plenty of danger. Dan Aguilar, a miner who put in fifteen years long after so-called department store mining had come into vogue, insisted, "I was lucky I survived." In one case, a seven-ton ore car into which he was loading material tipped on top of him—he was only just saved by a ladder wedged between him and the car. In another case, the cage he took underground every day dropped so fast it would have killed him had the operator not been so quick on his feet.[42] A nurse who worked in a Butte hospital ward dedicated solely to the treatment of miners explained, "There was kind of that fatalistic idea.... If you work in the mines, you're apt to get it."[43] Even the silicosis problem never went away. Defeating ballot initiatives and unfavorable legislation, Anaconda successfully ensured for decades that the state workers' compensation program would not cover silicosis.[44]

To minimize the danger below, almost all employees worked with partners—usually the same man every day. Each man would learn to "look out for each other."[45] According to miner Perle Watters, most miners refused to work with anyone else but their longtime partners. Austrians Joe Carleen and Joe Manduch "worked together so long they were just like machinery."[46] Shift bosses would often place greenhorn miners with longtime employees, so the partner system not only served as a safety precaution, but as a training system.[47]

The dangerous environment helped to create what miners saw as one of the most positive aspects of their working life: camaraderie. Historian Thomas Andrews has suggested that the harsh underground environment in Colorado's coal mines brought its miners together despite ethnic differences. While some might call this view environmental determinism, the situation in Butte matches Andrews's assessment.[48] In the early twentieth century, Butte's established migrant groups felt threatened by the seeming invasion of darker-skinned immigrants from southeastern Europe, partly because these new workers lacked the skills of longtime miners and hence affected workscape safety, and partly because newcomers sometimes acted as strikebreakers.[49] The danger of the workscape, however, slowly connected these groups. Miner Dennis Murphy told how a "Montenegro Yugoslavian" would embrace him underground, exclaiming "good kid son of a bitch—too bad you Irish."[50] The togetherness sometimes made its way into workplace hierarchy. Tensions could occur when the company promoted one miner, especially if he was younger, over men who had been working underground for years. Workers, however, were less likely to question the leadership of shift bosses who came from the ranks of the mine workforce, which meant that they understood the duties and troubles that the job entailed.[51]

The peculiarities of the underground workspace also resulted in freedom for many of Anaconda's laborers. Experienced workers became contract miners who were paid by the task, not by the hour like employees. They would either receive money for every cubic foot that they advanced in a drift or crosscut, or they would get paid by the amount of material they blasted and removed from a stope. Contract miners and the engineers who measured their progress sometimes bickered about the numbers, but, overall, miners liked the system, particularly because it meant little supervision.[52] Because engineers measured progress only after mining concluded, bosses stopped by only twice a day. Miners therefore "had to make a lot of decisions on their own," explained one engineer.[53]

Employees not on contract also had a fair amount of freedom simply because the miles of underground tunnel meant they worked so far away from other miners. Miner Louis Forsell thought that miners felt like they were "independent" on the job, just like "you don't have a boss."[54] On breaks, miners might turn off their headlamps and enjoy the complete darkness and silence. Because underground mining required so many workers to keep operations running, Butte miners also had a reputation for quitting at a moment's notice, only to rustle up a job at another mine the next day.[55] In their work, Butte's miners were so individualistic that writer Teresa Jordan called them the "cowboys of the underground."[56]

The independence that the workscape provided promoted the connection between masculinity and mining—a connection that had become normalized in the Americas and Europe by this time.[57] Butte miners saw themselves as paragons of masculinity due to the physicality of their work, which echoed the focus on male physique in the rise of team sports and bodybuilding at the turn of the century. The competitive nature of many miners' tasks also played into emerging ideas of masculinity. Most miners devoted a lot of energy into getting to the top of the contract board, which was the list of the most productive miners, so as to prove their mental and physical acuity. A good contract miner was multiskilled—not only an expert in drilling and explosives, but also in carpentry and even pipefitting.[58]

Miners also believed that they practiced great courage in performing their tough work under dangerous conditions. In 1939 one author described Butte's mining men as "no slaves of an assembly line: despite the great advance of metal mines technology since the war, their job is still personal and integrated." For each miner, he continued, "his is still a man's work, a man's contest with Nature jealous of her riches, a struggle against dust and heat and fire and gas and death."[59] Miners' thorough connections with each other, their job, and the dangerous earth thus made them true men. Employees who worked aboveground were called "softies" by underground miners, but both groups could still prove their worth by breadwinning for their families.[60]

## The Mining Landscape

As an industrial center that churned out copper, Butte seemed like a world apart to those in the predominantly rural surrounding region. Although some boosters marveled at Butte's sudden industrial power, most late-nineteenth- and early-twentieth-century descriptions of Butte added to its reputation as the "the ugliest town on earth," a label given by Helena resident Helen Fitzgerald Sanders.[61] Early mine operators used the process of open-heap roasting, which was a way to recover copper from the rock they dug out of the ground. Because roasting meant lighting stacks of ore-bearing rock on fire, noxious fumes wafted over Butte. Despite mine owners' assurances that the gasses were perfectly safe—even health inducing, according to copper baron William Clark—angry residents pushed operators to stop open-heap roasting. Mine owners moved their roasting operations to furnaces, trying a more effective means of conveying the smoke to stacks, intending to carry the smoke aloft. Residents again became upset at the smoke billowing out of new chimneys. These actions against smelter smoke show that some Butte residents agreed, at least to a degree, with outsiders' assessments that the area's environment was in need of care. Tired of complaints and looking for both more space and water resources, Butte's mining executives soon transferred all of their smelting to the Anaconda Copper Mining Company's smelter at Anaconda, about twenty-five miles west of the more heavily populated Butte.[62]

Saved from the stifling smoke, Butte residents were now able to avoid thinking about the environmentally nastiest parts of copper mining. By the 1920s outside reporters rejoiced over the greening of what they described as once treeless Butte. According to them, an area that allegedly could not grow anything green had experienced redemption.[63]

Of course, removing smelter smoke did not change every outsider's impression during the 1910s, 1920s, and 1930s. Even promotional literature admitted that Butte still contained "heat, dust, impure air, and microbes of nervousness."[64] It was also still full of immigrants who, some outside observers claimed, were slaves to duplicitous companies and union leaders. In their free time, workers supposedly engaged in only the seedy activities offered by mining towns, like gambling and prostitution. This view certainly showed up in the work of detective novelist Dashiell Hammett, who based his novel *Red Harvest* on Butte, nicknaming it Poisonville, while describing the grimy conditions of its environment and its labor relations.[65]

ENTERING BUTTE, MONTANA, 1939. Photograph by Arthur Rothstein. Photo courtesy of Library of Congress, the Farm Security Administration, Office of War Information Photograph Collection, Digital Id fsa 8a11191.

Butte's residents accepted what local Waino Nyland described as the turbulent, dirty, and barren nature of a mining area.[66] They held a more positive view of Butte's mining landscape and social environment, though, because those aspects, of course, meant employment. Their view was also not as rooted in the stereotypes of mining areas as outsiders' because Butte had transitioned from a mining camp to a more established city.[67] The company came to own all but one of the state's newspapers, so it did practice information management, but, despite its political power, Anaconda did not have as much control over the local populace as it would have had in the stereotypical company town. It did not control people's lives outside of work, including their important connections to religion—mostly Catholic in Butte—or their leisure. In between the two world wars, new leisure opportunities opened up for or were created by women, so the all-male mining town stereotype also failed to match reality.[68]

In other cases, Butte's residents accepted conditions common to mining communities. As the city that was said to be a mile high and a mile deep, residents thought of the underground mines as part of the city above, and not as a dangerous interloper. People lived by the rhythms of the mine whistle.[69] Work in the tunnels below Butte also left quite an impression on the people above. While in their homes, residents felt shock waves from blasting far below or experienced unusual noises.[70] Children played on mine waste dumps sprinkled throughout each neighborhood. Others swam in the copper waste ponds, full of acidic metals. Children might even climb into mine yards and abandoned shafts during their free time.[71] The stories told by residents in oral history interviews suggest the intimacy neighbors had with the mines.

Locals were thus often accepting of hazards, but to them the environment was more than just a mineral warehouse. Like residents of other industrial centers, many kept large gardens.[72] Others appreciated the nearby mountains. One miner came to work in Butte partly for the excellent fishing streams. Another explained that many took the day off simply to enjoy outdoor activities.[73]

## From Depression to War

Both residents and outsiders agreed on the downturn facing Butte's mining industry after World War I. Butte, which housed perhaps 85,000 during World War I, quickly lost population. By 1920 the city's population was around 40,000, the county's 60,000. Anaconda consolidated its power over the workforce by promoting interethnic rivalries in the 1920s.[74] As the Great Depression intensified, Butte suffered as much as, if not more than, the rest of Montana. Butte's residents became used to uncertainty. Every year, the total mining workforce might rise by thousands or fall by thousands, meaning that many of them became dependent on federal programs set up by Roosevelt's New Deal, especially the Works Progress Administration.[75]

Residents of 1930s and early 1940s Butte recognized how difficult the times were, though their oral histories tend to emphasize community togetherness. One major reason for this positive view came in the final piece of President Franklin Roosevelt's "Hundred Days" legislative session: the National Industrial Recovery Act, which gave companies the power to perform actions that previously would have fallen under antitrust laws. In exchange, those companies had to grant unionists a right to organize. In the summer of 1933, Anaconda's workers revived the old BMU, using a strike during the next spring to force the company to recognize its bargaining rights.[76]

On the back of this revival, a strong labor community reformed in Butte, closely tying workers and their families together. As part of the International Union of Mine, Mill, and Smelter Workers (IUMMSW; this was the child of the old WFM, and is also called Mine-Mill), the revived BMU became the leader among many other unions in Butte. As in other mining communities, the IUMMSW helped breadwinners and their families to gain a financial ladder upward.[77] By the end of the 1930s, the rate of home ownership by occupants had reached 75 percent, while the proportion of men to women became much more equal. Vibrant social activities in Butte's ethnic neighborhoods, in which most working-class residents lived, helped locals to make it through the Depression.[78] Women also played important roles in waged and unwaged work, helping to maintain the community, especially during strikes, when they often proved more militant than the men.[79]

World War II jumpstarted the sluggish mine industry in Butte. Once again the federal government considered Butte's minerals to be vital war matériel. Most men could gain draft deferment if they were employed in the essential copper industry, although Butte's surface employees, whose work sometimes required less training and usually garnered less-masculine status than those underground, did not receive industrial deferments. The company took advantage of this situation, claiming that every miner who did not work hard or fast enough "is guilty now of a traitorous act." Many men stayed to be a "soldier of production" in Butte, but others decided it was more patriotic to leave for war.[80] Better factory pay outside Butte led additional residents to relocate to the West Coast.[81] Some population loss coupled with a lack of attention to infrastructure led to one outside journalist's report that "whole neighborhoods are mouldy, whole streets are rotten and decaying."[82]

Needing more employees to feed the wartime boom, the Anaconda Company inadvertently challenged the culture created by area residents, which was dominated by white men with recent European roots. The big challenge came in 1942, when the company's workers in Butte refused to accept black men sent by the federal government to work in the mines. Despite pleading by the international union's leadership, eight thousand employees walked off the job in a wildcat strike. At the same time, the Montana war manpower director claimed that women would not be recruited for Butte's mines because they lacked physical prowess, but the Anaconda Company needed workers at the smelter town of Anaconda, so management manipulated employees. It claimed that the federal government would send Mexican Americans or African Americans unless the smeltermen accepted local women. Smeltermen gave in to the ploy, but only after they received assurances that women's labor would be temporary.[83]

When World War II ended, men returning from overseas and those still mining in Butte found that the Anaconda Company would offer only limited jobs and insisted on frozen wages. After contract negotiations failed, miners and their wives embarked on an unusually violent strike, which lasted only nine days, but included damage to property in which strikebreakers (scabs) lived.[84] World War II shook up Butte's social life, but no one forced radical changes. Anaconda still maintained its grip on the workforce and white male workers kept their elevated status, even if that status had taken some blows.

## Subsidence

Butte residents and their culture increasingly faced another challenge during the war: mine-induced subsidence. The Anaconda Company had to work harder to convince residents to be as accepting of subsidence as they had been of other hazards. Not since smelter smoke had the company encountered such a problem. Subsidence in Butte comes in two types. The first, chimney caving, happens when mine shafts that have either been partially filled or bulk-headed (covered over with timbers near the surface) suddenly open up, or when the ground caves in on underground workings in a small area. Caving looks like either a shallow depression or a hole that appears suddenly. Old, rotting timbers covering shafts or holding up mine workings that gave way were often the cause. In Butte, old shafts sometimes opened up when a person walked by. Cars slid into openings. Big depressions sank in backyards.[85]

The second kind of subsidence was more troubling. Trough subsidence involves a much larger area over collapsing mine tunnels. The whole area slides as the overburden becomes depressed and compressed. The result is often substantial horizontal movement along with vertical movement, like books on a shelf falling away from each other. The result was house walls pulling apart from each other, and not just sinking downwards. Trough subsidence occurred over the Emma and Travona Mine workings in central Butte, an older, working-class neighborhood to the south of the central business district (CBD). The Butte Copper and Zinc Company had joined with Anaconda to mine manganese under central Butte in 1915. During the 1940s and 1950s manganese became a vital war matériel, and was often used in the production of firearms; in a mad dash for minerals, workers tunneled faster and closer to the surface.[86]

The damage was extensive. Widespread settling of the neighborhood led to structural damage in many of the homes, while it also cracked the sidewalks, shifted the sewers, and broke water and gas lines. The area over which settling occurred eventually equaled 3,600 feet east to west, and 4,000 feet north to south. The ground on Gold, Platinum, Montana, and Colorado Streets sank up to twelve feet.[87] Settling lessened a great deal after 1960, but Anaconda still received subsidence complaints on 1,576 properties in the Butte area, mostly in central Butte.[88]

Anaconda settled approximately 45 percent of the complaints it received about holes, shafts, and subsidence. It likely settled with owners of 712 properties, either with money or in-kind. One major form of settlement simply involved filling up holes or depressions, an in-kind settlement that involved no admission of guilt or money attached. It settled 184 complaints in this way.[89] Anaconda also spent a great deal of money settling subsidence claims. The damage was obvious. As Chief Mining Engineer William A. O'Kelly put it, when the ground was shifting by almost a foot every four months, the "deformation of the surface is plainly visible."[90] It was hard for them to deny complaints outright.

By 1943 Anaconda had developed a method to deal with complaints. Anaconda's attorneys worked hard to settle cases outside court "rather than give the plaintiff's attorneys access to the damaging information" about such a large region.[91] To combat residents who wanted to pool their claims, Anaconda dealt with each case individually. In communications with the Legal Department, O'Kelly explained that Anaconda would be fine "admitting some liability in connection with the plane of movement and tilting of the surface, [but] it would not be advisable to discuss with the claimant the structural features of the movement, which might have widespread application."[92] Meanwhile, the company stopped stoping (removing ore) under the thickly populated district, and it began to fill in the drifts (horizontal tunnels) with tailings and wet sand.[93]

The company's one-by-one complaint system primarily involved its engineers—both civil and mining—the Legal Department, the Abstract Department, and sometimes the Land Department. When the company received a complaint, it would first send a team of engineers to the location. Usually a civil engineer and a mining engineer would meet with the owner, then look over the house or other building, and the nearby ground. After examining its underground mine maps, Anaconda's engineers sent a report to Legal. Legal would then discuss the report with the Mining Engineering Department, determine the liability, and negotiate a settlement, if need be. In the meantime, Abstract would investigate the property chain. Then contractors or realtors would appraise the building in question, and determine its worth and the cost to repair the damage.[94]

When Anaconda's Legal Department believed a money settlement was needed, which it did in 521 cases, it negotiated for a perpetual release and easement. In each case, the company would pay for the damage it had caused to the property, giving it a release from any liability for mining damage. It also gained an easement at the same time on the property, giving it "the perpetual right to extract ores and minerals from beneath the property, without incurring any liability to support the surface of said premises."[95] A perpetual release meant that no matter what kind of claim anyone made about mining damage in the future, the company could not be held liable. In one 1972 case, resident Despina Famos argued that she wanted a settlement over blasting damage from open-pit mining. Famos was surprised to learn an earlier subsidence settlement had already been made on her property in the

1940s. The previous owners' perpetual release and easement meant that the company would never have to settle any new damage claim, no matter the type.[96] Settling implied some admission of liability, but the company received a good deal by doing so.[97]

Nevertheless, the cost of these settlements was substantial. Even toward the end of the Emma's operations, in June 1959, Anaconda still faced $6,500 in lawsuits, $227,650 in active claims, $175,800 in inactive claims, and $115,000 in potential claims, for a grand total of $523,950.[98] Anaconda eventually made settlements on 28 percent of the structures in central Butte. On November 28, 1959, Anaconda bought out its partner in the Emma and Travona Mines, Butte Copper and Zinc Company, and shut down the Emma and its pumps, flooding the tunnels with water.[99]

Anaconda had essentially paid its way out of the situation, meaning that most Butte residents, even those living in sinking areas, became fairly accepting of subsidence, just as they had with previous industrial hazards. As with the smelter smoke controversy, locals believed that Anaconda had dealt fairly with their problems, even if many of the hazards were simply later faced by others, whether inhabitants of the town of Anaconda in the case of smelter smoke, or later residents of subsiding homes in the case of central Butte.

## Mass Mining

Apart from the subsidence issue, Butte residents experienced positive changes after World War II. "Rapid growth in the consumer economy [and] Cold War military demands" in the postwar period meant soaring demand for copper: Anaconda responded with renewed efforts in Butte.[100] The revitalized economic productivity of the mines seemed to revive the spirit of the town. Residents' oral histories and local news articles from the late 1940s and early 1950s reflect this optimism.[101]

Outside journalists agreed this was a new and better era for Butte. Progress in the mining landscape signified moral progress. Dorothy Kilgallen, for example, wrote in *Good Housekeeping* that, although Butte was not a "pretty town," it had become more refined and gained a "new found virtue." The company, residents, and city government had become increasingly willing to invest money into the town.[102] Edmund

Christophers reported a similar tale of renewed boom in the *Saturday Evening Post*. To Christophers, mine dumps and head frames meant that "it's impossible to escape the signs of the digging and refining that have kept this overgrown copper-mining camp alive," yet new Anaconda operations also meant replacing some of the dilapidated and run-down housing with new construction, especially in the beautiful suburban Flats below the town. A new community-wide group called the Butte Development Association dedicated itself to civic improvement and the building of many new homes.[103] Despite outsiders being a bit more concerned than locals with Butte's bawdy mining camp past, both groups spoke the same economic language of boom, just as they had with bust.

Residents were particularly positive in the immediate postwar period because Anaconda, riding high from wartime profits, planned to reinvest heavily in Butte. To do so, its leaders turned to new mining methods. At that point, management could rely on a half-century's efforts by engineers, metallurgists, and management to move hard-rock mining away from a reliance on high-grade ore and a large, highly skilled workforce. The best options appeared to be open-pit mining and underground block-caving.

Mining through the use of surface pits was both new and quite old. Mining on many, if not most, ranges throughout the United States and elsewhere began from the surface in small pits. Only when stripping layer after layer of ground became too costly and difficult did mining move underground. In early copper-mining regions, like Michigan's Upper Peninsula, American Indians had dug small copper pits to get at native copper.[104] European Americans replicated those practices on a larger scale when they began mining in metal regions of the Midwest during the nineteenth century. Open-pits opened on the Marquette range, now located on the Upper Peninsula, in 1854, only a decade after iron mining had begun in the Lake Superior basin. After breaking up the rock, miners pulled ore out in wheelbarrows and wagons.[105]

Out West, early exploitation of the massive silver deposit known as the Comstock Lode began with small surface pits, known as glory holes, in 1859. The Comstock's first big mine, the Ophir Pit, helped to make Virginia City, Nevada, one of the first major mining towns in the

West. In some ways, digging a pit was essentially an extension of the placer mining performed during early gold rushes: Placer deposits of gold were located in sands deposited by rivers. Since they were near the surface, miners typically would dig into sandbars a number of feet to release more gravel. Then they used rockers, which were wooden boxes, to wash sand away from the gold.[106]

None of the aforementioned methods could solve the economic puzzle that faced the mining industry in the late-nineteenth century, however. Many metal deposits across the country were often so diffuse that they had never been touched, no matter how massive they were. Mining and metallurgical engineers began to experiment with economy of scale as a way to lower production costs enough to make big deposits feasible. Improvements in drills made it easier to sample the ground. A large mine could now be mapped ahead of time and miners knew better where to place explosives. Growth in the explosives industry, which made more-reliable, more-powerful explosives available by the 1880s, also helped to make all types of mining more efficient (and sometimes more dangerous for workers).[107]

Attempts to go beyond simple human labor for the purposes of loading rock from the mines brought another important step toward open-pit mining. Otis Young's steam shovel, invented in the 1830s, became a key earthmover for railroad companies that sought to dig through hills and to perform grading work. The steam shovel found a home in the construction industry at the same time that some innovative mine operators were trying to increase mining's speed. Unlike hard-rock mining, where underground work predominated, surface strip mining was more common in coal, and so two coal operators in Kansas were the first in the industry to use a steam shovel.[108]

After their 1877 breakthrough, metal mines were not far behind. Captain Joseph Sellwood transported a steam shovel to the Gogebic iron range, which runs from Wisconsin into Michigan's peninsula, in 1884. By the end of the 1890s, steam shovels were becoming common in rapidly expanding activity on the Mesabi iron range in Minnesota. The shovels ran on rail tracks, as did the cars that men used to move, or tram, material out of the mine. Open-pit methods required large iron companies with a lot of financing, because machines like steam shovels were major investments and additional land was necessary for dumping the overburden, or non-ore-bearing rock, left over from stripping.[109]

Industry leaders did not believe that open-pit mining was the answer to all questions about efficiency. The Lake Superior Region became a hotbed for mass methods. Just as some managers began to create surface pits on a larger scale, others, in Michigan's Pewabic Mine, began a new underground mining method known as block-caving in 1895. For deposits that were too deep for a pit, but that could be mined economically on a mass scale, engineers would repeatedly define a large ore body underground, undercut it free of the waste rock, and then let gravity and rock stresses pull it to the floor of a tunnel. Miners would then haul the crushed material to a shaft, where elevators could lift everything out of the mine.[110]

Both forms of mass mining made a lot of sense to progressive-era businessmen. As a later report put it, block-caving grew in popularity because, like open-pit mining, it was "an efficient and low-cost mining method permitting scientific planning, centralized management, labor specialization, and modern mechanization."[111] The emphasis on scientific management, expert planning, and efficiency made these mines especially modern, fitting well into turn-of-the-century thought.[112]

These new nonselective mining techniques tantalized copper companies but, in order to make such methods pay, metallurgists had to figure out how to better separate ore from waste *after* its extraction from the ground. Both open-pit mining and block-caving brought far more waste out of the ground than typical underground methods, in which skilled miners tried to take out only ore, with as little waste as possible. Engineers started to consider the concentration process as a way to separate ore from waste *after* extraction. Mills originally crushed ore-bearing rock, then used water and gravity to separate less-dense from more-dense particles, hence concentrating the ore. Unfortunately, much of the ore in this process was lost into a sludgy mess called slimes. James Colquhoun was the first to create a more successful concentration process for porphyry copper ores; the term *porphyry copper ores* applied to widely disseminated, but massive, deposits of low-grade ore. Other engineers pioneered an even better method, known as flotation, which used the selective attraction of oils to metals as a way to increase concentration. Improvements in the concentration process therefore allowed miners to be far less selective in mining. It could now be economical to dig up very low-grade ore.[113]

Anaconda considered both block-caving and open-pit mining during this period. In the fall of 1900 Henry H. Rogers, part of the new company ownership group, asked an experienced mining engineer in his employ, David W. Brunton, to travel to Minnesota's iron mines. Rogers knew that iron mines in the area had been using both caving and the new open-pit methods. He wanted Brunton to see if Butte could similarly be made more economical. Brunton spent some time meeting mine superintendents and examining Minnesota's new extraction practices, and eventually decided that open-pit mining could not work for Butte. As he wrote, "The mode of the occurrence of the ore bodies and the numerous surface improvements at Butte entirely prohibited the use of the more economical methods of iron mining." Looking back on this decision in 1915, Brunton seemed sad that he could not put Rogers's ideas into action, because such "far-sighted views on the subject were entirely correct." He noted that surface methods "have since come into general use in many of the large copper mines in the United States, Chile, and Spain," even if they could not work in Butte.[114]

Instead of Butte, the major developments toward nonselective copper mining came together in Bingham Canyon, Utah, on a claim that had supposedly been offered to both William Andrews Clark and Anaconda Company leaders. At about the same time that Rogers and Brunton were examining the iron range, Daniel Jackling and Robert Gemmel were developing a comprehensive analysis of Bingham, home to one of the world's great porphyry copper deposits. In 1899 they coauthored an audacious plan to mine a mountain's worth of low-grade copper using steam shovels, trains, and an enormous concentrating plant. Entrepreneurs and investors balked at the idea that an area with only 2 percent copper could be made profitable, so the plan was abandoned for a number of years.[115]

By the early 1900s an important part of Jackling's proposal, steam shovels, had come west via Rockefeller's Colorado Fuel and Iron Company (CF&I). CF&I was using them to dig iron ore at Sunrise, Wyoming.

Samuel Newhouse, who ran a mine above Jackling's favored spot, examined CF&I's operation and determined that steam shovels could make disseminated copper profitable. His first attempt, in February 1906, was unsuccessful because the rock was so hard, but his second, later that spring, did quite well because he had moved down into Bingham Canyon where the ore was softer, and because he made better use of explosives to loosen material first. He quickly reduced his mining costs to 10 cents a ton, far below the previous record in copper mining of 37.5 cents per ton.[116]

In the meantime, Daniel Jackling's Utah Copper Company had gained financial backing from the Guggenheims. Jackling joined up with Gemmel again, and they revisited their old plan to make low-grade copper pay. They both visited the Mesabi iron mines and hired a Minnesota man, J. D. Shilling, as their mine superintendent. Back in Utah, Jackling and Gemmel, already embarrassed by Newhouse's success, began their Bingham Canyon Mine that same spring, in 1906. Mining historian Richard Lingenfelter has recently argued that Jackling, long lauded as the Father of the Porphyrys, was often a very poor, and sometimes dishonest, financial manager of the property, and that he invested far too much in a failed attempt to block-cave, but his big surface mine was incredibly productive.[117] In 1910 the Utah Copper Company's Bingham Pit produced nearly 85 million pounds of copper, good enough to make it the third-largest copper producer in the nation, behind only Anaconda and Phelps Dodge. By the mid-1920s, it had eclipsed these early numbers with annual production figures in the $200 millions.[118]

Lingenfelter claims that Jackling did not gain respect during early Bingham operations partly because he brought in foreign workers from around the world and gave them only "starvation wages."[119] Jackling, however, was doing exactly what engineers and mine managers had been trying to do for decades. In an era in which order and efficiency were the call words, one point of mass mining was to decrease reliance on a large skilled workforce. Scientific management of the workforce, as progressive-era professionals would put it, was thus important to mass mining's growth. In both block-caving and open-pit mining, engineers and managers could train workers for specific tasks, instead of the long-standing system of relying on a jack-of-all-trades miner who could locate and then deliberately, yet dangerously, extract high-grade ore.[120]

Two writers on this subject, Logan Hovis and Jeremy Mouat, argue that mass methods like open-pit mining therefore meant more than just the introduction of machinery: they also meant the "abandonment of low-volume, high-value, selective mining techniques and the adoption of higher-volume, nonselective methods that emphasized the quantity rather than the quality of the ore brought to the surface."[121] Focusing on quantity meant that speed was a key element of mass mining, because it was important to dig up as much as possible in as little time as possible. The enormous scale and speed at which men and machines ripped earth into tiny pieces is one reason why historian Timothy LeCain calls open-pit work mass destruction and not mass production. These methods could be considered very efficient largely because energy sources for big machines were so cheap in twentieth-century America. There is also some irony to calling mass mining "efficient" because it produced such massive amounts of waste, but land for dumping waste seemed unlimited to early mass mining practitioners.[122]

Following the maxims of speed and power, new copper porphyry mines sprang up around the country. John Murchison Sully conceived of mass mining at around the same time as Jackling began his operations. After observing iron mining in Alabama, he proposed a steam-shovel-driven porphyry copper mine in 1906, which became the Chino Mine in Santa Rita, New Mexico, a few years later. Other porphyry copper mines throughout the West followed, especially in Arizona, which soon overtook Butte as the largest copper-producing region in the United States. In the 1910s Americans funded by the Guggenheims founded Chuquicamata, which was the largest of the many rich porphyries in Chile. Anaconda later purchased the property, which featured low labor costs (in comparison to the United States) and fairly high-grade ore for a porphyry mine. By the early 1920s workers were producing copper at an incredibly low 6.2 cents per pound, far below the previous cost of 17 cents per pound just a few years earlier.[123]

Block-caving also moved west and south. Familiar with block-caving methods in Michigan, superintendent Felix McDonald tried caving at the Ohio Copper Company mine that he was running in Bingham, Utah, in 1906. He later used a similar system at the Inspiration Mine in Arizona. Others adopted block-caving in Nevada, in much of the Rockies, and far south, in Chilean copper mines. Eugene Delos Gardner, supervising engineer for the U.S. Bureau of Mines' Southwest Experiment Station, explained that, when engineers encountered a large, low-grade copper mass that could either be a block cave or an open pit, "the thickness of the capping is the deciding factor." If there was a lot of non-ore-bearing rock (called capping or overburden) on top of the ore body, then engineers chose to block-cave. If there was little capping, which tended to be more common, engineers preferred the open pit. Mass mining was therefore the key idea, and it did not simply mean big pits.[124]

Copper thus benefitted from what historian Jeffrey T. Manuel calls a "close collaboration" with iron mining, but, in the early days, this collaboration looked mostly like copper borrowing from iron, and the borrowing went far beyond Jackling.[125] Similar to Jackling, Gemmel, Newhouse, Sully, and McDonald, the operators of open-pit or block-caving copper mines often examined, trained, or had already worked in iron first. John C. Greenway, for example, helped to open the western end of the Mesabi iron range. He left Minnesota in 1910 to become general manager of the Calumet and Arizona Copper Company. Looking for a "mine where he could utilize his Minnesota experience and open a pit with steam shovels," in one author's words, he began a major open-pit mine at Ajo, Arizona.[126]

## Mass Mining Comes to Butte

Mining in Butte began with big veins of copper sulfides rather than porphyry copper deposits. Rich ore built the Anaconda Company into a giant, while supporting a large, though fluctuating, group of workers. As mining continued, however, many big veins eventually played out, forcing miners to dig deeper and deeper. The problem was, as one expert later put it, "The deeper you went the more disseminated the veins got."[127] Anaconda's management began to worry that Butte's days as a major mining center would soon be over.

Technological innovation was often the first and most important step taken by mining regions trying to stave off decline.[128] Anaconda president Cornelius Francis "Con" Kelley thus introduced the Greater Butte Project in 1947 as a way to guarantee continued mining in Butte and hence his hometown's future. The project's primary focus was a highly publicized block-caving operation, the first nonselective operation in the area. Anaconda geologists had found that, unlike the western part of the Butte hill where big veins of ore were widely spaced—one hundred to several hundred feet apart—the eastern part of the hill contained small veins that were very close to each other. Smaller veins would make mass-mining economically viable. Clyde E. Weed, Anaconda's vice president in charge of mining operations, had worked at Arizona's Inspiration Mine when some of the first innovations in block-caving were developed. Based on this early experience, he and his assistant, Richard S. Newlin, originated the idea of block-caving in Butte.[129]

Kelley bought into this suggestion, telling the people of Butte that the Greater Butte Project "is the most conclusive answer that can be given of Anaconda's faith in Butte and its confidence in the future."[130] A man with both a mining and a legal background, Kelley first made his name with Anaconda as its representation in Apex Law and smelter smoke litigation. He served as President John D. Ryan's lieutenant before later coming into the presidency himself in 1918.[131] In introducing the Greater Butte Project to the city's chamber of commerce, he was effusive: "It is therefore of the utmost importance that the feeling of insecurity—a natural heritage from the failure of so many mining camps in the past—be banished and, instead, a conviction of faith in the future and a courage, justified by demonstrable facts, be instilled that will lead this community on to its destiny." The mining of low-grade ore, he explained, meant that Butte still had a natural treasure below. Surely, he proclaimed, the project would make the city "a more desirable and a better place in which to live—Here's to the future of a greater Butte!"[132]

PORTRAIT OF CORNELIUS F. KELLEY, 1935. Kelley was president of the Anaconda Company when it undertook the Greater Butte Project. Photograph by Kaiden Kazanjian Studios. Photo courtesy of the Montana Historical Society, Catalog #PAc 75-28.M4.

By investing in the Greater Butte Project, Con Kelley was perhaps the father of Butte's open-pit mining because it developed out of block-caving, but the Geology Department, particularly Vincent "Vin" Perry and Ed "Eddie" Shea, should likely get more of the credit. Both Perry and Shea had long careers with Anaconda before their involvement with open-pit work in Butte. Upon his graduation with a master's degree in geology from Columbia University, Perry interviewed with the Anaconda Company's chief geologist, Reno Sales, whose work during litigation cases against Anaconda helped to make the Geology Department more important than it had previously been in mining companies. No mine development work could proceed without its recommendation.[133] Perry joined the Butte department in 1924. The next years took him south, setting up a Butte-style Geology Department at Cananea in Mexico and exploring the Sonora for metals, before he ended up as the head of western U.S. exploration in 1937. As one of

Sales's favored students and his confidant, Perry became chief geologist upon his mentor's retirement in 1948. Shea had also worked in Anaconda's Geology Department for a long time, learning from Sales and Perry. By the early 1950s, Shea had worked his way up to the position of chief geologist for Montana operations.[134]

It fell to both Perry and Shea to make a bigger picture out of Butte exploration data. Shea, Sales, and Charles "Chuck" Meyer, another of Sales's mentees and later a professor of geology at the University of California Berkeley, had all long recognized the existence of widely disseminated copper, but it was not until the Greater Butte Project that bulk mining techniques were considered.[135] Aiming to expand that project, the Geology Department initiated a program in 1949 to take large samples of the mineralization between and within as many of Butte's old east-side mines as possible. Many of the upper mines had become inaccessible, but the department's leaders were able to combine data from old geological records, churn drill samples from above ground, and take mineral samples from old crosscuts and drifts. According to the department, many remnants and ore veinlets had been left behind during old operations. There was also a thick, but disseminated, blanket of copper-bearing material that had been deposited by a process known as secondary enrichment, just to the east of the new block-caving operation. Butte had never been a porphyry copper body, but as Anaconda executive Robert P. Corbett explained, "Because of the high grade veinlets and everything it has the same effect of a continuously homogenous low grade deposit."[136]

Members of Anaconda's Geology Department began to think that maybe Brunton had ultimately been wrong back in the early 1900s. Open-pit methods developed for porphyry deposits might, almost magically, turn leftover waste rock into lucrative ore. Volume could now make up for Butte's declining ore grade.[137] Besides, earthmoving technology had greatly improved since the early 1900s. Electricity or diesel fuel now ran shovels that previously used steam. Tractors developed for the lumber, road-building, and construction industries had also increased in power and maneuverability. World War II brought more profits to companies that made earthmovers, as the war demanded the creation of landing strips and temporary bases across Europe and, especially, the South Pacific. Companies like Caterpillar, LeTourneau, and International Harvester had created new machines and large trucks for the U.S. military; now these vehicles could serve industry.[138]

By the 1950s the Anaconda Company also had more experience with mass mining. Its talented geologists and engineers had developed and managed open-pit mines around the Americas, including Chuquicamata, Chile, which served as the training ground for Butte's first pit superintendent and its first pit engineer. Anaconda's experience in Yerrington, Nevada, also showed the way. The company began mining a small low-grade copper deposit there because big haul trucks suddenly made such operations fast and flexible.[139]

A thoughtful geologist, Vin Perry was not necessarily a fan of open-pit mining in every case. In the early 1950s he actually argued that Anaconda should shift its Chuquicamata operation from open-pit to underground block-caving, but he was overruled.[140] On the other hand, he was vitally interested in continued Butte operations. Like his mentor, he loved Butte. When he transferred to the head office in New York, he even asked Sales to send Kodachrome prints of Butte back to him so that he could blow up a picture to put on his wall.[141]

Perry therefore did not need too much pushing to get him to focus on Butte, but it was actually his Montana associate, Edward Shea, who made the first official suggestion that Anaconda go into open-pit mining in Butte. In January 1952 Shea wrote a letter to Edward S. McGlone, who ran Montana operations and who would become executive vice president that year. He clearly laid out the collected data, arguing that "serious consideration should be given to possible open pit operations on the Butte Hill."[142] Perry agreed that a "major change" was necessary.[143] As chief geologist, he convinced the rest of Anaconda's top management to move forward with open-pit mining by arguing that, without doing so, Butte's operations might die out. He claimed, "Butte has been a high-grade, selective mining operation for 75 years, during which time mounting labor and supply costs plus the inevitable deeper level engineering problems have encroached inexorably on the profit margin.... Fortunately, there is strong evidence that the problem can be solved." He suggested that Anaconda needed to go beyond the Greater Butte Project, which he called only "one step in the right direction." He also went farther than Shea, envisioning a massive open pit that would produce more tonnage than the underground.[144] As both the visionary and the person who made the pit happen, Perry is correct in later recollecting, "In 1952 I initiated a plan to change Butte from underground to open pit mining."[145] By 1953 engineers had initiated tests to determine the viability of a pit and Shea became the lead in the geologic part of its design.[146]

Anaconda appreciated the work that Shea and Perry were doing. Shea continued with the Geology Department, serving as head of Montana geology until his retirement in February of 1967.[147] Just two years after the Berkeley Pit opened, Perry became a company vice president. Professional accolades followed. In 1961 Perry was lauded by the American Institute of Metallurgical and Petroleum Engineers and the Society for Mining, Metallurgy, and Exploration, especially for his leadership in "the finding and developing of mines." They gave Perry their top medal for "significant contributions to technical progress," known fittingly as the Daniel C. Jackling Award.[148]

Due partly to Perry's work, Butte would continue for many years as a major copper producer, but Butte itself would never be the same. Just as Anaconda was putting a test pit into action in 1955, Perry claimed that further examination might show that "under conditions of large volume, low-cost production, open pit and glory hole operations may be extended on a much larger scale than planned now." Somewhat ominously, he suggested to management that "the possibilities for future pit ore, both laterally and at depth are so impressive" that Anaconda needed to contemplate the "solution of related operational concerns," including the "acquisition of additional surface property rights" for the mine itself and the waste from the mine.[149]

Perry foresaw something that Butte residents would never have guessed—mass mining would soon encounter people's neighborhoods—but he did not foresee that Butte's mass mining would also upend workers' lives and test residents' resilience. By moving smelter smoke out of the city, Butte residents had managed to avoid the worst effects of mine development; open-pit mining meant that industrial work would again intrude into people's lives at home.

## Notes

1. Rodman Wilson Paul, *Mining Frontiers of the Far West, 1848–1880*, ed. Elliott West (Albuquerque, NM: University of New Mexico Press, 2001, original 1963), 11; Frederick Jackson Turner, "The Significance of the Frontier in American Life," in *Frederick Jackson Turner: Wisconsin's Historian of the Frontier*, ed. Martin Ridge (Madison: State Historical Society of Wisconsin, 1986), 26–47.

2. Kent Curtis, "Producing a Gold Rush: National Ambitions and the Northern Rocky Mountains, 1853–1863," *Western Historical Quarterly* 40, no. 3 (2009): 275–97, population on 284.

3. Curtis, "Producing a Gold Rush"; Susan Lee Johnson, *Roaring Camp: The Social World of the California Gold Rush* (New York: W. W. Norton, 2000); Kent Curtis, "An Ecology of Industry: Mining and Nature in Western Montana" (PhD diss., University of Kansas, Lawrence, 2001), 61–67.

4. Michael P. Malone, Richard B. Roeder, and William L. Lang, *Montana: A History of Two Centuries* (Seattle: University of Washington, 1991), 68–71, 92–97, population on 68. Clark C. Spence, *Territorial Politics and Government in Montana 1864–89* (Urbana: University of Illinois Press, 1975), 4–19.

5. Malone, *Battle for Butte*, 3–10; Isaac F. Marcosson, *Anaconda* (New York: Dodd, Mead, 1957), 18–21.

6. Malone, *Battle for Butte*, 10–18; Curtis, "An Ecology of Industry," 107–9, 115–16, 126–27.

7. Malone, *Battle for Butte*, 12–24, 82–130; Emmons, *Butte Irish*, 99–102, 205–9; David M. Emmons, "The Orange and the Green in Montana: A Reconsideration of the Origins of the Clark-Daly Feud," *Arizona and the West* 28 (1986): 225–45.

8. Marcosson, *Anaconda*, 33–49; C. B. Glasscock, *The War of the Copper Kings* (New York: Grosset and Dunlap, 1935), 45–46; Works Progress Administration, *Copper Camp: Stories of the World's Greatest Mining Town, Butte, Montana* (New York: Hastings House, 1943), 31–47.

9. Kent A. Curtis, *Gambling on Ore: The Nature of Metal Mining in the United States, 1860–1910* (Boulder: University Press of Colorado, 2013), 127–43.

10. LeCain, *Mass Destruction*, 26–30; Malone, *Battle for Butte*, 34–39.

11. Emmons, *Butte Irish*, 13, 20–21, 190–94; Malone, Roeder, and Lang, *Montana*, 202–11; K. Ross Toole, "Marcus Daly: A Study of Business in Politics" (master's thesis, University of Montana, Missoula, 1948).

12. Dave Walter, "Montana's Venomous 'Capital Fight,' 1894," *Montana Magazine* 124 (Mar.–Apr. 1994): 38–45; Brian Leech, "Corporate Tyranny and Foreign Labor: Populism, Nativism, and Montana's 1894 State Capital Election" (master's thesis, University of Wisconsin, Madison, 2007); Dennis L. Swibold, *Copper Chorus: Mining, Politics, and the Montana Press, 1889–1959* (Helena: Montana Historical Society Press, 2006), 39–59; Emmons, "The Orange and the Green," 225–45; K. Ross Toole, "A History of the Anaconda Copper Mining Company: A Study in the Relationships between a State and Its People and a Corporation, 1880–1950" (PhD diss., University of California, Los Angeles, 1954); Forrest L. Foor, "The Senatorial Aspirations of William A. Clark, 1898–1901: A Study in Montana Politics" (PhD diss., University of California, Berkeley, 1941).

13. Malone, *Battle for Butte*, 141–89; Works Progress Administration, *Copper Camp*, 31–47.

14. Malone, *Battle for Butte*, 182–200; Swibold, *Copper Chorus*, 74–87; Eric Nystrom, "Underground Mine Maps and the Development of the Butte System at the Turn of the Twentieth Century," *IA: The Journal of the Society for Industrial Archaeology* 37, no. 1–2 (2011): 97–113.

15. F. Ernest Richter, "The Amalgamated Copper Company: A Closed Chapter in Corporation Finance," *Quarterly Journal of Economics* 30, no. 2 (Feb. 1916): 387–407; Richard E. Lingenfelter, *Bonanzas and Borrascas: Copper Kings and Stock Frenzies, 1885–1918* (Norman: Arthur H. Clark Company, University of Oklahoma Press, 2012), 70–92; Malone, *Battle for Butte*, 42–45, 131–89.

16. Jerry W. Calvert, *The Gibraltar: Socialism and Labor in Butte, Montana, 1895–1920* (Helena: Montana Historical Society Press, 1988); Emmons, *Butte Irish*, 340–97.

17. Michael Punke, *Fire and Brimstone: The North Butte Mining Disaster of 1917* (New York: Hyperion, 2006); Clemens P. Work, *Darkest Before Dawn: Sedition and Free Speech in the American West* (Albuquerque: University of New Mexico Press, 2005); Emmons, *Butte Irish*, 255–91, 340–97.

18. Clarence Adami Wendel, "Mining in Butte 40 Years Ago," *Montana: The Magazine of Western History* 31, no. 3 (Summer, 1981): 60–66, quote on 60.

19. Matthew Basso and William Allen Burke, "Another Look at Burke's Butte: The Great Depression and William Allen Burke's 'Greenhorn Miner,'" *Montana: The Magazine of Western History* 56, no. 4 (Winter 2006): 18–31, 97–98; Wendel, "Mining in Butte," 60–61; *Anaconda Copper Mining Company: Its Properties and Operations in Montana*, n.d., 4–7, VF 1390, BSBPA; Dominick Bontempo, interview by Judy Cusick, Nov. 26, 1982, Butte, MT, transcript, Oral History Collection (hereafter OHC), BSBPA; Leonard Hendrickson, interview by Ronald D. Hendrickson, Apr. 17, 2005, Butte, MT, transcript, OHC, BSBPA; J. J. Carrigan, "Mining in the Butte District," Talk Before the Butte Rotary Club, Mar. 16, 1939, in Butte Chamber of Commerce, *Seven Talks about Mines*, n.d. (1939–40), 20–30, VF 1423, BSBPA.

20. Wendel, "Mining in Butte," 63–64; Basso and Burke, "Another Look at Burke's Butte," 25–27; Carrigan, "Mining in the Butte District," 21–23; Frank M. Monninger testimony, Nov. 23, 1974, U.S. Senate Subcommittee on Minerals, Materials, and Fuels of the Committee on Interior and Insular Affairs, *Closure of Underground Copper Mines: Hearing to Review the Anaconda Company's Plans to Close Its Underground Copper Mines in Butte, Montana*, 93rd Cong., 2nd sess., Nov. 25, 1974, 30; John Mihelich, "The Richest Hill on Earth: An Ethnographic Account of Industrial Capitalism, Religion, and Community" (PhD diss, Washington State University, 1999), 118–27.

21. Wendel, "Mining in Butte," 61–62, 64; Basso and Burke, "Another Look at Burke's Butte," 25–27; *Anaconda Copper Mining Company*, 4–7, VF 1390, BSBPA; Bontempo interview; Mihelich, "The Richest Hill on Earth," 118–27; "The Anode in Retrospection," *Anode* (Butte, MT) 21, no. 1 (Jan. 1935): 1–2.

22. Monninger testimony, Nov. 23, 1974, 29.

23. Martin K. Hannifan, "Merits of Open Pit Vs. Underground Mining in Butte," *Mining Congress Journal* 58, no. 4 (Apr. 1972): 119–23; Edward R. Borcherdt, "Improvements in Mining Practice in the Butte District," *Mining Technology* 5, Technical

Papers 1339 (July 1941): 1–11; L. F. Bishop, "Vertical Slice and Slot Stoping at Butte," *Transactions of the American Institute of Mining and Metallurgical Engineers: Mining Practice* 5 (1945): 262–73; Monninger testimony, 29–30; Carrigan, "Mining in the Butte District," 22–24.

24. Anaconda Company Bureau of Safety, *Anaconda Safety Rules: Mining Operations*, n.d. (1956?), Fldr. 1, Box 1, Butte Miners Union #1/United Steelworkers Labor History Collection 12 (hereafter LH12), BSBPA, 12–16. See also Frank M. Monninger, "The Three Greatest Causes of Accidents in Mines," *Anode* 21, no. 4 (Oct. 1935): 1–2.

25. Wendel, "Mining in Butte," 64–66; Basso and Burke, "Another Look at Burke's Butte," 27–31; *Anaconda Copper Mining Company*, 4–7, VF 1390, BSBPA; Bontempo interview.

26. On envirotechnical landscapes, see Martin Reuss and Stephen H. Cutcliffe, eds., *The Illusory Boundary: Environment and Technology in History* (Charlottesville: University of Virginia Press, 2010). Applying that concept to underground mines is LeCain, *Mass Destruction*, 24–63, 129–37. Use of the term *workscape* for underground mines is in Andrews, *Killing for Coal*, 125. See also Jonathan S. Franklin, "Working-Class Themes in the New Environmental History," *Labor: Studies in Working-Class History of the Americas* 9, no. 1 (Spring 2012): 119–31.

27. "She's Talking to You," *Anode*, Apr. 1936, 7; *Mining World*, July 19, 1916.

28. George Tarrant, interview by Adam Lawrence, Apr. 9, 2002, Butte, MT, transcript, OHC, BSBPA. See also Hendrickson interview.

29. On the underground climate, see "Digging Ore . . . a Mile Below Surface," *Anaconda Company Trailsman* (Butte, MT) 7, no. 1 (June 1965): 3, VF "Anaconda Company Trailsman," Montana Historical Society Library and Archives, Helena, MT (hereafter MHS); Bontempo interview; Al Beavis, interview by Teresa Jordan, Dec. 3, 1985, Walkerville, MT, partial transcript and notes, OHC, BSBPA; Art Braach, interview by Teresa Jordan, July 13, 1984, Butte, MT, partial transcript, OHC, BSBPA; Tom Holter, interview by Brian Leech, Oct. 22, 2009, Butte, MT, transcript and digital recording, OHC, BSBPA.

30. Daniel Harrington and A. J. Lanza, "Miners' Consumption in the Mines of Butte, Montana," U.S. Bureau of Mines Technical Paper No. 260 (Washington, DC: Government Printing Office, 1921), 10–12. See also David Rosner and Gerald Markowitz, *Deadly Dust: Silicosis and the On-Going Struggle to Protect Workers' Health*, rev. ed. (Ann Arbor: University of Michigan Press, 2006).

31. John Connors, interview by Laurie Mercier, Dec. 9, 1982, Butte, MT, transcript, OHC, BSBPA; Jim Carden, interview by Joe McCarthy, Apr. 7, 2001, Butte, MT, partial transcript and notes, OHC, BSBPA; Anaconda Company Bureau of Safety, *Anaconda Safety Rules*, 61.

32. Brian Shovers, "The Perils of Working in the Butte Underground: Industrial Fatalities in the Copper Mines, 1880–1920," *Montana: The Magazine of Western History* (Spring 1987): 26–39.

33. Floyd Bossard, interview by Dawn Ingersoll, Nov. 29, 1999, Butte, MT, transcript, OHC, BSBPA. See also Ernest Robert Ayres, interview by Brenda Ayres, Oct. 26, 1998, Butte, MT, transcript, OHC, BSBPA; George Wold, interview by Alexis Wold, Apr. 7, 2002, Butte, MT, transcript, OHC, BSBPA ; Allen L. Rowe, interview by Sheila Franey,

Mar. 5, 2002, Butte, MT, transcript, OHC, BSBPA; Robert P. Corbett, interview by Laurie Mercier, Apr. 17, 1984, Butte, MT, Oral History 674, cassette tapes, OHC, MHS.

34. "Digging Ore... a Mile Below Surface," *Anaconda Company Trailsman* 7, no. 1 (June 1965): 3, VF "Anaconda Company Trailsman," MHS.

35. LeCain, *Mass Destruction*, 44.

36. "Pump Station 3900 Feet Underground," *Anaconda Company Trailsman* 14, no. 1 (Aug. 1974): 3, VF "Anaconda Company Trailsman," MHS.

37. "The Underground," *Anaconda Company Trailsman* 13, no. 1 (July 1973): 3, VF "ACM: General Mining," BSBPA.

38. Murphy, *Mining Cultures*, 125–27. See good examples of coverage in the *Anode*: John L. Boardman, "Field Day and the Reason," and "Bad Practice—The Result," and "Remorse," 4, no. 6 (Aug. 1918): 1–5, 8–9; "Great Disaster Takes Toll of Human Life," and "Game Fish Swim Upstream," 20, no. 1 (Jan. 1934): 1–3; "Butte First Aid Men Recognized," 22, no. 4 (Oct. 1936): 13; "Honor Roll," 25, no. 1 (Jan. 1939): 9–10; "The Safest Crew in Butte," *Anode* 25, no. 3 (July 1939): 8; "The C. F. Kelley Safety Award," 26, no. 3 (July 1940): 1–2.

39. On salt tablets, see C. P. Anderson, "Mining—Then and Now," *Anode* (Dec. 1940): 5–6; Basso and Burke, "Another Look at Burke's Butte," 25. Butte miners used candles, then gas lamps, then electric lights. See "Butte Mines Adopt Modern Equipment," *Anode* 20, no. 2–4 (Apr. 1934): 1–2; "The Anode in Retrospection," *Anode* 21, no. 1 (Jan. 1935): 1–2. On headlamps, see Paul J. Bartos, "A Light in the Darkness: U.S. Mine Lamps, the Early Years—Candlesticks, Oil Lamps, and Safety Lamps," *Mining History Journal* 16 (2009): 45–63; Paul J. Bartos, "A Light in the Darkness: Modern U.S. Mine Lamps—Carbide and Electric Cap Lamps," *Mining History Journal* 17 (2010): 90–102.

40. Herbert A. Wendel, interview by Ray F. Calkins, July 1979, Butte, MT, transcript, OHC, BSBPA.

41. Anderson, "Mining—Then and Now," 5.

42. Dan Aguilar, interview by Teresa Jordan, Apr. 10, 1986, Butte, MT, partial transcript and notes, OHC, BSBPA. For similar sentiment, see Nick Gevock, "Historian in the Making," *Mont. Std.*, Aug. 11, 2011.

43. Joyce Coupal, interview by John Terreo, Apr. 3, 1990, Noxon, MT, tapes and transcript, OHC, BSBPA.

44. Bradley Dean Snow, "From the Sixth Floor to the Copper Dome: 'The Company's' Political Influence in Montana, 1920–1959" (master's thesis, Montana State University, Bozeman, 2003), 161–75; Mercier, *Anaconda*, 102–7.

45. Gevock, "Historian in the Making." Exceptions to the partner rule included shift bosses and "nippers," who transported tools underground. See Ayres interview.

46. Perle Watters, interview by Laurie Mercier, Mar. 25, 1983, Butte, MT, Oral History 484, cassette tapes, OHC, MHS. See also John Onkalo, interview by Ray Calkins, Feb. 1980, Butte, MT, transcript, OHC, BSBPA.

47. Ayres interview.

48. Andrews, *Killing for Coal*, 125–56; the critique of determinism is in Chad Montrie, review of *Killing for Coal: America's Deadliest Labor War*, by Thomas G. Andrews, *Journal of American History* 114, no. 3 (June 2009): 772–73.

49. Emmons, *Butte Irish*, 367–73; Basso, *Meet Joe Copper*, 32–36.

50. Dennis J. Murphy, interview by Teresa Jordan, May 30, 1986, Butte, MT, partial

transcript and notes, OHC, BSBPA. See also Basso, *Meet Joe Copper*, 152–53; Murphy, *Mining Cultures*, 108–11.

51. John Frank Gardiner, interview by Brian Leech, Aug. 3, 2010, Wise River, MT, transcript and digital recording, OHC, BSBPA; Bob McCarthy, interview by Brian Leech, Oct. 26, 2009, Butte, MT, transcript and digital recording, OHC, BSBPA.

52. Hovis and Mouat, "Miners, Engineers," especially 453–54; Gardiner interview; George Vidrich, interview by Brian Leech, Aug. 2, 2010, Butte, MT, transcript and digital recording, OHC, BSBPA.

53. Gardiner interview. See also Corbett interview, by Mercier; Tarrant interview.

54. Louis Forsell, interview by Teresa Jordan, Jan. 2, 1986, Clancy, MT, partial transcript and notes, OHC, BSBPA. See also Victor Giuseppe Segna and Louise Zanchi, interview by Russ Magnaghi, May 3, 1983, Butte, MT, cassette tapes and transcript, Oral History 579, OHC, MHS.

55. Tarrant interview. See also Larry Hoffman, interview by Teresa Jordan, Apr. 12, 1988, Butte, MT, partial transcript and notes, OHC, BSBPA; Gardiner interview; Watters interview.

56. Rick Foote, interview by Teresa Jordan, Jan. 7, 1987, Butte, MT, partial transcript and notes, OHC, BSBPA.

57. Jaclyn J. Gier and Laurie Mercier, ed., *Mining Women: Gender in the Development of a Global Industry, 1670–2005* (New York: Palgrave Macmillan, 2006), 1–6.

58. Basso, *Meet Joe Copper*, 23–29; Murphy, *Mining Cultures*, 108–11; Gardiner interview; Monninger testimony, 30. On understandings of men during the early twentieth century, see Gail Bederman, *Manliness and Civilization: A Cultural History of Gender and Race in the United States, 1880–1917* (Chicago: University of Chicago Press, 1996); David S. Churchill, "Making Broad Shoulders: Body-Building and Physical Culture in Chicago 1890–1920," *History of Education Quarterly* 48, no. 3 (Fall 2008): 341–70; Roberta J. Park, "'Soldiers May Fall but Athletes Never!': Sport as an Antidote to Nervous Diseases and National Decline in America, 1865–1905," *International Journal of the History of Sport* 29, no. 6 (Apr. 2012): 792–812.

59. Joseph Kinsey Howard, "Boisterous Butte," *Survey Graphic* 28, no. 5 (May 1939): 316–20, 348–51, quote on 318. See discussion of this quote in Basso, *Meet Joe Copper*, 28.

60. Mercier, *Anaconda*; Basso, *Meet Joe Copper*, 110–28.

61. Helen Fitzgerald Sanders, "Redeeming the Ugliest Town on Earth," *Craftsman* (June 1907): 315–27. See also Mary MacLane, *The Story of Mary MacLane by Herself* (Chicago: Herbert S. Stone, 1902), 10; Carroll Van West, "The Pit and the Stack: Decoding the Southwestern Montana Landscape," *Drumlummon Views* 3, no. 1 (Spring 2009): 21–26.

62. Fredric L. Quivik, "Smoke and Tailings: An Environmental History of Copper Smelting Technologies in Montana, 1880–1930" (PhD diss., University of Pennsylvania, Philadelphia, 1998), 204–69; Donald MacMillan, *Smoke Wars: Anaconda Copper, Montana Air Pollution, and the Courts, 1890–1920* (Helena: Montana Historical Society Press, 2000).

63. F. J. Clifford, "Greenery Hunger: and How It Brought About the Transformation of the 'Richest Hill on Earth,'" *American Forests* 8 (1930): 523–24; Jessie C. Knox, "'The Redemption of 'Barren Butte' by Flowers and a Woman's Faith," *Garden Magazine and Home Builders Magazine* 5 (1925): 256–57.

64. Adolph H. Heilbronner, *Sights and Scenes and a Brief History of Columbia Gardens: Butte's Only Pleasure Resort* (Butte, MT: Butte Miner Co., 1902), 6, in VF 1041, BSBPA.

65. Dashiell Hammett, *Red Harvest* (New York: Alfred A. Knopf, 1929).

66. Waino S. Nyland, "A Famous Mining Camp in Literature, Written in 1926 While a Student at the University of Colorado," Nyland File Fldr., BSBPA.

67. Arthur O'Dane (Reuben Maury), "Butte Circa 1925: Hymn to an Oasis," *American Mercury* (Oct. 1925): 190–96; Waino Nyland, "Western Mining Town," *Scribner's Magazine* (May 1934): 365–69.

68. Mihelich, "The Richest Hill on Earth," 1–6, 150–91, 228–308; Swibold, *Copper Chorus*, 125–310; Murphy, *Mining Cultures*, 42–199.

69. Onkalo interview; Marie Cassidy, "Mine Whistle as Good as Clock," *Mont. Std.*, July 1, 1999.

70. Joseph J. Picarelli to Nelson Jones, "Possible Underground Movement—Walkerville," Sept. 21, 1973, Fldr. "Bonner, Mrs. Delia," Box "Cabinet 1, Drawer 3 (Box 1)"; F. C. Jaccard to F. A. Linforth, June 10, 1941, Fldr. "Dwyer, Richard," Box 4B, Dave Piper Collection, Manuscript Collection 388 (hereafter MC388), BSBPA; Angelo Petroni and Virginia Petroni, interview by Brian Leech, Oct. 21, 2009, Butte, MT, digital recording and transcript, OHC, BSBPA; Fritz Daily, interview by Brian Leech, Aug. 4, 2010, Butte, MT, interview and digital recording, OHC, BSBPA.

71. Jim Street, interview by Craig M. Chatriand, Nov. 25, 1998, Butte, MT, transcript, OHC, BSBPA; Frances Dow, interview by Ray F. Calkins, Feb. 1980, Butte, MT, transcript, OHC, BSBPA; Debbie Peterson, interview by Jori Liva, Apr. 19, 2005, Butte, MT, transcript, OHC, BSBPA; Jim Matteuchi, interview by Brian Leech, Oct. 19, 2009, Butte, MT, transcript and digital recording, OHC, BSBPA; McCarthy interview, by Leech; Finn, *Mining Childhood*, 64–71.

72. Joan M. Filpula, *Meaderville Revisited* (Butte, MT: Pit Printers, 2008), 106–7; Tracy Thornton, "'Little Italy': Reunion Set to Remember Meaderville," *Mont. Std.*, Sept. 3, 2006. Keeping gardens to supplement mine income was common in many mining areas. See Benjamin Herber Johnson, "Conservation, Subsistence, and Class at the Birth of Superior National Forest," *Environmental History* 4, no. 1 (Jan. 1999): 80–99.

73. Wendel, "Mining in Butte," 60–61; Watters interview.

74. Basso, *Meet Joe Copper*, 21; Malone, Roeder, and Lang, *Montana*, 280–85.

75. Malone, Roeder, and Lang, *Montana*, 292–95; Mary Murphy, *Hope in Hard Times: New Deal Photographs of Montana, 1936–1942* (Helena: Montana Historical Society Press, 2003); "Anaconda and Copper," *Time* 26, no. 15 (Oct. 7, 1935): 35–36; Brenda Pentland, "Letters from World's End," *Montana: the Magazine of Western History* 53, no. 4 (Dec. 2003): 37–49; "The Outlook," *Anode* 19, no. 5–7 (July 1933): 1–2; Basso, *Meet Joe Copper*, 41–42.

76. Janet Ore, "Labor and the New Deal in Butte, Montana: The International Union of Mine, Mill and Smelter Workers' Strike of 1934" (master's thesis, Washington State University, Pullman, 1987).

77. Christopher J. Huggard and Terrence M. Humble, *Santa Rita Del Cobre: A Copper Mining Community in New Mexico* (Boulder: University Press of Colorado, 2012), especially 100–1.

78. On social life, see Basso, *Meet Joe Copper*, 21, 40; Finn, *Tracing the Veins*. On the depression in Butte, see Murphy, *Mining Cultures*, 200–31.

79. Janet L. Finn,"A Penny for Your Thoughts: Stories of Women, Copper, and Community," *Frontiers: A Journal of Women Studies* 19, no. 2 (1998): 231–49; Laurie Mercier, "Bordering on Equality: Women Miners in North America," in *Gendering the Field: Towards Sustainable Livelihoods for Mining Communities*, Asia-Pacific Monograph 6, ed. Kuntala Lahiri-Dutt (Canberra: Australian National University E-Press, 2011), 33–48.

80. Quotes in "A Sensible View of the War," *Anode* 28, no. 2 (Apr. 1942): 16. See also Basso, *Meet Joe Copper*, 95–130; Amanda Graham, "Copper Commando and the Anaconda Company's Wartime Production," *Montana: The Magazine of Western History* 59, no. 4 (Winter 2009): 67–72.

81. Joseph Kinsey Howard, "What Happened in Butte," *Harper's Magazine*, Aug. 1948; Basso, *Meet Joe Copper*, 129–31.

82. John Gunther, "Butte Is the Black Heart of Montana," *Inside USA* (New York: Harper & Brothers, 1947): 163–82; quote on 164.

83. Basso, *Meet Joe Copper*, 129–56, 159–94, 231–65; Mercier, "Bordering on Equality," 33–48.

84. Finn, *Tracing the Veins*, 45–46.

85. Portage Environmental, "Belmont Subsidence Evaluation and Redevelopment Project," Prepared for the City and County of Butte-Silver Bow, Jan. 2002, in Butte Urban Redevelopment Assistance Office, Butte, MT, 2–3; LeCain, *Mass Destruction*, 35–36. Sample subsidence complaints are in P. C. Jaccard to P. J. Berget, Aug. 4, 1943, Fldr. "Strom, Gust, 618 South Dakota," Box 2, MC388, BSBPA; W. A. O'Kelly to P. J. Berget, Oct. 9, 1944, Fldr. "Aluminum," Box 3A, MC388, BSBPA.

86. John Byrne, "Pieces from the Past: The Central Butte Neighborhood of Butte, Montana," 1982, VF 1609, "Central Business District," BSBPA; Butte Community Union, "Preservation of a Neighborhood: A Neighborhood Preservation Plan for the Central Butte Neighborhood" (Butte, MT: Nov. 1985), 1–11; Bob J. McCarthy, "Re-Claiming Butte: The Doctrine of Subjacent Support," *Montana Law Review* 49, no. 2 (Summer 1988): 267–83.

87. Woodrow Roble to Anaconda Company Legal Department, "Property Damage Due to Underground Mining & Blasting," Aug. 6, 1973, Fldr. "Roble, Woodrow," Box 8, MC388, BSBPA; Butte Community Union, "Preservation of a Neighborhood," 2–3.

88. Butte Community Union, "Preservation of a Neighborhood," 1–11; McCarthy, "Re-Claiming Butte," 267–83.

89. See an example in William A. O'Kelly to P. J. Berget, "Inspection of Shaft Under Sundberg Building," Nov. 2, 1950, Fldr. "Park St.," Box 5B, MC388, BSBPA.

90. William A. O'Kelly to Edward S. McGlone, Feb. 1, 1944, Fldr. "Subsidence Reports," Box 1, MC388, BSBPA.

91. Roy H. Glover to W. H. Hoover, May 11, 1943, Fldr. "Property Damage Miscellaneous Files," Box "Cabinet 1, Drawer 4 (Box 1)," MC388, BSBPA.

92. William A. O'Kelly to P. J. Berget, July 29, 1944, Fldr. "Creden, W. L.," Box "Cabinet 1, Drawer 3 (Box 1)," MC388, BSBPA. See also Roy H. Glover to W. H. Hoover, June 3, 1943, Fldr. "Property Damage Miscellaneous Files," Box "Cabinet 1, Drawer 4 (Box 1),

MC388, BSBPA; William A. O'Kelly to P. J. Berget, Oct. 30, 1951, Fldr. "Travona St.," Box 5B, MC388, BSBPA.

93. Edward S. McGlone to Clyde E. Weed, Feb. 10, 1944; and Edward S. McGlone to Clyde E. Weed, Mar. 16, 1944; both in Fldr. "Property Damage Miscellaneous Files," Box "Cabinet 1, Drawer 4 (Box 1)," MC388, BSBPA.

94. Doug Dasinger to Krest Cyr, James Robischon, Dave Piper, and Nelson Jones, "Subsidence Claims and Procedures," Apr. 21, 1969, Fldr. "Subsidence Reports," Box 1; Con J. Kelly Memorandum, May 5, 1944 and May 18, 1944, Fldr. "Blair, Ann," Box "Cabinet 1, Drawer 3 (Box 1)," MC388, BSBPA; Butte Board of Realtors to John Frankovich, Jan. 28, 1964, Fldr. "Bert Martin," Box "Cabinet 1, Drawer 2 (Box 2)," MC388, BSBPA.

95. P. J. Berget to A. A. Shelare, Aug. 5, 1944, Fldr. "Kall, Frank," Box "Cabinet 1, Drawer 2 (Box 2)," MC388, BSBPA.

96. P. W. Murray to W. P. Mufich, "Despina Famos Property," Apr. 26, 1972, Fldr. "Famos, George," Box "Cabinet 1, Drawer 3 (Box 2)," MC388, BSBPA; P. J. Berget to H. A. Kenck, Mar. 14, 1945, Fldr. "Johnson, George," Box "Cabinet 1, Drawer 2 (Box 2)," MC388, BSBPA.

97. Woodrow Roble to Anaconda Co. Legal Department, "Property Damage Due."

98. "Subsidence—Emma Mine Area—Butte, Montana, June 11, 1959," Fldr. "Butte Copper & Zinc Reports," Box 2C, MC388, BSBPA.

99. Butte Community Union, "Preservation of a Neighborhood," 10; Frank Quinn, "Anaconda Buys Mine Company: Butte Copper and Zinc Firm Holdings Sold," *Mont. Std.*, Nov. 28, 1959.

100. Timothy J. LeCain, "'See America the Bountiful': Butte's Berkeley Pit and the American Culture of Consumption," *Montana: The Magazine of Western History* 56, no. 4, (Winter 2006): 5–17, quote on 10.

101. Rex J. Henningsen, interview by Teresa Jordan, Apr. 12, 1988, Butte, MT, partial transcript and notes, OHC, BSBPA; Bernie Lazzari, interview by Matt Moore, Nov. 27, 2005, Butte, MT, transcript, OHC, BSBPA; Luigi Pasini and Norma Pasini, interview by Russ Managhi, May 4, 1983, Butte, MT, cassette tapes, Oral History 582, OHC, MHS; "Anaconda Company to Spend Huge Sum in Providing Vital Copper, Board Chairman Says in Press Convention Talk," *Butte Daily Post*, Aug. 19, 1956; "Anaconda Developing Metal Resources, C. H. Steele Reports," *Mont. Std.*, Dec. 7, 1953.

102. Dorothy Kilgallen, "A Big Night in Butte," *Good Housekeeping* (1953): 28, 250–54; quotes on 28.

103. Edmund Christophers, "Butte," *Saturday Evening Post*, Dec. 8, 1951, 40-41, 124, 127, 129-130; quotes on 40, 41, 124. See also Edward B. Reynolds, "When Copper Is King," *Christian Science Monitor* 5, no. 23 (1942): 7.

104. Susan R. Martin, *Wonderful Power: The Story of Ancient Copper Workings in the Lake Superior Basin* (Detroit, MI: Wayne State University Press, 1999), 81–214.

105. Dwight E. Woodbridge, "The Mesabi Iron-Ore Range: 4," *The Engineering and Mining Journal* 79 (Feb. 9, 1905): 266–68.

106. Otis E. Young, *Western Mining: An Informal Account of Precious-Metals Prospecting, Placering, Lode Mining, and Milling of the American Frontier from Spanish Times to 1893* (Norman: University of Oklahoma Press, 1970), 58–65, 102–45, 234–47; James, *The Roar and the Silence*, 255.

107. Eric Twitty, *Blown to Bits in the Mine: A History of Mining and Explosives in the United States* (Ouray, CO: Western Reflections, 2001), 137–64; LeCain, *Mass Destruction*, 137–46.

108. E. D. Gardner and McHenry Mosier, *Open-Cut Metal Mining*, Bulletin 433 (Washington, DC: Bureau of Mines, 1941), 112, 130; LeCain, *Mass Destruction*, 146–59.

109. David A. Walker, *Iron Frontier: The Discovery and Early Development of Minnesota's Three Ranges* (Minneapolis: Minnesota Historical Society Press, 1979), 132–38; Woodbridge, "Mesabi Iron-Ore Range: 4," 266; LeCain, *Mass Destruction*, 152–57.

110. A. J. Myers, "The Block Caving System at Pewabic Mine," *Proceedings of the Lake Superior Mining Institute* 21 (Ishpeming, MI: Lake Superior Mining Institute, 1917), 25–28; William Hawes, "A Brief History of Block-Caving," *Mining History News*, July 2008, 5.

111. C. C. Popoff, *Block-Caving at Kelley Mine, The Anaconda Company, Butte, Mont.*, Bureau of Mines Info. Cir. 7758 (Washington, DC: Department of the Interior, 1956), 6. See also A. B. Parsons, *The Porphyry Coppers* (New York: The American Institute of Mining and Metallurgical Engineers, 1933), 323–428.

112. On efficiency and scientific management, see Samuel Haber, *Efficiency and Uplift: Scientific Management in the Progressive Era 1890–1920* (Chicago: University of Chicago Press, 1964); Samuel P. Hays, *Conservation and the Gospel of Efficiency: The Progressive Conservation Movement, 1890–1920* (Cambridge, MA: Harvard University Press, 1959); Jonathan Zimmerman, "Simplified Spelling and the Cult of Efficiency in the 'Progressive' Era," *Journal of the Gilded Age and Progressive Era* 9, no. 3 (July 2010): 365–94.

113. James Colquhoun, *The Story of the Birth of the Porphyry Coppers* (London: William Clowes and Sons, 1933), 1–31; Hovis and Mouat, "Miners, Engineers," 429–56; LeCain, *Mass Destruction*, 160–68.

114. D. W. Brunton, "Technical Reminiscences," *Mining and Scientific Press* 111, no. 22 (Nov. 27, 1915): 811–18, quotes on 817. On Brunton, see T. A. Rickard, *Interviews with Mining Engineers* (San Francisco: Mining and Scientific Press, 1922), 69–96; William Columbus Ferril, *Sketches of Colorado*, vol. 1 (Denver, CO: Western Press Bureau Company, 1911), 221.

115. LeCain, *Mass Destruction*, 111–29; Parsons, *Porphyry Coppers*, 53–64.

116. Lingenfelter, *Bonanzas and Borrascas*, 382–86, 390–91; Parsons, *Porphyry Coppers*, 71–77; Leonard J. Arrington and Gary B. Hansen, *"The Richest Hole on Earth": A History of the Bingham Copper Mine* (Logan: Utah State University Press, 1963), 41–46.

117. Lingenfelter, *Bonanzas and Borrascas*, 382–86, 386–403; Parsons, *Porphyry Coppers*, 71–77; Arrington and Hansen, *"Richest Hole on Earth,"* 41–46.

118. Arrington and Hansen, *"Richest Hole on Earth,"* 90.

119. Lingenfelter, *Bonanzas and Borrascas*, 403.

120. Huggard and Humble, *Santa Rita Del Cobre*, 61; LeCain, *Mass Destruction*, 129–37.

121. Hovis and Mouat, "Miners, Engineers," 435.

122. LeCain, *Mass Destruction*, 129–37, 206–8.

123. Parsons, *Porphyry Coppers*, 208–9, 282-3; Huggard and Humble, *Santa Rita Del Cobre*, 62.

124. E. D. Gardner, *Undercut Block-Caving Method of Mining in Western Copper Mines*, Info. Cir. 6350 (Washington, DC: U.S. Bureau of Mines, 1930), 1–6, quote on 4; Parsons,

*Porphyry Coppers*, 393–96; Hovis and Mouat, "Miners, Engineers," 443–45; E. M. J. Alenius, *A Brief History of the United Verde Open Pit, Jerome, Arizona*, Arizona Bureau of Mines Bulletin 178 (Tucson: University of Arizona, 1968), 1.

125. Jeffrey T. Manuel, "Mr. Taconite: Edward W. Davis and the Promotion of Low-Grade Iron Ore, 1913–1955," *Technology and Culture* 54, no. 2 (Apr. 2013): 317–45, quote on 323.

126. Donald L. Boese, *John C. Greenway and the Opening of the Western Mesabi* (Bovey-Coleraine Joint Bi-Centennial Commission, 1975), 188. See also Ira B. Joralemon, *Copper: The Encompassing Story of Mankind's First Metal* (Berkeley, CA: Howell-North, 1973), 189–98.

127. Tom Finch, interview by Darby Stacey, Mar. 14, 2002, Butte, MT, OHC, BSBPA.

128. Manuel, *Taconite Dreams*, ix–32.

129. Marcosson, *Anaconda*, 259–60, 272–82; Joralemon, *Copper*, 343–44; Gardner, *Undercut Block*, 2–3.

130. Kelley quoted in Marcosson, *Anaconda*, 278.

131. Ibid., 73–65, 101–6, 129–31.

132. "Greater Butte Project," MAM Newsletter, Sept. 1947, Fldr. 2, Box 335, Manuscript Collection 169, Anaconda Copper Mining Company Records (hereafter MC169), MHS.

133. Richard N. Miller, "Production History of the Butte District and Geological Function, Past and Present," n.d., F-8, Fldr. "Butte Mine Operations 1965–1985," File 29906.01, Anaconda Geological Documents Collection, American Heritage Center, University of Wyoming, Laramie, WY (hereafter AGDC, AHC); Edward P. Shea, "The Role of the Geologist at Butte," *Mining Engineering* 13, no. 3 (1961): 264–68.

134. John P. Hunt, "Vincent Denis Perry—An Appreciation," *Mining Engineering* (Dec. 1997): 95–96; Miller, "Production History."

135. Paul Lindberg, e-mail to the author, July 30, 2012; Leo Brewer et al., "Charles Meyer, 1915–1987," Earth and Planetary Science, University of California, Berkeley, eps.berkeley.edu/content/charles-meyer.

136. Corbett interview, by Mercier. See also Marcosson, *Anaconda*, 275–80; Vincent D. Perry to Clyde E. Weed, Dec. 18, 1953, Fldr. "Butte Mine Operations, Prospect Data and Correspondence," File 30005.02, AGDC, AHC. On Butte's geology, see Charles Meyer et al., "Ore Deposits of Butte, Montana," in *Ore Deposits of the United States, 1933–1967*, Graton-Sales, vol. 2 (New York: AIME, 1968), 1363–416.

137. Bossard interview.

138. William Haycraft, *Yellow Steel: The Story of the Earthmoving Equipment Industry* (Urbana and Chicago: University of Illinois Press, 2000), 25–31, 100–7; Ted P. Flynn, "From Bulls to Bulldozers," ed. by Joseph A. Miller, *Forest History* (Fall 1963): 14–17; LeCain, *Mass Destruction*, 146–59; Robin McCulloch, interview by Adam Lawrence, Nov. 26, 2002, Butte, Montana, transcript, OHC, BSBPA.

139. On the Chuquicamata background of George Parker, Butte's first pit superintendent, and John Dougherty, the first pit engineer, see "Anaconda Maps Greatest Expansion," *Mining World*, Nov. 1956, 56–61. On the Chilean experience of the Kelley Mine's manager, Martin Hannifan, see Thomas S. Jenks, "An Interview with Anaconda Co.'s Montana Chief: Martin Hannifan Likes the Looks of Butte's Future," *Mont. Std.*,

Dec. 14, 1969. On Anaconda's use of knowledge about open-pit mining elsewhere in Butte, see John R. McWilliams, "Mining Methods and Costs at the Anaconda Company Berkeley Pit, Butte, Mont." (Bureau of Mines, Info. Cir. 7888, 1959), 8. On trucks in open-pit mines, see "Trucks Make Marginal Open-Pit Mining Pay," *Business Week*, Apr. 11, 1953.

140. Vincent D. Perry, "A Half Century as Mining and Exploration Geologist with the Anaconda Company," an oral history conducted in 1990 by Eleanor Swent, Carmel, California, Regional Oral History Office, The Bancroft Library, University of California, Berkeley (hereafter ROHO), 1991, 102–3.

141. Vincent D. Perry to Reno H. Sales, Oct. 29, 1957, Fldr. "Reno Sales Correspondence," File 3107.01, AGDC, AHC.

142. Edward P. Shea to Edward S. McGlone, "Possible Open Pit Operations on the Butte Hill," Jan. 15, 1952, Fldr. "Butte Mine Operations," File 29801.13, AGDC, AHC. On McGlone, see Marcosson, *Anaconda*, 283.

143. Vincent D. Perry to Frank M. Monninger, Dec. 5, 1975, File 30005.02, AGDC, AHC.

144. Vincent D. Perry, "Future Copper Possibilities at Butte," Apr. 10, 1953, document 12, Fldr. "Butte, 1937–1979 Mine Operations, Silver Bow County," File 29705.12, AGDC, AHC.

145. Vincent D. Perry to Frank M. Monninger, Dec. 5, 1975, File 30005.02, AGDC, AHC.

146. Miller, "Production History"; Edward P. Shea, Memorandum no. 8, "Re: Butte District Special Development," May 20, 1955, Fldr. "Progress Reports, Special Development Memos nos. 1 to 22 (Mar. 31, 1955 to Aug. 26, 1955)," File #29704.05, AGDC, AHC. See also Paul Lindberg e-mail.

147. "Frank Trask Succeeds Charles Goddard Here," *Mont. Std.*, Jan. 10, 1971.

148. "The 1961 Daniel C. Jackling Award, Awarded to Vincent D. Perry," Society for Mining, Metallurgy, and Exploration, accessed June 1, 2013, test.smenet.org/membership/awards/division-awards/mining-exploration-(m-e)/daniel-c-jackling-award. See also Hunt, "Vincent Denis Perry," 95–96.

149. Vincent D. Perry, "Future Exploration and Development of Copper Ore at Butte, Montana," Apr. 1955, Fldr. "Butte Mine Operations, 1937–1979," File 29705, AGDC, AHC, 4–5.

CHAPTER 2

# NEIGHBORHOODS

DURING THE BOOMING YEARS of the late 1940s and early 1950s, growth in Butte was occurring in an area of Summit Valley known as the Flats. Movement in that direction had begun in the 1890s and early 1900s, when streetcar suburbs began to grow along the eastern part of the hill and the valley down below, connected to the rest of the area by the thirty-eight miles of rail transit in the county. By the time the streetcar system met its demise in 1937, cars and buses had become common, allowing the city to spread farther out on the valley floor.[1]

Above the valley, Butte's social geography remained much as Mary Murphy described it in *Mining Cultures,* a study rooted in the early twentieth century. The hill on the northern section of the city, which overlooked the valley, remained the center of economic and social life. Businesses clustered in a large central district in the middle of the hill. Residential areas, each with its own distinct social world, sat on each side of the CBD. Skilled white-collar workers lived on the western side of the hill, "an attractive neighborhood of rambling Victorian houses, Queen Anne cottages, and craftsman bungalows." A few older mansions built for wealthy beneficiaries of Butte's booms were also situated on the West Side. Miners, their families, and other working-class residents made their homes elsewhere on the hill. Many lived in workers' cottages north of the CBD while others found homes on the East Side, which was "crammed with walk-up flats, fourplexes and rambling boardinghouses."[2] Past the East Side proper were three neighborhoods for working-class white ethnics: Meaderville, McQueen, and East Butte.

A wide economic gap existed between the West Side and other areas on the Butte hill. Former McQueen resident Jim Michelotti recalls that the differences were evident even in the uniforms worn by teams in the parochial baseball league. Comparing his McQueen team with that of St. Patrick's Church on the West Side, "We didn't have very good uniforms. McQueen—they sort of had hand hand-me-downs where these neighborhoods up here had warm-ups and everything, you know—they were considered the rich kids."[3] Underground mining took place all around the hill, but neighborhoods on the eastern and northern parts of the hill had to cope with far more of it.

Outside of the oddity of mining structures and refuse, Butte's working-class neighborhoods were reminiscent of those in larger industrial cities in the United States. Butte was a far smaller city than New York or Boston, but its industrial history and pattern of immigration had nurtured the ethnic enclaves common in mining settlements.[4] Mineral wealth lasted longer in Butte than in most mining towns, however, and so its enclaves turned into true neighborhoods, just as they did in eastern cities.

By the 1930s the younger members of Butte's ethnic groups were adopting and adapting American customs. Unions and mine work continued to connect residents across the city, forming a strong oppositional culture to the company and providing some copper workers with interethnic solidarities. The threat of outsiders coming into Butte, including African Americans, because of labor shortages during World War II, also brought white ethnics together.[5]

Despite these challenges to each neighborhood's insularity, each place actually maintained much of what made it stand out, including ethnic culture.[6] The extent of their intricately woven community culture strongly bonded residents of eastern neighborhoods like the East Side, Meaderville, McQueen, and East Butte, and residents of northern areas, like Dublin Gulch and Walkerville, to each other and their respective neighborhoods.

## THE EASTERN SIDE OF THE HILL

People in Meaderville, McQueen, East Butte, and the East Side lived in communities full of complicated social connections and cultural vibrancy. Each neighborhood had its own distinctive characteristics. As working-class enclaves, however, the four had just as many elements in common.

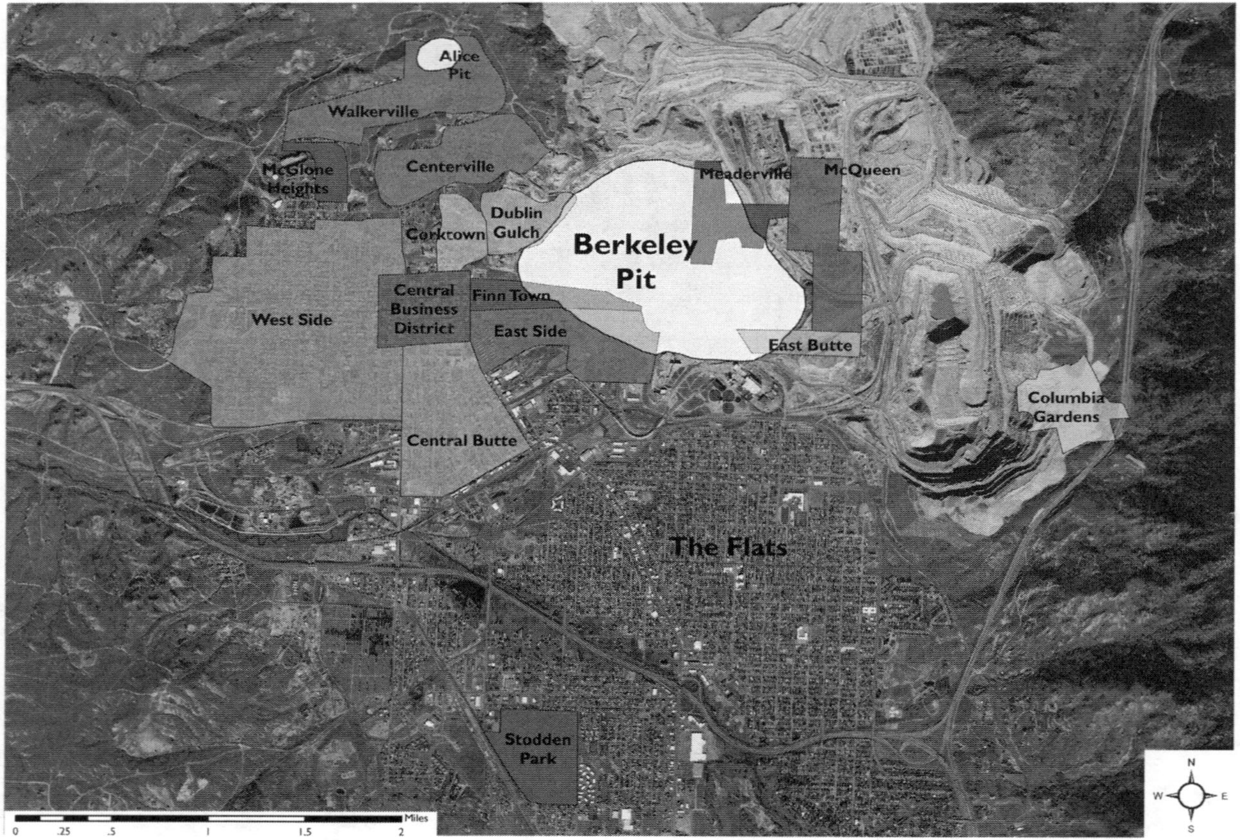


Butte's Social Geography. This map displays the Butte neighborhoods most impacted by mining. It also features both the city's most controversial open-pit mines (the Alice and the Berkeley) and its two most important parks (the Columbia Gardens and Stodden Park), which themselves became centers of relocation debates during the 1970s. Map created by geographer Matthew Fockler, 2017.

Consisting of approximately two hundred homes, Meaderville formed relatively early in Butte's history. The original inhabitants were Cornish miners in the northern end of the neighborhood and later-arriving Italians and Austrians in the south.[7] The McQueen Addition and mountains rising up to the Continental Divide bordered it to the south and east. Originally known as Gunderson, Meaderville sat about one mile from Butte's CBD. The highway from Butte to Helena, the state capital, was the sole major road connecting Butte to Meaderville. Charles T. Meader, who built the Bell smelter in Meaderville in 1881, later claimed that he put his facility there to move acidic smoke, which plagued the city in the early days of mining, away from Butte's city center. By the end of the nineteenth century, the community had taken his name.[8]

Meaderville developed haphazardly around the smelter works and the Leonard Mine, originally run by the Boston and Montana Consolidated Silver and Copper Mining Company. Its seventeen streets jutted out in many directions. From above, the town appeared awkwardly elongated as it followed the highway and the often-flooding and always-polluted Silver Bow Creek. Meaderville's layout was a stark contrast to that of the later-developing, strictly organized suburban Flats lying to the south. It also housed many businesses as well as residences.[9]

Anaconda came to own most of the ground beneath residents' homes in Meaderville, as the company bought out other mining companies and other mining claims during the late nineteenth and early twentieth centuries. The Leonard Mine was one of its prizes. In a classic display of welfare capitalism, the Anaconda Company sought to keep its workforce happy by supporting neighborhood activities. The gymnasium in the Leonard's fenced mine yard became Meaderville's athletic center, hosting basketball games and boxing bouts. The company also supported undertakings like the baseball teams that residents of Meaderville and Butte enjoyed as players or fans.[10] Meaderville, however, did not become a true company town. Anaconda did not develop company-owned stores, housing, or other amenities in Meaderville. For the most part, the company left Meaderville's people to construct their own social and cultural world.[11]

After Meaderville lost its smelting operations in the early twentieth century, most of Meaderville's men subsequently worked in the underground mines that surrounded the community. Many of the men who originally moved into the area were single and lived in boardinghouses. As families became more common, residents built cabins, many of which later became remodeled into more substantial structures. In the 1930s the New Deal's Works Progress Administration installed a sewerage system in the area.

By the middle of the twentieth century, Meaderville had aspects of a suburban settlement, but its rough-hewn appearance, high proportion of businesses to residential structures, proximity to the mines, and more-organic development meant that it never felt like a true suburb.[12] Nor did Meaderville fit the definition of a town. It never incorporated as a separate place, nor did Butte incorporate Meaderville as part of the city. Despite its fairly dense settlement, Meaderville, with a haphazard settlement pattern, remained somewhat rural. Many residents clearly remember enjoying the open land near Meaderville, raising animals, and picnicking next to the mountains.[13] Some residents appreciated Meaderville's anomalous status, which let them enjoy the feeling of being out of the city; their situation was also more legally vulnerable, though, due to the community's lack of incorporated status.[14]

McQueen and East Butte were more typically suburban than Meaderville, with orderly street grids and most of their space dedicated to housing. From the beginning, the two neighborhoods served primarily as centers for single-family residences. They contained fewer boardinghouses and large apartment buildings than other neighborhoods on the eastern side of the hill. East Butte featured small boardinghouses originally created for Northern Pacific Railroad workers, but McQueen had none. Likewise, East Butte and McQueen lacked Meaderville's concentration of businesses, although they had enough to function self-sufficiently, which was important because they were south and east of Meaderville and hence far from the CBD. McQueen had two groceries, a barber shop, an ice cream shop, and a few other businesses near a highway that ran across Woodville Hill.[15]

East Butte came into being in 1890 in connection with the Pittsmont Mine, located at the bottom of Woodville Hill. In 1891 McQueen, named for a nearby placer claim, began right next door. In 1901 it solely contained fifteen log homes, but it quickly grew into an important Butte suburb so that, by the 1950s, it hosted 267 homes and 1,390 people. Residents of both towns drew from the same well in the early years, but later federal aid improved the infrastructure. Through the federal Works Progress Administration, McQueen gained sidewalks and a sewerage system during the 1930s.[16]

Residents of the McQueen/East Butte area were similar in social class and employment to their neighbors in Meaderville. Former East Butte resident Ann Bone remarked, with some exaggeration, "Everybody worked in the mines. That was just the way it was." No doubt, working-class jobs, primarily for the Anaconda Company, were the norm. The homogeneity of employment could be troubling. "Most of the families in East Butte lived day-to-day, paycheck-to-paycheck," Bone commented. "And when there was a strike at the pit or at the mines, I mean, it was just heart wrenching to everybody."[17]

Descriptions of childhood leisure in McQueen/East Butte resembled those for Meaderville. Most McQueen natives remember playing on the dumps near the Pittsmont Mine. They also participated in more definitively rural activities, like hiking in the foothills leading up to the Continental Divide, fishing, hopping on trains, and using railroad tracks to flatten nails.[18]

Butte's East Side shared a similar history with those locales. In 1880, the Leggat Brothers from St. Louis laid out the plats. The original inhabitants were single men who flocked to boardinghouses near the Parrot and Northwest mines. The presence nearby of other high-producing mines, especially the St. Lawrence, Anaconda, and Mountain View Mines, made the East Side a popular destination. Miners felt it was important to be within walking distance of their workplace. At the turn of the twentieth century, an influx of immigrants from Eastern Europe transformed the growing East Side. Many of the new arrivals worked in the mines, but others worked in blue-collar jobs as carpenters, teamsters, and clerks.[19]

Sitting just east of the CBD, the East Side was far more urban than either Meaderville or McQueen. Historians Mary Murphy and Brian Shovers claim that its mix of commercial buildings, boardinghouses for single men, ethnic enclaves, and common blue-collar occupations initially gave the East Side an identity "distinct from the business district."[20] During a boom period in Butte, from 1916 to 1918, the link between the East Side and the CBD became stronger as brick edifices replaced wood-frame buildings. Duplexes, fourplexes, and walk-up flats joined boardinghouses as the most significant forms of housing. Some single-family Queen Anne and small, gable-fronted workers' cottages existed in the area, along with a shantytown known as the Cabbage Patch. A New Deal project called the Silver Bow Homes replaced the Cabbage Patch in 1940, making the East Side more densely packed than nearby neighborhoods.[21] East Siders, unlike residents of Meaderville or McQueen, rarely had yards. Onetime resident Aili Maria Goldberg recalled, "There wasn't a blade of grass there. Nobody had a piece of lawn or anything. You just lived with the old mine dumps for a backyard up there."[22]

The variety and number of businesses added to its urban quality. A drug store, bakery, hardware store, casket company, butcher shop, and dairy company joined a laundry, multiple grocery stores, the Harrington Ice Cream Parlor, Wayrynen Shoe Repair, hotels, and gas stations.[23] Nevertheless, the people of the East Side shared with the residents of Meaderville, East Butte, and McQueen working-class backgrounds, the experience of living in the midst of mining operations, and the satisfaction that each of their living areas provided most of the services they needed.

## Ethnic Enclaves

These neighborhoods also shared the fact that they became centers for southern and eastern European immigrants. These immigrants were a substantial part of the approximately 18 million immigrants arriving in the United States between 1890 and the 1920s. Many became members of the American industrial workforce, including those who came to Butte's mining operations, which had been dominated by Irish and Cornish immigrants up to that time. According to historian David Emmons, the arrival of new immigrants destabilized and challenged Butte's Irish labor aristocracy and gave the mine operators greater control over the workforce than ever before.[24] Butte's new immigrants were inexperienced miners, which became a major problem at a time of rapid technological change. The inability of the new groups to communicate with each other and the rivalries that sometimes existed among them helped to create an increasingly unsafe underground environment during the 1910s.[25] Some of the newcomers left when the mining industry experienced hard times in the 1920s and 1930s, but many became a permanent part of the city's workforce, often remaining in their ethnic enclaves on the eastern side of Butte's hill.

No neighborhood better showcased how important ethnic ties and traditions became to Butte's eastern communities than Meaderville. Immigrants from Italy dominated its business and social scene by the early 1900s. Those Italians who did not live in Meaderville often lived nearby. Italian-speaking Swiss, for example, operated dairies in nearby Elk Park. Many Meaderville Italians hailed from northern Italy, particularly the Piedmont region, but central areas, including Tuscany, were also a common source of new immigrants. While men often labored in the mines, women operated small laundry services, worked in small businesses, or cooked, baked, and waited tables in the food service industry.[26]

Meaderville's Italian heritage made it close-knit and distinct from Butte's other neighborhoods. Although they spoke different regional dialects when they arrived in Butte, interactions among the Italians led to the emergence of a common form of Italian.[27] Butte hosted three generations of Italian Americans in the 1950s and 1960s. One-time Meaderville resident Pauline deBarathy recalls that the first generation—her grandfather's—spoke little English. Members of the second and third generations tended to speak little Italian outside of the neighborhood, but the language was still common at Meaderville's shops and family gatherings, especially among the elderly.[28]

Whether speaking in Italian or English, Meaderville's residents maintained an active social life. During the neighborhood's early years, immigrants created the Italian-American Club, but it did not last long: later-arriving Italian men instead joined the Cristoforo Colombo Club. The latter lodge functioned as a benefit society to compensate families for the injuries and deaths that men suffered, usually while mining. Due to state requirements, the Anaconda Company became increasingly likely to provide some financial assistance for mine-related injuries, but the club still served its death-benefit function in the 1950s and 1960s.

The club maintained an Italian identity. Members had to have an Italian name. By that criterion, being considered Italian in the tightly bonded, homosocial environment was a phenomenon of male descent. The club's minutes switched from Italian to English in the 1930s, although even as late as the 1950s and 1960s members continued to conduct much of their oral business in Italian; indeed, some older members did so just to spite younger upstarts. The club, however, opened up to outsiders several times a year. It hosted dinners, picnics, Columbus Day events, and fundraising activities like dances, all of which helped to connect the area's Italian Americans.[29]

Italian food traditions were a major force in creating interconnections for the Meaderville neighborhood. They provided a strong sense of an Italian ethnic identity for residents and gave the place recognition within the larger region. In the early days, freight cars brought grapes to Meaderville so that residents could make wine. Once the homemade beverage became popular in Butte, Italians sold it to the wider community. Five small groceries, including the Meaderville Mercantile and the Guidi Brothers' grocery, provided favorites—the salty fish *salicina*, sausages, salamis, olives, and cheese. Some of the items were imported. What residents did not find at the store, they grew. The smelters had originally killed much of the vegetation around Meaderville, but by the 1920s and 1930s, much of the foliage had revived. According to resident Jim Troglia, every house had a garden as well as a dirt cellar to store canned goods, wines, and sausages. Some residents took the garden concept further, raising chickens and cows, while other locals hunted for additional meat.[30]

Large family dinners and other informal gatherings centered around traditional dishes, and Meaderville's lively restaurant and club scene showcased the neighborhood's Italian food, which became an important link between Italians and non-Italians. The first of many Meaderville clubs opened in the 1920s, and most of them served elaborate Italian meals as a way to attract business. Their enormous meals consisted of antipastos with caviar, crudités, salami, smoked salmon, anchovies, and breadsticks; a second course with salads, spaghetti, and raviolis; and a third course with a piece of chicken or an enormous steak.[31]

Historian Harvey Levenstein claims that Italian food was unpopular in the United States before World War I.[32] Americans believed that vegetables and fruits contained too much water to be of nutritional value. A general distaste for intense spices as well as distrust of the supposedly dirty and disloyal immigrants coming from southern and eastern Europe added to the notion. Wartime conditions, however, put a greater value on vegetables, and the status of Italy as an ally improved the status of Italians in the United States. Likewise, the introduction of strict immigration quotas in the 1920s removed the stigma many Italians carried as sojourners, whose high rate of return immigration implied to many Americans that they were not truly loyal residents.

Prohibition also helped to trigger the popularity of Italian food, both in Meaderville and across America. Little Italies in many cities became havens for drinkers.[33] When immigration quotas reduced the arrivals of single men, the women who had run boardinghouses for them stepped up their production of wine and grappa to attract new customers. Meaderville's Italians became especially well known during Prohibition for making large batches of bootleg grappa. Of course, people coming to Meaderville and other Italian communities for libations desired food as well. The women who ran boardinghouses had long cooked local specialties for the men they housed; now, they were serving this Italian food to newcomers.

By the 1930s and 1940s, non-Italians had become used to going to Italian restaurants. Meaderville became locally famous for its food and drink. The new nightclubs called on local female cooks to run their kitchens. Teddy Traparish's Rocky Mountain Café became the most renowned of Meaderville's clubs, and its steak dinners even managed to gain national press. The Café employed the mother of former resident Marlene Holayter as its head cook. Her own mother had been widowed by a mine accident. Only in her thirties and with her relatives still in Italy's Piedmont, Holayter's grandmother opened a boardinghouse, where her daughter honed her skills.[34] Food offered a fairly easy path into the business world for Italian immigrants like these women, just as it did for other migrants moving to a marketplace environment from rural areas with strong food traditions.[35]

The 1955 *Butte City Directory* shows that Meaderville's club life was still going strong when open-pit mining began. On Main Street, the International Order of Odd Fellows' Colusa Bar was at 36, the Brass Rail at 37, Rex's at 40, Sam's at 44, the Vegas Club at 45, the Pioneer Lounge at 48, the Big Wheel at 51, the Aro at 65, and the Rocky Mountain Café at 78. Rounding out the roster of major bars and restaurants was Club 45, which sat on nearby Leatherwood Street.[36]

Food traditions promoted by the clubs also helped to connect Meaderville's residents to each other. Meaderville's Italians selected some food traditions from Italy over others. In the process, they created a new Italian food tailored to their new lives in America, while also helping to build and maintain a sense of identity for Meaderville's residents. What Italians ate in Italy varied due both to regional variations in the availability of crops and to social class. In northern Italy, the most common staple was the stiff cornmeal of polenta, though rice, butter, and a diversity of other foods were also available. In the south of the country, pasta, tomatoes, beans, some cheeses, and fish reigned supreme. As historian Hasia Diner makes clear, social class also came into play. The rich often ate meat, macaroni, and other delicacies, while laborers and farmers generally made due on very meager diets, with small amounts of bread, oil, and salt the most common elements. Despite the origins of its working-class population in northern Italy, Meaderville's traditional Italian food was mostly southern Italy's pasta, olive oil, and tomato sauce coupled with the upper-class's chicken, beef, and veal.[37]

Why this transformation, which generally occurred across the United States, took place is not entirely clear, though a number reasons are likely. Like most immigrants, Italians wanted to enjoy the abundance that America offered, and that included high-status items like beef, veal, and chicken. Under pressure from people who wanted to Americanize immigrants, the second generation also wanted to hew more closely to American tastes. Southern Italy staples like dried pasta, beans, tomato sauce, greens, and olive oil were also relatively inexpensive in America.[38]

What is clear is that Italians with roots from diverse regions came to rally around this selected cuisine in America, not Italy. Historian Hasia Diner notes, "Feasting upon dishes once the sole preserve of their social and economic superiors enabled them to mold an Italian identity in America around food."[39] For Meaderville's Italians, divergent food traditions like polenta and blood sausage continued in many homes, but residents eventually rallied around the southern and upper-class foods that restaurants sold as representing the now singular Italian identity of Meaderville. Even non-Italians living in Meaderville often ate this food. Frank Shone, for instance, relates that his mother hid his Choctaw Indian background from him while he was growing up in Meaderville, partly by serving spaghetties from the area at mealtime and not foods related to his own heritage.[40]

The clubs' role as social centers for the area's residents went beyond food and drink. Known across the region as little Las Vegas, Meaderville offered men and women additional ways to socialize. As long as customers kept up a collection to pay off the police, there was open gambling across Butte. Whatever part of the gambling took place outside Butte's CBD usually found its way to the backrooms of Meaderville's clubs.[41] Mary Murphy suggests that Meaderville's clubs also stood out from the super-masculine, rough-and-tumble bars that had originally dotted the Butte landscape. The new clubs were far more open to women, who came to sit in closed-off booths and to dance, while the supposedly seedier elements were relegated to back rooms.[42] Indeed, dancing was so popular that most clubs hosted extensive live music. An eleven-person band played on most nights at the Rocky Mountain Café. The clubs featured additional ways to socialize, including long outdoor courts for the Italian game of bocce ball.[43] Meaderville's clubs and cuisine therefore helped to build ties between Italians from different parts of Italy, Italians and non-Italians, and even men and women.

Ethnic traditions also held together the other neighborhoods on the eastern side of the hill. No single ethnic group dominated cultural life in either McQueen or East Butte. English, French, and Germans were among the early settlers of McQueen, but much of the population consisted of people from the former Austro-Hungarian Empire and additional Eastern European Slavic immigrants. Slovenes, Croats, and Poles joined many German-speaking immigrants and Italians in early-twentieth-century McQueen. In nearby East Butte, Swedes and English dominated the north end, while the rest of the area housed Croats and Slovenes.[44] The term *Bohunk* popularly served to label the primarily Catholic Slavic immigrants and their offspring. Although sometimes used affectionately, the word originated as a slur.[45]

Many second- and third-generation residents of both McQueen and East Butte remained in their original neighborhoods and maintained strong connections with other families from the same place. Former McQueen resident Ivan Stepan, for instance, described the ten Slovenian families who lived near him in McQueen as "a close-knit group."[46] A few of those ethnic groups created newspapers in the 1910s and, as they became better established, set up formal clubs, as had occurred in Meaderville. Like the Cristoforo Colombo Club, the Croatian Club and East Butte's Slovenian Hall sponsored dances, banquets, and picnics.[47] Louis Govednik, who arrived in McQueen from the Kingdom of Yugoslavia in 1940, found a fully formed ethnic community whose traditions gave him an immediate sense of belonging.[48]

Even by midcentury, languages other than English were still spoken in both McQueen and East Butte, though primarily by members of the older generation. George Vidrich, for instance, grew up just outside of East Butte. His father had immigrated from Austria and so he spent hours speaking with his Austrian neighbors in German. As in Meaderville, the children of immigrants sometimes could not learn English at home, but they tended to transition to it as they continued through the school system.[49]

More suburban than Meaderville, McQueen and East Butte lacked its richness of food businesses and nightlife, but they had groceries and eight saloons. Truly neighborhood bars, they packed men, women, and even children together within their walls. The last two to survive were the long-functioning Spehar's Bar and Hotel in East Butte and the McQueen Athletic Club, which housed an auxiliary room for community meetings.[50] As Tom Holter, a former officer of the Club, describes it, the McQueen Athletic Club is "kind of like a bar and it's like a social club." Men would participate in bowling leagues through the club and social events after every meeting. Of the leaders at the club, Holter says simply, "We took care of the neighborhood."[51] Although the McQueen Athletic Club served more functions than most drinking establishments, it is a perfect example of how Butte's bars operated as community centers.

Life in McQueen and East Butte also involved intricate home food traditions. Because McQueen and East Butte were so close to the foothills, hunting was very popular, as was the gathering of chokecherries. In the 1950s and 1960s, every house had a large vegetable garden. Despite Butte's short growing season and frequent raids by neighborhood children, the gardens were productive. Women with roots in the old Austro-Hungarian Empire made their own sauerkraut from cabbage in their gardens. They also canned fruits and vegetables to feed their families through Butte's long winters. Like inhabitants of Meaderville, they often made wine as well. Ethnic traditions affected the processing of the chickens, hogs, and rabbits that some residents raised to add to the food supply. Hams, strips of pork, and kielbasas went through backyard smokehouses. As former resident Ivan Stepan explains, "They would buy pigs and they'd use, I always say, every bit of that pig except the eyeballs. They smoked it then stored it for the winter."[52] At home, common specialties included *povatica* (a walnut bread identified with Croatians), blood sausage, cheese strudel, and *sarma* (an eastern and central European dish that usually involves sauerkraut leaves, ground or smoked meat, and rice).[53]

Strong ethnic identifications sometimes led to tensions, especially in the early days. Croats and Slovenes contended with each other in East Butte. According to former East Butte resident Carl Gegurich, the Slovenian Hall hosted so many ethnic-driven fights that it became known as the Bucket of Blood.[54] By midcentury, however, the second generation had forgotten most disputes, so much so that one longtime McQueen resident only remembered, "They were really a nice group of families that migrated from Europe.... They each swapped recipes and it was all good."[55] Although nostalgia perhaps clouds their memories, most interviewees agree that by the 1950s and 1960s the tension had dissolved into a love-fest, where Croats, Serbs, Austrians, and Irish all "got along beautifully."[56] Many ethnic-related traditions, including accordion and polka music, became shared experiences for residents of the two neighborhoods. Mesopust, a Croatian version of Carnival, was also popular.[57]

Butte's large East Side neighborhood was more diverse than the rest of the city. Ethnic enclaves in the East Side were more numerous than those in McQueen or East Butte. The East Side included Austrians and Eastern European Slavs—Croats, Slovenes, Montenegrins, and Serbs—along with a smattering of Mexicans, Swedes, and Irish. Slavic organizations founded during the early twentieth century, including the Montenegrin Literary Society, the Serb-Yugoslav Unity Lodge, the Serbian Montenegrin Federation, and the Circle of Serbian Sisters, became important to people on the East Side.[58] Residents also celebrated many of their own traditions. One Croatian resident, Bob Pavlovich, for example, spoke Croatian and quipped that he "celebrated two Christmases—one designated by Christians and one designated by Serbians."[59] One exceptional group to the East Side was the Lebanese, known as Syrians in the early twentieth century. Most of them lived between Ohio and Oklahoma Streets, toward the center of the neighborhood. Many of the Lebanese made their living as merchants of one kind or another. In 1908 they formed the first Syrian Peace Society in the country, building a lodge to house it on East Galena Street.[60]

Although the enclaves had scattered to some extent by the 1950s, the place remained relatively diverse. One woman who grew up on the East Side during the 1960s, for example, remembered living across from a Native American family, down the street from a Mexican American family, and right next door to Tom Powers, the Irish American mayor of Butte.[61]

Finns dominated the northeast corner of the neighborhood. The earliest-arriving Finns had scattered around Butte at first, but by the 1920s Finntown had become their social and cultural hub.[62] As historian Mary Murphy makes clear, Finns were not the only people in what became known as Finntown, but their cultural contributions "flavored the neighborhood."[63] With Broadway Street as its epicenter, Finntown took up the area east of Arizona Street, including Granite, Broadway, and Park Streets. As resident Aili Maria Goldberg joked, locals claimed that Broadway Street "was like a fish, too, because it had Finns on both sides."[64]

Displaced from their farms by the growing market economy, the Finns had first moved to European cities in search of work. When their urban adventures failed, a large number embarked for the United States.[65] Although some of Butte's Finns arrived straight from Finland, Butte represented a second or third stop for others. Many came from the upper peninsula of Michigan, where they had learned the trade of copper mining. As Butte displaced the Upper Peninsula as the largest copper producer at the turn of the twentieth century, many Finns set out for Montana. Ray Wayrynen, an East Side Finlander, businessman, and politician, was born in Michigan and came to Butte in 1932 at the age of ten. The aforementioned Aili Maria Goldberg also found Butte via Michigan, arriving in 1920 when her widowed mother could not find work in Ishpeming.[66] Some of the Finns who eventually made their way to Finntown, including Ida Myklebust as well as Laina and Sylvia Maki, started in Red Lodge, a coal mining area in south-central Montana, before arriving in Finntown.[67]

With eighteen establishments by 1916, the area had more boardinghouses for single male workers than any other in Butte. The most locally famous houses were Mrs. Rippi's, Mrs. Suominen's, Fanny Tuomala's, Kauhanens' Central House, the Kingston House, the Belmont, and the Florence (also known as the Big Ship). Local lore claims that a piston drill sat in the basement of the Big Ship so that new arrivals could learn its operation before hustling a job in the mines.

As in Meaderville, women ran most of the boardinghouses. These women employed other women and girls to serve as housekeepers, waitresses, and cooks. Like those in Meaderville, the Finntown boardinghouses gained a reputation for excellent food. Sometimes the women made Finnish stews or other specialties that featured whole meal products, potatoes, and roasts. The boardinghouses were gone by the time the Berkeley Pit arrived in the 1950s. The last of them, Mrs. Rippi's, also known as the Broadway Dining Room, closed in 1948 or 1949. Aili Maria Goldberg, whose mother ran Mrs. Rippi's at the time, explains that, after World War II, "it seemed like we got more steady families and less of the floating population, so there wasn't really the demand for boarding houses anymore."[68] Despite their absence in postwar Butte, boardinghouses had made it possible for many women to keep their families afloat during hard times. They also helped to establish Finnish traditions as important identifiers for the area.

The first major effects of the Finnish presence in Butte were contentious, but they did showcase the compact nature of the community and Finns' distinct sense of themselves. The newly arrived Finns were more politically radical than the conservative Irishmen who ran the town at that time. The Anaconda Company fired at least five hundred of them from the mines in 1912 for their leftist beliefs. The BMU, run by Irishmen who viewed the Finns as troublesome, refused to support these Finns.[69]

Finnlander Hall, built in 1902 on the 300 block of North Wyoming, hosted socialist-tinged plays throughout the 1910s, 1920s, and 1930s. During those years, the Hall also served as the headquarters for the IWW, a radical workers' group intent on forming one big union for all workers and abolishing wage labor entirely.[70] Before members sold the hall to Anaconda at midcentury, its function had changed from supporting an alternative political culture to less-controversial social activities, including hosting weddings, dances, and basketball games between the Finnish American Athletic Club and outside groups.[71]

Historian Mary Murphy rightly suggests that changes like these show the decrease in insularity for Butte's ethnic groups during the 1930s and 1940s, but Finns in the area showed remarkable resilience in protecting cultural traditions. Former resident Aili Maria Goldberg claims that her mother's refusal to learn English showed just how unwilling Butte's Finns were to give up their old ways. Others told similar stories. Neither John Latinen nor Agnes Cramer spoke any English before they attended grade school in the 1920s and 1930s. Agnes Cramer's mother wanted her to speak English every day after school, but Agnes refused, because she was worried about forgetting Finnish.[72] Even in the 1950s and 1960s, older locals still used Finnish in their daily lives. Bar owner and funeral house director Ray Wayrynen convinced sheriff Bill Dallino that he should hire his chief bartender, Lenny Bay, because it would be a great boon to have a deputy who could understand Finnish, especially during a time when Anaconda was buying up Finntown property for open-pit operations.[73]

Ethnic traditions became particularly important in Finntown's saloons. Just as in Meaderville, McQueen, and East Butte, bars served as social and cultural hubs. Up to twelve saloons existed in or near Finntown at one time, including Victor Kontola's Broadway Bar, the Board of Trade, the Pacific, the Corner Bar (which became the still-extant Helsinki Bar), Laitkainen's Alaska Bar, the Moonlight Bar, Oscar Tuomi's Alley Bar, Kirk's, the COD, the Atlas, and the Wayrynen's Yellowstone Bar. Like other Butte bars, many of them were open twenty-four hours a day, and they offered drink specials for miners when their shifts ended.[74]

The music of choice in Finntown's bars was usually polka, which describes a variety of music and dance of Central European origin. After spreading through Eastern and Northern Europe, polka music arrived in the United States, then blended with North American folk songs and jazz. This hybridized polka music gained popularity throughout the United States from 1930 until about 1960. It then fell off the mainstream pop music charts but still thrived within white ethnic communities.[75]

Polka remained the most popular music and dance in Finntown bars up until the 1960s and early 1970s, when Anaconda bought much of the area's property. The ethnically ambiguous tradition of polka became a means for the neighborhood's residents to celebrate together around a shared tradition. When the Broadway Bar celebrated its last night in 1975, Johnny's Hot Shots, featuring Johnny and Einer Kivela, played to the packed house, just as they had for twenty years.[76]

Finntown housed multiple saunas, which were more obviously Finnish in origin than polka. When Finns first arrived in the United States, most saunas were one-room log-cabin structures with a firebox, but no chimney. Finns believed that the accumulating smoke was itself was a cleansing force. The later, now more familiar, ventilated saunas around the country are plank-walled structures with two rooms—one for the sauna steam bath, in which water is placed on hot rocks, and one for the dressing room. Saunas are removed from daily life—usually dark, hot, quiet, and relaxing—and people who enter them usually are naked.

For Finns, the sauna was not simply a bath. Finns traditionally were born in the sauna, ritually cleaned for weddings in the sauna, and even prepared for burial there. By the 1960s and 1970s, the ritual functions had certainly changed. One folklorist claimed in the 1970s that "in contemporary North America the primary social function of the sauna has become manifestation and reassertion of Finnish-American ethnic identity."[77]

The strong sense of identity and ritual associated with the sauna made it a common and lasting feature in areas across the upper-Midwest and the Pacific Northwest where Finns settled. As late as 1962, two researchers traveling across the upper-Midwest found that 77 percent of rural Finnish families in northern Minnesota and 88 percent of rural upper-Michigan Finns all had saunas on their property.[78] The same held true for Butte's Finns. According to historians Mary Murphy and Brian Shovers, the existence of saunas helped to make Butte's Finntown a "viable ethnic enclave."[79] The aforementioned Ray Wayrynen split wood for one steam bath on Covert Street when he was growing up. Isa (meaning "father") Maki ran a particularly popular sauna in the area as well. The continued existence of saunas likely helped Butte's Finnish Americans to reassert and reconfirm their own identity as Finns.[80]

Ethnic-based traditions were still thriving and ethnic ties were still important for much of the population in Finntown, as in McQueen, East Butte, and Meaderville. Across the nation, Americans in the 1920s and 1930s had focused on acculturating the nation's European immigrants. The 1924 Johnson-Reed Act instituted quotas that severely limited the number of immigrants allowed into the United States, which meant fewer reinforcements for the new immigrant groups.[81] Yet the people in Butte's eastern neighborhoods had actually retained many of their ethnic landmarks and traditions during the immediate postwar period, which reinforced their already-strong bonds to each other. Moreover, a number of institutions, including religious ones, strengthened the ties across ethnic groups.

## Institutions and Connections

Churches, local schools, the volunteer fire department, and sports teams helped to maintain civic vitality in Meaderville, McQueen, East Butte, and the East Side. For most residents of those eastern neighborhoods, the Catholic Church and its attached parochial grade school were among the most important forces bonding people both to their neighborhoods and to their neighbors. In 1958 the city directory listed ten Catholic parishes. Each parish save one had its own grade school, which fed into two Catholic high schools. As sociologist John Mihelich states, "[In Butte] the neighborhood parish acted as a meeting ground and unifying force within the neighborhood." Some Butte Catholics attended Mass every day, but most attended every Sunday at least, with social activities often following the service.[82]

Meaderville's residents were tied particularly closely to their parish church. St. Helena's, built in the 1920s to give Italians "a church of their own," served as the social center for all of Butte's Italians, and not just those living in Meaderville. The church provided an all-men's choir and a women's altar society. Holy Savior Church sat two blocks away and housed services for non-Italian Catholic residents of both Meaderville and McQueen. Some of Meaderville's children went to the public Franklin Elementary. Other Meaderville children attended the parochial Holy Savior School, located on the border between Meaderville and McQueen.[83]

Religion could bridge ethnic differences. Lorraine Lazzari moved to Meaderville with an Italian husband whose family lived in the area. After a nervous lunch with the "ladies" of the "Meaderville group," one of her husband's relatives told everyone to accept her "even though I wasn't Italian." Lazzari explains that this relative "gave in because I was Catholic."[84]

Almost everyone in McQueen was also Catholic. Holy Savior Church and its attendant grade school served McQueen and East Butte, as well as the northeastern section of the Flats. It opened in 1902, and was replaced with a larger building in 1927. In addition to space for religious services, Holy Savior had a large hall to host community dances, potlucks, and celebrations. The few in the neighborhood who were not Catholic were likely to go up the road to Unity Methodist in Meaderville.[85]

WASHINGTON SCHOOL CHILDREN IN FINN TOWN, C. 1942. Children celebrating their scrap metal collection efforts during World War II. Photo courtesy of Butte-Silver Bow Public Archives, Riipi Collection, DIG019.

To the west, Sacred Heart parish worked to fulfill the educational and spiritual needs of the East Side's Catholics. The church was formed in 1901, and burned to the ground in 1912. Sacred Heart Catholic Church was rebuilt in 1913 with unusual Spanish-style architecture.[86] Besides Sacred Heart grade school, other schools served the neighborhood, including the presidentially named Washington, Grant, and Jefferson Schools. Of course, the large group of Finns in the neighborhood were likely to attend the public schools and unlikely to join the Catholic Church. Many of them attended Trinity Methodist Church on East Park Street. There was also a Finnish Lutheran Church and an evangelical Finnish Mission Church in Butte at one time.[87]

Small schools in each of the neighborhoods meant that every neighborhood child eventually knew every other neighborhood child. It also meant that children spent almost all of their hours in their respective neighborhoods. According to East Butte resident Ann Bone, her home was close enough to her grade school that she returned for lunch every day.[88]

Volunteer fire departments often provided key hubs for residents' social lives. The City of Butte maintained its own professional fire department, but volunteer fire departments still played an important role in many areas. That role often went far beyond firefighting. According to one recent study of volunteer fire departments, many "local communities in rural regions are structured around VFDs [volunteer fire departments]."[89] That was certainly true for Meaderville's fire department during the 1950s and early 1960s. The department hosted many events, including the community's summer picnic, where nearly a thousand residents usually joined together in Butte's treasured Columbia Gardens. Department members also constructed elaborate floats for all of the local parades and hosted big parties for young men drafted into World War II. The department was especially noteworthy for its Christmas traditions. Each year, members drove Butte's nuns around on a Christmas lights tour and hosted a Christmas party for four hundred Meaderville children. They also created enormous, moving Christmas displays that drew hundreds to Meaderville each winter.[90]

For men in Meaderville's volunteer fire department, the social aspects of membership were as important as the job itself. Their approach matched what one historian describes for early volunteer firefighters in the United States; she claims they "viewed their membership as a masculine brotherhood."[91] Like the Cristoforo Colombo Club, the volunteer fire department served as a gathering place for Meaderville's young men. Firemen would help each other (and each other's families) out during long strikes at the mines and other rough economic spells. Former fire chief Angelo Petroni and his wife Virginia recalled the many times that fellow firemen assisted their young family, helping to build a house after Angelo had been hurt fighting a fire, bringing salami and bread during strikes, and even providing their children with toys during one particularly tough Christmas.[92]

The volunteer fire department also mattered mightily for the residents of McQueen and East Butte. As Jim Michelotti recalled about his childhood, "The volunteer fire department would show movies on Friday nights, and then they would give us treats." Men in the fire department also maintained a winter ice skating rink and even built a neighborhood playground on Anaconda Company land in 1952.[93]

Games and sports served as an additional unifying factor for people living in Butte's eastern neighborhoods. In addition to functioning as a meeting place and bar, the McQueen Athletic Club and its 282 members sponsored many neighborhood sports teams, but baseball reigned supreme. According to one 1948 newspaper article, the "McQueen Addition is to Butte What Brooklyn is to New York" in relation to the "same devotion" each place bestowed on its baseball team.[94] A state league functioned until World War II, but amateur baseball continued in Butte for the next couple of decades, first through a city league and then through the Copper League, which drew over 136,000 fans in 1948.[95] Community sports teams thus formed another important institution for Butte's working-class neighbors to express allegiance to their neighborhood and to each other.

Residents of these three neighborhoods therefore had a full social and cultural world contained in each community. Since each neighborhood housed most of the services residents required, people rarely needed to leave it. Simply staying so close to one another for such a long time increased the strength of their social ties. Local historian Pat Kearney claims that "McQueen's isolation forced residents to be self-sufficient," a condition that likely held for the other neighborhoods as well. As East Butte resident Rose Marie Ralph recalls, "It was almost like we had our own little town."[96] Indeed, former residents who spent their childhoods there claimed that they largely stayed in their own neighborhoods day-in and day-out until their teenage years. Seldom did any child from McQueen, East Butte, or Meaderville go uptown to the business district; former McQueen resident Jim Michelotti says that going "uptown was viewed almost like going to Seattle.... You just didn't leave. I mean, there was no reason to."[97] For many children, the only reasons to leave the neighborhood were to buy clothes or to see a movie.[98]

Outside of those few activities, most children's leisure took place within the area, particularly at the Columbia Gardens. Its origins came in 1899, when William A. Clark purchased twenty-one acres, tucked up against the mountains, just a few miles east of Butte's CBD.

He hoped the move would win him local support against rival copper baron Marcus Daly. Clark, who had been disqualified that year from taking up a U.S. Senate seat by that body's own Committee on Privileges and Elections, wanted to generate enthusiasm for his next effort to join that millionaire's club. His plan was to take the land, which had solely hosted men's events like horse racing and cockfights, and turn it into a grand public playground, meant especially for local children.[99]

Clark enlisted a trusted associate, Jesse Wharton, who managed the Butte Electric Railway Company for Clark, to spend more than $100,000 of Clark's wealth on the Gardens. Wharton built a grand pavilion with a large dance floor, a children's playground, Montana's first fish hatchery, sporting grounds, and a zoo that contained bison, elk, mountain lions, and a bear. A lake hosted swimming, boating, and a long water slide. The real showpieces, however, were the meticulously landscaped gardens, which featured enormous flowerbeds and floral pieces such as harpsichords, butterflies, and the BMU insignia. The Gardens employed a sixteen-person maintenance crew who tended to the gardens full time.

The Columbia Gardens became readily accessible to all of Butte's residents when Wharton added a streetcar track to it in 1902. Clark later bought the surrounding land, building up sixty-eight acres. An alcohol ban ensured that the Columbia Gardens remained a family park. Free transportation for those under sixteen years of age packed the Gardens with as many as 25,000 youngsters on children's days. In an attempt to show how much Clark supported working families, the park also hosted miner's union day celebrations, which included drilling and mucking (shoveling) contests. A small neighborhood formed around the park.[100]

Wharton's Butte Electric Railway operated the Columbia Gardens until 1928. Clark died in 1925, and in 1928 the Anaconda Company, which increasingly held a monopoly on the area's mines, bought Clark's holdings from his heirs. Although the company closed the zoo and did not refill the artificial lake, it installed a new roller coaster and carousel, as well as fresh playground equipment, including cowboy swings. Caretakers also continued the tradition of weekly children's days, lush flower displays, and picnic spots, while adding big band dances with the hottest touring groups.[101]

Leisure activities beyond the gardens typically happened even closer to home. While men had bonding experiences in places like the volunteer fire department, the Cristoforo Colombo Club, or the McQueen Athletic Club, women had their own homosocial opportunities. In Meaderville, women's social world included scheduled get-togethers and informal visits when they would share salami, cheese, and coffee.[102]

Former residents believe that proximity reinforced their social attachments. On the East Side, for example, buildings stood very close to each other. According to John Onkalo, on East Park Street "the houses were almost skin to skin.... In those days, if you wanted to see somebody, you'd go in front of his house and you'd call him."[103] Many residents did not have cars because work, especially for the Anaconda Company, was so close by. Street cars ran through the neighborhoods until the 1930s, when a bus system replaced them. Vince Crnich, a resident of McQueen for almost seventy years, believed that McQueen had lost some of its closeness when he was interviewed in the 1970s: "When people began relying more on autos, they started leaving the neighborhood."[104] Nevertheless, most people's stories from the neighborhoods' later decades suggest that socializing still took place because of consistent on-foot traffic. Residents frequently stopped to chat over a fence, or popped in on each other to enjoy impromptu meals, pinochle nights, and other social gatherings.[105]

In contrast to newer, more-spread-out suburban neighborhoods on the Flats below Butte's hill, residents on the hill maintained a strong attachment to their neighborhood and their neighbors. Almost all former residents of the vanished neighborhoods interviewed in the past four decades made statements similar to Ann Marie Quilici Reynolds from McQueen. These were close-knit communities where everybody was good friends who had lived next to each other for decades, some for over half a century. Some of this closeness may be exaggerated, since interviewees certainly have reason to look back fondly on places that no longer exist. Yet, in interviews today, decades after these neighborhoods disappeared, residents can still list off the top of their heads who once lived in each residence down their street.[106] Many statements about neighborhood life mirror those of Bob Pavlovich, who claims that "there was never a door locked on the East Side." Indeed, "at supper time, where you were, that's where you ate."[107] Former East Butte resident Rose Marie Ralph expressed a similar sentiment: "Everyone seemed to take care of each other."[108]

Those who were not friends were usually relatives. In interviews, residents often list various family members living right next to each other—aunts and uncles next door, cousins down the street. Former McQueen resident Jim Michelotti, for instance, remembered, "My grandparents both lived half a block or a block away—aunts and uncles throughout the neighborhood."[109] Other McQueen residents, like Fran Corcoran and Arlene Stepan, agree that most family members stayed near to each other up until Anaconda began neighborhood relocation.[110]

Of course, this close-knit situation meant that "everybody was in everybody's business," in the words of Meaderville resident Bernie Lazzari. According to former residents, few could keep a secret in Meaderville, McQueen, or the East Side. Residents looked after each other's children; youngsters were not only safe at play, but also under a supervision that curtailed mischief.[111]

Of course, these working-class neighborhoods did exhibit tensions and divisions. In McQueen, some families did not get along. Some remember that strict divisions existed between McQueen folk and the nearby East Butte folk, but others deny that.[112] Some remember less pleasant aspects of the East Side's diversity. One lady who grew up there during the 1960s recalled, "Because of the mixture of ethnic backgrounds it was a tough neighborhood to grow up in. You hear that statement from many who grew up there.... I remember coming home from popular Thompson Park picnic area and seeing a swastika painted on our front door or hearing from time to time prejudiced remarks about folks from the south or different races."[113] Children who were newcomers to the neighborhood often ended up in fights with the close-knit kids around them.[114]

Most residents nevertheless look back positively rather than negatively on a compact, unified neighborhood life. One man recalls that even violence could bring together a neighborhood, as it did when his East Side neighbor children gathered together "regardless of nationality" to fist-fight with children from the Dublin Gulch neighborhood that sat above them.[115] Although nostalgia for now-lost places clearly plays a role in residents' memories, it is hard to deny how strongly ethnicity, religion, class, leisure, and proximity had bonded them together. Their strong ties made it especially tragic when these neighborhoods encountered expanding open-pit operations.

## Up North

Despite the fact that eastern neighborhoods would eventually bear the brunt of open-pit mining, the first neighborhoods to encounter mass mining were actually north, not east, of the CBD. In that area, clusters of houses continued up a very steep grade. Because most of Butte's original lode mines were located in the area, northern neighborhoods like Dublin Gulch and Walkerville had been established longer than eastern neighborhoods like McQueen and East Butte. They therefore tended to house people with heritage from the area's longer-standing ethnic groups, like the Cornish and Irish, rather than the southern and eastern European ethnics who predominated in the eastern neighborhoods by the 1920s.

Dublin Gulch sat just north and east of the CBD. Once known as Town Gulch, Dublin Gulch had actually hosted Butte's first permanent, non–American Indian inhabitants in the 1860s. Many of Butte's richest mines, like the Neversweat, Anaconda, St. Lawrence, High Ore, Steward, and Parrot, were in this area, making it an easy trip to work for residents. Houses followed Anaconda Road to the east, right up to the mine gate. Like most of Butte's eastern neighborhoods, Dublin Gulch sat outside of the city limits, possibly because the Anaconda Company held so many mining properties there and thus did not want to pay city taxes on them.[116]

Many of Butte's Irish American families made the area home, as is evident in both the neighborhood's name and the name of the subcommunity nearby, Corktown, which bore the moniker of the Irish home county for its first inhabitants. There were originally a few boardinghouses, but most residents lived in single-family homes. In

1934 one writer called the gulch's residents "shanty" Irish because their small wooden workers' cottages delineated them from the "brick house" Irish in Centerville, which sat above the gulch. Dublin Gulch also served as an entry point for southern and eastern Europeans in the 1910s and 1920s, but, facing outward hostility from the Irish, many of them later relocated. Unlike some other neighborhoods, residents did not have too much of a business district, although the neighborhood store and a few saloons served as community centers.[117]

Like the eastern neighborhoods, Dublin Gulch was close-knit. One resident described its hundred plus homes as housing "neighbors who were all caring for each other."[118] It seemed like everyone in the neighborhood was Catholic and hence went to St. Mary's Church in Corktown. Most children attended St. Mary's grade school. Socioeconomic class also bonded residents. If someone perceived a resident as being "a little richer than they were," they quickly decided "they wouldn't associate with you," in one former resident's words.[119] Thus, members of the gulch community remained staunchly working-class in culture if not in fact.

No matter how poor they were, the residents of Irish heritage benefitted socially by being part of the most celebrated ethnic group in Butte, a place where St. Patrick's Day had long been a major holiday. Continuing the work of a Gaelic cultural revival overseas, St. Mary's Church offered Irish history and Gaelic language classes. On the day after Easter, children would wear cardboard lilies to remind everyone of the 1916 Easter Rebellion, thus expressing solidarity with the cause of Irish independence. Gaelic poems and tales of banshees show up in oral histories with men and women raised in the area. Although some traditions gradually disappeared, others remained. Traditional Irish wakes for the dead, for instance, continued to the middle of the twentieth century, even if participants' adherence to customs like keening, which happened when women poetically lamented the death through crying and wailing, faded.[120]

Above Dublin Gulch and Centerville, at the top of "the richest hill on earth," was Walkerville. The town bears the name of Utah silver-mining capitalists David, Joseph, Matthew, and Samuel Walker. The brothers had purchased the Alice claim on the advice of their agent, the future copper baron (and Anaconda Company founder) Marcus Daly, in 1876. Popular boardinghouses opened, and stone and brick commercial buildings became fixtures on the main thoroughfare, later to become known as Daly Street, while small wooden miners' cottages sprang up on narrow, winding roads on the slopes and gullies around the Lexington and Alice Mines. By the early 1900s, almost 2,500 people lived in Walkerville, and more than half of the residents worked in the copper mines. In 1890 Walkerville incorporated and elected a mayor and city council.[121] At 6,172 feet high, it held the distinction of being the highest city in Montana. Although its haphazard early development meant less city planning at first, Walkerville joined McQueen in gaining a sewerage system through the federal Works Progress Administration in the 1940s.[122]

Although an incorporated town, Walkerville shared a number of similarities with working-class communities in eastern Butte. In these neighborhoods, ethnic enclaves were important to community culture. In Walkerville's case, the Cornish predominated during the earliest wave of immigration, with a number of Irish joining them later, despite the long-standing enmity between the two groups. A few eastern and Central Europeans followed in the 1910s and 1920s. Social life in Butte's working-class communities revolved around local schools and churches. In Walkerville this meant Mount Bethel Methodist Church for those with Cornish heritage and St. Lawrence O'Toole Catholic Church for everyone else. The volunteer fire department served as a neighborhood focal point and the source of much of its leadership.[123]

As in Butte's eastern neighborhoods, Walkerville's residents had few reasons to leave their close-knit community. Resident Dolores Barsanti recalled that mid-twentieth century Walkerville had a theater, a barbershop, and five bars that hummed with activity. The Manza Market, a grocery, served the community for more than seventy years. Livina Richards, a longtime resident, reminisced, "It really was a thriving town."[124] During the 1940s and 1950s, some Walkerville residents worked in these businesses, but many others continued to work for the Anaconda Company.

Chain migration, a desire to live next to their workplace, and long-standing economic inequality had forged deep social bonds in Butte's working-class neighborhoods, and many cultural activities had nurtured those bonds over the decades. Little did these residents know that the same industry that had provided for these communities' growth was poised to change them forever. First Dublin Gulch and Walkerville, then the East Side, Meaderville, McQueen, and East Butte, would face mass mining's new hazards and hunger for land.

## Notes

1. Ira L. Swett, *Montana's Trolley's 2: Butte Anaconda BAP, Interurbans Magazine, Special* 50: 26, no. 4 (Winter 1969): 12–34; Murphy, *Mining Cultures*, 6.

2. Murphy, *Mining Cultures*, 10–15; quotes on 13.

3. Jim Michelotti, interview by Brian Leech, July 28, 2010, Butte, MT, transcript and digital recording, OHC, BSBPA.

4. On ethnic enclaves in mining communities, see Francaviglia, *Hard Places*, 103–12.

5. On immigrants' adaptation in 1930s Butte, see Murphy, *Mining Cultures*, 93, 151–52. On white interethnic alliances and racial challenges, see Basso, *Meet Joe Copper*, 159–93. On the union movement's ability to connect residents, see Mihelich, "The Richest Hill on Earth."

6. For a suggestion that ethnic identity still played a major role in the lives of white ethnics during the 1950s and 1960s, see Margaret Brennan Lee, "Ethnic Limits: Negotiating Ethnicity, Race, and Community Boundaries in Suburban Chicago" (PhD Diss., University of Wisconsin, Madison, 2010). The view that whites shed their ethnic identity in favor of a racial one after World War II can be found in Eric Avila, *Popular Culture in the Age of White Flight: Fear and Fantasy in Suburban Los Angeles* (Berkeley and Los Angeles: University of California Press, 2004); David R. Roediger, *Working Toward Whiteness: How America's Immigrants Became White, The Strange Journey from Ellis Island to the Suburbs* (New York: Basic Books, 2005).

7. Murphy, *Mining Cultures*, 14, 58–60; Filpula, *Meaderville Revisited*, 1; Esther Franconi, Maurice Maffei, and Angelo Petroni, Butte Historical Society Meeting, Topic: Meaderville Italians, Nov. 19, 1992, Butte, MT, cassette tapes, OHC, BSBPA; Al Hooper, interview by Marc J. Lee, Dec. 1, 1999, Butte, MT, transcript, OHC, BSBPA.

8. Quivik, "Smoke and Tailings," 144–48; Meader Family Association, "Charles Thain Meader and Meaderville, Montana," in *Meaderville Reunion Book* (Butte, MT: 2006).

9. "Silver Bow County and City of Butte Civil Defense Evacuation Map, 1962," VF 1361 "Butte Street Maps," BSBPA; Butte Chamber of Commerce, "Detailed Map of Butte," 1957, VF 1361 "Butte Street Maps," BSBPA; "Butte Map, County Surveyor," 1956; Butte Chamber of Commerce, "Butte," 1939, VF 1361 "Butte Street Maps," BSBPA. See also Filpula, *Meaderville Revisited*, 104–5.

10. Petroni and Petroni interview; John Orizotti, interview by Cody Maldonado, Apr. 22, 2005, Butte, MT, transcript, OHC, BSBPA; Franconi, Maffei, and Petroni, BHS meeting; Holter interview, by Leech. See also "Meaderville Volunteer Fire

Department: Fifty Years of Community Service," 1960, VF 1117, BSBPA; John Mihelich, "Baseball Was Our Life: Amateur Baseball in Butte, Montana, 1920–1960," *Montana: the Magazine of Western History* 59, no. 2 (Summer 2009): 62–73.

11. On company towns, see James B. Allen, *The Company Town in the American West* (Norman: University of Oklahoma Press, 1966); Oliver J. Dinius and Angela Vergara, ed. *Company Towns in the Americas: Landscape, Power, and Working-Class Communities* (Athens: University of Georgia Press, 2011).

12. Segna and Zanchi interview; Camille Maffei and Lucy Maffei, interview by Russ Magnaghi, May 4, 1983, Butte, MT, Oral History 583, cassette tapes, OHC, MHS; Angelo Petroni, interview by Russ Magnaghi, May 4, 1983, Butte, MT, Oral History 581, cassette tapes, OHC, MHS; Franconi, Maffei, and Petroni, BHS meeting; Matteuchi interview.

13. Franconi, Maffei, and Petroni, BHS meeting.

14. Petroni and Petroni interview.

15. Pat Kearney, *Butte Voices: Mining, Neighborhoods, People* (Butte, MT: Skyhigh Communications, 1998), 209. Ivan Stepan and Arlene Stepan, interview by Brian Leech, Aug. 5, 2010, Butte, MT, transcript and digital recording, OHC, BSBPA; Ann Bone, interview by Brian Leech, July 30, 2010, Butte, MT, transcript and digital recording, OHC, BSBPA; Michelotti interview.

16. Andrea Ciabattari, "McQueen Spirit: Ageless, Endless," *Mont. Std.*, July 29, 1973; Stephen James Luft, "Virtual Reconstruction: A Primer to the Digital Reconstruction of McQueen" (master's thesis, Montana Tech, 2004), 15; Kearney, *Butte Voices*, 186–87; Rudolph Petritz, interview by Keri Petritz, Nov. 15, 1998, Butte, MT, transcript, OHC, BSBPA; Stepan and Stepan interview.

17. Bone interview.

18. Ibid.; Michelotti interview; Kearney, *Butte Voices*, 188.

19. Mary Murphy and Brian Shovers, "The Decline of a Neighborhood: Butte's Eastside," Butte Historical Society, 1986, VF 1615, BSBPA, 2–3, 7; Kearney, *Butte Voices*, 189; Andrea McCormick, "East Park Survivors Go It Alone," *Mont. Std.*, Oct. 21, 1979.

20. Murphy and Shovers, "The Decline," 1.

21. Ibid., 7–9.

22. Aili Maria Goldberg, interview by Mary Murphy, Feb. 29, 1980, Butte, MT, transcript, OHC, BSBPA.

23. Cheryl Lynn Madison, "Butte's 'East Side': Memories and History of the 1960s," Oct. 19, 1999, page 5, VF 1614, BSBPA; Ray Wayrynen, interview by Nick Badovinac, n.d., Butte, MT, Box 1, Fldr. 17, Personal Collection 096 "Nick Badovinac—East Side" (hereafter PC96), BSBPA; Andrea McCormick, "Old Drug Store Bottles Up Years of Memories," *Mont. Std.*, Oct. 21, 1979; McCormick, "East Park Survivors."

24. Emmons, *Butte Irish*, 13, 255–91.

25. Shovers, "The Perils of Working in the Butte Underground."

26. Father Sarsfield O'Sullivan, interview by Teresa Jordan, Feb. 10, 1985, Butte, MT, notes, OHC, BSBPA; Ann Marie Quilici Reynolds, interview by Roxanna Lynn Reynolds, Oct. 30, 2005, Butte, MT, transcript, OHC, BSBPA; Marlene Holayter, interview by Brian Leech, July 29, 2010, Anaconda, MT, transcript and digital recording, OHC, BSBPA; Peter Zanchi, interview by Brian Leech, July 30, 2010, Butte, MT,

transcript and digital recording, OHC, BSBPA; Orizotti interview; Matteuchi interview; Pasini and Pasini interview.

27. Segna and Zanchi interview.

28. deBarathy is quoted in Thornton, "'Little Italy.'" See also Petroni interview by Magnaghi; Segna and Zanchi interview; Matteuchi interview; Pat Bartoletti, interview by Camie Hoppe, Nov. 18, 2005, Butte, MT, transcript, OHC, BSBPA; Petroni and Petroni interview; Peter Zanchi interview; Holayter interview.

29. Petroni interview, by Magnaghi; Maffei and Maffei interview; Petroni and Petroni interview; "Meaderville Volunteer Fire Department," VF 1117, BSBPA.

30. Thornton, "'Little Italy.'" On food and wine in Meaderville, see Petroni interview by Magnaghi; Maffei and Maffei interview; Segna and Zanchi interview; Frank Jursnick, interview by Ray F. Calkins, June 6, 1979, Butte, MT, transcript, OHC, BSBPA; B. Lazzari interview; Orizotti interview; Franconi, Maffei, and Petroni, BHS meeting; Ann Pentilla, interview by Caroline Smithson and Ray Calkins, Apr. 27, 1979, Butte, MT, transcript, OHC, BSBPA; Peter Zanchi interview; Paula J. McGarvey, "Mining City Christmas Memories," *Mont. Std.*, Dec. 25, 2003; Filpula, *Meaderville Revisited*, 106–7.

31. On food and drink at Meaderville clubs, see Minnie Piazzola, interview by Laurie Mercier, Mar. 11, 1982, Butte, MT, cassette tapes, Oral History 454, OHC, MHS; Petroni interview, by Magnaghi; Segna and Zanchi interview; Segna and Zanchi interview; Maffei and Maffei interview; Tracy Thornton, "King of Fine Dining," *Mont. Std.*, Dec. 12, 2004; Thornton, "'Little Italy'"; Tom Holter, interview by Teresa Hayward, Nov. 12, 1998, Butte, MT, transcript, OHC, BSBPA; John Scofienza, interview by Ray F. Calkins, June 27, 1979, Butte, MT, transcript, OHC, BSBPA; Jursnick interview; Petroni and Petroni interview; Holayter interview.

32. Harvey Levenstein, "The American Response to Italian Food, 1880–1930," in *Food in the USA*, ed. Carole M. Counihan (New York: Routledge, 2002), 75–90.

33. Ibid.

34. Thornton, "King of Fine Dining"; Murphy, *Mining Cultures*, 59; Holayter interview.

35. Tracy Nicole Poe, "Food, Culture, and Entrepreneurship Among African-Americans, Italians, and Swedes in Chicago" (PhD diss., Harvard University, 1999); Levenstein, "American Response."

36. *Butte City Polk Directory*, 1955, 711–16.

37. Tracy Nicole Poe, "The Labour and Leisure of Food Production as a Mode of Ethnic Identity Building Among Italians in Chicago, 1890–1940," *Rethinking History* 5, no. 1 (2001): 131–48; Hasia R. Diner, *Hungering for America: Italian, Irish, and Jewish Foodways in the Age of Migration* (Cambridge, MA: Harvard University Press, 2001).

38. Diner, *Hungering for America*; Levenstein, "American Response."

39. Diner, *Hungering for America*, 54.

40. Frank Shone, interview by John Terreo, Apr. 3, 1989, Helena, MT, cassette tape, Oral History 1229, OHC, MHS. On food traditions that differed from restaurant menus, see Maffei and Maffei interview; Segna and Zanchi interview.

41. On gambling at the clubs, see Manila Murphy, interview by Ray F. Calkins, Jan. 1980, Butte, MT, transcript, OHC, BSBPA; William Antonioli, interview by Chris Antonioli, Mar. 27, 2002, Butte, MT, transcript, OHC, BSBPA; Shone interview.

42. Murphy, *Mining Cultures*, 58–63.

43. Franconi, Maffei, and Petroni, BHS meeting; Peter Zanchi interview; Thornton, "King of Fine Dining."

44. Vidrich interview; Bob Kovacich, interview by Teresa Jordan, Mar. 19, 1986, Butte, MT, partial transcript and notes, OHC, BSBPA; Rose Marie Ralph, interview by Brian Leech, Aug. 2, 2010, Butte, MT, transcript and digital recording, OHC, BSBPA; Holter interview, by Leech; Stepan and Stepan interview; Fran Corcoran in Trux Fisher and Fran Corcoran, interview by Brian Leech, Oct. 21, 2009, Butte, MT, OHC, BSBPA; Murphy, *Mining Cultures*, 14; Kearney, *Butte Voices*, 187.

45. For early, aggressive uses of this term, see Emmons, *Butte Irish*, 256; Basso, *Meet Joe Copper*, 22–36. On its later, more-friendly use, see Basso, *Meet Joe Copper*, 44–45; Nancy Rae and Ronald Rae, interview by Teresa Jordan, June 21, 1986, transcript, OHC, BSBPA.

46. Ivan Stepan in Stepan and Stepan interview.

47. Emmons, *Butte Irish*, 255–56; Andrea McCormick, "Given a Choice, They Never Would Leave," *Mont. Std.*, July 2, 1989; Ciabattari, "McQueen Spirit."

48. McCormick, "Given a Choice."

49. Vidrich interview; Fran Corcoran in Fisher and Corcoran interview; Michelotti interview.

50. Margaret R. Lemm, interview by Mary Murphy, May 9, 1980, partial transcript and notes, OHC, BSBPA; Ciabattari, "McQueen Spirit"; Kearney, *Butte Voices*, 187.

51. Holter interview, by Leech. See also Michelotti interview.

52. Ivan Stepan in Stepan and Stepan interview. See also McCormick, "Given a Choice"; Ciabattari, "McQueen Spirit."

53. Ciabattari, "McQueen Spirit"; Bone interview; Ralph interview; Michelotti interview; Fran Corcoran in Fisher and Corcoran interview; Kearney, *Butte Voices*, 209.

54. Carl Gegurich quoted in Kearney, *Butte Voices*, 187.

55. Ciabattari, "McQueen Spirit."

56. Lemm interview. See also Ivan Stepan in Stepan and Stepan interview.

57. Ralph interview; Murphy, *Mining Cultures*, 14.

58. Murphy, *Mining Cultures*, 14, 151; Murphy and Shovers, "The Decline," 3; Thad Kelling, "Finntown, East Side Residents Gather to Reminisce," *Mont. Std.*, Aug. 21, 2005; Kearney, *Butte Voices*, 189; Basso, *Meet Joe Copper*, 152–53.

59. Bob Pavlovich quoted in Kelling, "Finntown." See also Stepan and Stepan interview.

60. Murphy, *Mining Cultures*, 12; Murphy and Shovers, "The Decline," 7; Onkalo interview.

61. Madison, "Butte's 'East Side,'" 3.

62. Murphy and Shovers, "The Decline," 4; Kathleen Cook, "Finn Town Era Is Past," *Mont. Std.*, July 5, 1976.

63. Murphy, *Mining Cultures*, 13.

64. Goldberg interview. On the social geography of Finntown, see also Onkalo interview; Murphy and Shovers, "The Decline," 7.

65. Thomas J. Archdeacon, *Becoming American: An Ethnic History* (New York: The Free Press, 1983), 128.

66. Wayrynen interview; Goldberg interview; Murphy and Shovers, "The Decline," 4–6.

67. Andrea McCormick, "Finn Stories Still Flourishing in Old East Park Apartments," *Mont. Std.*, Oct. 21, 1979; Bonnie Christensen, *Red Lodge and the Mythic West: Coal Miners to Cowboys* (Lawrence: University Press of Kansas, 2002), 59–70.

68. Goldberg interview. On boarding houses in Finntown, see Murphy and Shovers, "The Decline," 6–7; Murphy, *Mining Cultures*, 49–50; Cook, "Finn Town Era Is Past"; McCormick, "Finn Stories"; John Laitinen to Nick Badiovinac, Dec. 13, 1989, Box 1, Fldr. 4, PC96, BSBPA; Agnes Cramer, interview by Mary Murphy, Mar. 28, 1980, Butte, MT, transcript, OHC, BSBPA; Goldberg interview; Wayrynen interview.

69. Emmons, *Butte Irish*, 255–91.

70. Calvert, *The Gibraltar;* John P. Enyeart, *The Quest for "Just and Pure Law": Rocky Mountain Workers and American Social Democracy, 1870–1924* (Stanford, CA: Stanford University Press, 2009).

71. Murphy, *Mining Cultures*, 13, 150–51; John Laitinen to Nick Badiovinac, PC96, BSBPA; McCormick, "Finn Stories"; Goldberg interview; Onkalo interview.

72. John Laitinen to Nick Badiovinac, PC96, BSBPA; Goldberg interview; Cramer interview.

73. Wayrynen interview.

74. Ibid.; McCormick, "East Park Survivors"; Cook, "Finn Town Era Is Past"; Madison, "Butte's 'East Side,'" 3–5; Kearney, *Butte Voices*, 190.

75. Anne Hetzl Gunkel, *Polish American Studies* 61, no. 2 (Autumn 2004): 5–8; Richard March, "Polka and Tamburitza: Ethnic Music and Dance Traditions in the Upper Midwest," in *Ethnic and Border Music*, Norm Cohen, ed. (Westport, CT: Greenwood Press, 2007), 139–70.

76. Rick Foote, "Mining City Hears Another Sad Song," *Mont. Std.*, May 24, 1975. See also Kelling, "Finntown."

77. Yvonne R. Lockwood, "The Sauna: An Expression of Finnish-American Identity," *Western Folklore* 36, no. 1 (Jan. 1977): 71–84, quote on 75. See also Aili K. Johnson, "Lore of the Finnish-American Sauna," *Midwest Folklore* 1, no. 1 (Apr. 1951): 33–39.

78. Cotton Mather and Matti Kaups, "The Finnish Sauna: A Cultural Index to Settlement," *Annals of the Association of American Geographers* 53, no. 4 (Dec. 1963): 494–504. On saunas in the Pacific Northwest, see Jon T. Kilpinen, "Finnish Cultural Landscapes in the Pacific Northwest," *Pacific Northwest Quarterly* 86, no. 1 (Dec. 1994): 25–34.

79. Murphy and Shovers, "The Decline," 6.

80. Wayrynen interview; John Laitinen to Nick Badiovinac, PC96, BSBPA; Goldberg interview.

81. Archdeacon, *Becoming American*, 204–7; Will Herberg, *Protestant–Catholic–Jew* (New York: Doubleday, 1955).

82. Mihelich, "The Richest Hill on Earth," 201–24, quote on 202.

83. Quote is Camille Maffei from Maffei and Maffei interview. See also Petroni interview, by Magnaghi; Pasini and Pasini interview; Michelotti interview; Orizotti interview; A. M. Q. Reynolds interview; *Butte City Polk Directory*, 1955, 711–16; Cindy Gaffney, "Glimpse of Heaven," *Mont. Std.*, Dec. 25, 2008; Kearney, *Butte Voices*, 216.

84. Lorraine Lazzari, interview by Callie Boyle, Nov. 27, 2005, Butte, MT, transcript, OHC, BSBPA.

85. Michelotti interview; Bone interview; Stepan and Stepan interview; Ciabattari, "McQueen Spirit."

86. "Sacred Heart Church Closes Doors," *Mont. Std.*, July 2, 1970; Madison, "Butte's 'East Side,'" 4.

87. Murphy and Shovers, "The Decline," 7; Madison, "Butte's 'East Side,'" 4; Murphy, *Mining Cultures*, 14; Madison, "Butte's 'East Side,'" 5; Kearney, *Butte Voices*, 189.

88. Bone interview. See also Ralph interview; Holter interview, by Hayward; Stepan and Stepan interview.

89. Charles R. Simpson, "A Fraternity of Danger: Volunteer Fire Companies and the Contradictions of Modernization," *American Journal of Economics and Sociology* 55, no. 1 (Jan. 1996): 17–34, quote on 17.

90. Petroni and Petroni interview; "Meaderville Volunteer Fire Department," VF 1117, BSBPA; A. J. Boyle to Meaderville Fire Department, Feb. 2, 1947, in VF 1117, BSBPA; McGarvey, "Mining City Christmas Memories."

91. Amy S. Greenberg, *Cause for Alarm: The Volunteer Fire Department in the Nineteenth-Century City* (Princeton, NJ: Princeton University Press, 1998), 4.

92. Petroni and Petroni interview. See also Pauline deBarathy interview by Brian Leech, July 20, 2010, transcript and digital recording, OHC, BSBPA; Kearney, *Butte Voices*, 211–21.

93. Michelotti interview. See also Stepan and Stepan interview; Holter interview, by Leech; Charlie Staples, Anaconda Company Rent Dept., to McQueen Fire Dept., May 13, 1952, Box 3, Fldr. 15 "McQueen Contracts and Land Agreements," Organizational Collection 19 (hereafter OC19), BSBPA; McCormick, "Given a Choice."

94. "McQueen Addition Is to Butte What Brooklyn Is to New York," *Mont. Std.*, Mar. 24, 1948. See also Kearney, *Butte Voices*, 187; Holter interview by Leech; Michelotti interview; Bone interview; Stepan and Stepan interview.

95. Mihelich, "Baseball Was Our Life," 62–73.

96. Kearney, *Butte Voices*, 210; Ralph interview.

97. Michelotti interview.

98. Stepan and Stepan interview; McCormick, "Given a Choice."

99. "Purchased the Gardens," *Daily Inter-Mountain* (Butte, MT), Mar. 29, 1899; Dave Walter, "Butte Paradise: Columbia Gardens," *Montana Magazine* 73, no. 5 (Sept.-Oct. 1985): 22–27.

100. Dave Walter, "Butte Paradise," 22–27; "Beautiful Columbia Gardens: 'Senator Clark's Masterpiece,'" 1899, VF 1045, BSBPA; Carl Englebach, "Remembrances of Carl Englebach," in *Interurbans Magazine, Special 50: Montana's Trolley's 2: Butte Anaconda BAP*, ed. Ira L. Swett, 26, no. 4 (Winter 1969): 36; Virginia Bugni, interview by Brian Leech, Oct. 22, 2009, Butte, MT, digital recording and transcript, OHC, BSBPA.

101. "Columbia Gardens," *Mont. Std.*, Aug. 2, 1964; Ted Beech, interview by Ray Calkins, Oct. 27, 1979, Butte, MT, transcript, OHC, BSBPA; Fisher and Corcoran interview; Frank Panisco, interview by Ray F. Calkins, May 6, 1980, Butte, MT, transcript, OHC, BSBPA.

102. deBarathy interview; Holayter interview; Michelotti interview.

103. Onkalo interview. See also Kelling, "Finntown."

104. Crnich quoted in Ciabattari, "McQueen Spirit." See also Stepan and Stepan interview.

105. Fisher and Corcoran interview; deBarathy interview; Holayter interview; McCormick, "Given a Choice."

106. A. M. Q. Reynolds interview. See also L. Lazzari interview; Franconi, Maffei, and Petroni, BHS meeting; Orizotti interview; Manila Murphy interview; Scofienza interview; Holayter interview; Peter Zanchi interview; deBarathy interview; Petroni interview by Magnaghi; Pasini and Pasini interview.

107. Pavlovich quoted in Kelling, "Finntown."

108. Ralph interview. See also Michelotti interview; Kearney, *Butte Voices,* 209.

109. Michelotti interview.

110. Fisher and Corcoran interview; Stepan and Stepan interview; Bone interview.

111. B. Lazzari interview. See also Bartoletti interview; Petroni and Petroni interview; Matteuchi interview.

112. Kearney, *Butte Voices,* 209; Ralph interview; Bone interview; Michelotti interview.

113. Madison, "Butte's 'East Side,'" 2–3.

114. Kearney, *Butte Voices,* 189.

115. John Laitinen to Nick Badiovinac, PC96, BSBPA. See also Goldberg interview; Edward Jursnich, *Butte's East Side: Gone, But Not Forgotten* (Denver: Outskirts Press, 2009).

116. Mary Murphy and Brian Shovers, "St. Mary's Neighborhood: A Bedroom to Butte's Miners," Butte Historical Society, 1986, VF 1629, BSBPA, 8–10; Mary Murphy, "Neighborhoods in Decline: An Inventory of Centerville and Butte's East Side," Butte Historical Society, 1984, VF 1606, BSBPA, 2–3.

117. Murphy and Shovers, "St. Mary's Neighborhood," 8-10; Kearney, *Butte Voices,* 185–86.

118. Jule McHugh, "The Gulch and I," VF 1610, BSBPA, 11.

119. Catherine Hoy, interview by Caroline Smithson and Ray F. Calkins, May 11, 1979, Butte, MT, transcript, OHC, BSBPA; Finn, *Mining Childhood,* 87–91, 180, 192–93.

120. Finn, *Mining Childhood,* 87-91; Hoy interview; Alma E. Hileman, interview by Ray F. Calkins, June 27, 1980, Butte, MT, transcript, OHC, BSBPA; Kearney, *Butte Voices,* 181.

121. Malone, *Battle for Butte,* 11–24; Brian Shovers, "Walkerville: Butte's Sister Silver Town," *Report for the Butte Historic Landmark District Project,* June 1986, 5, 2, VF 1637 "Neighborhoods: Walkerville Inventory 1984 David Cassidy," BSBPA; "What's in a Name," *Butte Miner* (Butte, MT), Sept. 4, 1877; Beverly J. Brothers, *Sketches of Walkerville* (Butte, MT: Ashton Printing, 1973), 61.

122. Thomas S. Jenks, "Walkerville's Jimmy Shea: Trustee for a Family," *Mont. Std.,* Apr. 18, 1971; Jim Tracy, "Walkerville Now a Town," *Mont. Std.,* July 16, 1981.

123. Shovers, "Walkerville"; Kearney, *Butte Voices,* 230–34; Clayton W. Maxwell, "Hill City Group Asks Death for Mining Bill," *Mont. Std.,* Jan. 23, 1961.

124. Dolores Barsanti, interview by Cliff Rickey, Apr. 15, 2005, Butte, MT, transcript, OHC, BSBPA; Dan Carter, "Walkerville Won't Say Die," *Mont. Std.,* Mar. 31, 1985.

# PART II

# WORKING THE PIT, 1955–75

## CHAPTER 3

# PROTEST

A STRONG CHALLENGE to the Anaconda Company's authority was almost unprecedented in Montana history, but that is exactly what Butte's labor community gave it in the late 1950s and early 1960s.[1] The challenge came from union workers, who feared a loss of power over their working lives, as well as from Butte-area families, who worried about their homes and lifestyle. The major issue for all involved was a new form of mining: the open pit. Protestors were having real success. The first neighborhood to suffer from open-pit operations, Walkerville, used legal means to amazingly halt production in a major mine. Meanwhile, a resilient workforce shut down most of the company's other operations with a massive strike. Company leaders feared the worst, as a coalition began to form between the two groups. Walkerville residents offered support to miners. The city council censored Anaconda's national leaders for firing 212 Butte workers during the strike.[2] Miners similarly supported Walkerville's cause. During the autumn and winter of 1960, their union newspaper ran news about Walkerville's fight against the encroaching Alice Pit and they posted photos of the neighborhood's damaged landscape in the union hall.[3]

Perhaps Anaconda should have foreseen this challenge. One report suggests that many Butte residential districts were already targeted for demolition almost a decade before open-pit mining would begin to eat away at them. In May 1955, just months after it began open-pit operations on Butte's East Side, Anaconda was churn-drilling (a way to discover possible ore) just forty feet east of Main Street in the Meaderville neighborhood. A few months later, plans suggested the inclusion of parts of the East Side neighborhood within proposed pit boundaries.

By the autumn, a test pit for the main pit in this vicinity had already meant the rather expensive relocation of the Great Northern Railroad's tracks.[4] Three years into open-pit operations, the company's chief geologist was arguing that Butte should not plan any future developments—whether it be roads or residences—anywhere near the mining area on the eastern side of Butte's hill. As he explained to the company's national chairman, "Butte's future mine requirements demand absolute and incontestable priority over any other surface features designed to serve the community."[5]

Seeing their goals as vital to the city, Anaconda's management jumped into Butte's open-pit era with a number of small surface operations. Many of them were test pits to examine the ore grade near what would become the main pit, an East Side operation later known as the Berkeley Pit. Others were simply small pits throughout the area that the company planned as short-term enterprises. By October 1955, Anaconda had placed 167 workers on its special development program for the Berkeley, Gagnon, Skyrme, Syndicate, and East Colusa Pits. Soon Anaconda would add to this list the controversial Alice Pit north of Butte.[6] Each of these pits became successful in the short term, with decent ore grades of valuable metals like copper, silver, gold, zinc, and lead, but the Alice and Berkeley were the only pits to continue operating into the 1960s.

The growth of each pit led to as much community anger as it did enthusiasm. The 1950s meant job reduction in the traditional underground mining sector that had long been Butte's bread and butter. Certainly big job losses would have caused concern, but the open pit was such a dramatically different way to work that it quickly became a symbol of mechanization's social consequences. The wide-open spaces of the open pit made possible the use of bigger and bigger machines. Those machines' greater share of the mining workload became obvious to workers and community members, all of whom could see the growing open pits from their backyards. The pits' success also created industrial hazards that angered many Butte residents, and their expansion threatened the very existence of closely bonded neighborhoods. In other words, industrial work intruded into neighborhoods, challenging residents' lifestyles in ways that mining never had before. Panic and protest were the result.

Butte's grassroots protests against open-pit mining might surprise others who study the postwar workplace. In detailing the changes that mass production made to working life after World War II, most scholars discussing similar shifts have overlooked just how strongly many opposed these changes.[7] Although the specifics of Butte's development ultimately resulted in the acceptance of open-pit technologies, the initial protests against open-pit mining by Butte's labor community provide a different picture. Because the open pit exposed the mechanization of the workplace in such a dramatic fashion, Butte-area workers stood against mechanization despite Cold War pressures. Butte communities showed that they valued their social environment over jobs, and not all accepted open-pit mining as the best method for extracting ore, despite its efficiency.

That this labor community put up any fight is particularly surprising in light of the Anaconda Company's power at midcentury. The company typically had some of its own candidates in the city government, it controlled Montana's daily newspapers, and it held sway over lawmakers at the state capital.[8] Yet by 1959 the company still faced an angry, and rather resilient, protest by Butte workers, residents, and an important local politician. To explore this unexpected protest, this chapter will examine Butte's geography of labor—the connections between work, landscape, neighborhood life, and social and environmental action.[9]

## The Unions Meet Mass Mining

Some union members had already become worried when Anaconda announced a massive block-caving project called the Kelley Mine in 1947. This project, introduced at the optimistic height of the postwar boom in metal mining, was meant to modernize Butte's operations. Like open-pit mining, block-caving was a highly mechanized form of mass mining. It involved the creation of large blocks of rock hanging on the top of an underground tunnel. Miners would then use a blast to trigger a gravity-induced fall of this material onto the floor of a tunnel, or drift. They could then load the mass of low-grade material onto carts and ship it up to the surface to be processed. Block-caving required semiskilled workers to move the massive amount of material out of the mine, but it needed far fewer skilled workers than a typical underground mine. The Kelley Mine was thus the first attempt by Anaconda to use nonselective, mass-mining methods in Butte. A corporate historian hired by Anaconda called the project "a triumph of new ideas and mechanical progress."[10]

When the company introduced the project, both Butte's residents and national publications rejoiced. Anaconda was now going to concentrate on its birthplace, Butte, and not on its operations in Chile, Mexico, or the U.S. Southwest. The local newspaper praised the "faith of the Anaconda Copper Mining Company in Butte, her natural resources, and her people."[11]

Almost immediately, though, all was not well. The company placed the Kelley Mine yard right in the center of Dublin Gulch, one of Butte's key Irish American neighborhoods. As a result, Anaconda had to buy a fair number of homes, though without spending too much money, because it owned the ground on which those homes sat. By the end of the Kelley's first year of operation in 1953, subsidence around the mine had commenced. Ground above was sinking into the large empty cavities below.[12] Anaconda had predicted this event, so it had already moved its own carpentry, machine, and boiler shops out of the way. The company then purchased even more property in the Dublin Gulch neighborhood, clearing away additional houses. Although a number of homes remained, an area that had long served as the heart of Butte's Irish working class had swiftly been disrupted and put on the path toward erasure. Displaced locals, who could no longer find old community landmarks, expressed great sadness, but no record of a significant communal protest exists.[13] They seem instead to have believed that relocation was a justifiable step to ensure the continued success of Butte mining. Besides, subsidence was a familiar problem for Butte's residents.

Although they were a bit tentative, Butte's unions expressed more concerns than other locals about block-caving, focusing on its increased use of mechanization. Some members thought that mechanization had renewed the silicosis problem underground, because more dust was being churned up and safety practices were being disregarded in favor of speedier production. Others feared block-caving techniques because they appeared to be the latest in a line of technology that deemphasized the role of workers' skills in the mining process. Their biggest worry, though, was that the Kelley meant a new corporate direction toward nonselective mining. Rumors circulated among the rank and file that a more disruptive method might arrive in Butte: the open pit. Miners were aware of a dramatic increase in the number of U.S. open-pit mines in the immediate postwar period.[14]

Union leader Joe Mason tried to allay members' fears. In the June 13, 1953, edition of the local union newspaper, Mason joined the Anaconda Company in its celebration of postwar success, including a great rise in employment numbers. He addressed so-called misunderstandings about the Kelley Mine project: "It is well known that the Kelley ore is low grade, averaging only about twenty pounds of copper to the ton. This has given rise to the thought that Butte's high grade ores are playing out and that it must now resort to the mining of low grade remnants. Nothing could be further from the truth.... There is some apprehension that the Butte Hill is to become a huge open pit, but that is not really going to happen. Although the Greater Butte ore bodies are very large, they do not constitute an all-embracing mass."[15]

Mason and his compatriots should have suspected something; during bargaining one year earlier, one of Anaconda's major arguments insisted that mining in Butte's deep underground was high cost when compared to other corporations' open pits.[16]

Anaconda indeed proved Mason wrong when, using geological data from the Greater Butte Project, it began test pits for surface mining in late 1954. At first, the BMU saw open-pit mining as another attempt by the company to lessen union power over Butte's operations. Anaconda subcontracted the F&S Contracting Company to carry out its initial open-pit operations.[17] To the BMU, such an action represented a devious way to maneuver around the bargaining contract's requirement that Anaconda had to hire its members. The BMU held no jurisdiction over the blasting and earth-moving being done by F&S. The BMU's infuriated leadership also had no prior knowledge of the

F&S Contracting Company, which made them quite suspicious. Local leaders turned to Bernard "Bernie" Stern, who directed the research division for the IUMMSW.

Robert "Bob" Schrank, a Butte resident and international representative for the union, told Stern, "It has been my feeling since I have been here that this open pit operation is a real threat to our members in Butte. The company has taken the position that it has nothing to do with this contracting outfit. We are convinced here that this is just a ruse. Actually, the F&S Contracting Company is just a front for the Anaconda Copper Mining Company." The evidence for this allegation was that F&S, which formed in 1946, had filed only two officers with the state of Montana. One of these officers, J. T. Finlen, was Anaconda's western operations attorney and likely an Anaconda Company board member.[18] Stern's research division agreed with local union members: all they could find was a credit rating for F&S. Stern concluded, "It is pretty clear that this outfit is certainly an Anaconda 'stooge' corporation if not an actual dummy controlled by the Company. You will note, for example, that their main regular source of income is hauling for Anaconda Copper."[19]

Concerned union leadership started an open pit committee to investigate what this new form of mining might mean to Butte's workers. As bargaining began for the next contract, local president Ernest Sjoman appointed three stalwart BMU members to talk to the Anaconda Company: Ray Lee, Sunny Powers, and Earl Gregory.[20] From the autumn of 1955 to the spring of 1956, committee members met with company representatives about the growing pits around Butte. The committee's stated goal was to ensure that the BMU gained jurisdiction over pit operations. Anaconda's initial response was simply that it could not tell F&S whom to hire. Much to the BMU's dismay, the operating engineers instead gained jurisdiction over many workers in the new open pits.[21] The BMU turned to its lawyer for advice; its leaders wanted to file unfair labor charges against Anaconda, but it was not clear whether the union could make a strong enough case.

Faced with BMU allegations that the company was using the pit "to drive a wedge in the labor movement," however, company representatives went farther than the BMU had expected. During meetings,

Anaconda's representatives issued an ominous threat: "If the price of copper were to drop the open pit would probably be the only operation working in Butte, and the Butte miner would be out on the street."[22] Meanwhile, the United Steelworkers of America (USWA), trying to raid the local unions, riled up BMU members, claiming in one instance that, if workers failed to "wake up, . . . more machines all the time" would continue to displace miners.[23]

Underground block-caving should have been just as threatening as this new form of mass mining, but mechanization and job loss had become much scarier to everyone when they could see the growing open pits. The threat of mass mining's reduced labor needs had become truly visible in the form of open pits.

## The Unions Fight Each Other

The amount of attention Butte's locals devoted to open-pit mining is notable given the controversy that was engulfing unions nationwide: accusations of communism. Midcentury Butte was a close labor community with strong, often aggressive, unions and a constant distrust of company actions.[24] Its mine workers were the backbone (and founders) of the fiercely independent IUMMSW. The national labor movement had attracted leftists of all types during the 1930s and 1940s, but many unionists came to worry about the radicals they had accepted into their midst when anticommunism became a ubiquitous feature of Cold War America. In 1950 the Congress of Industrial Organizations (CIO) expelled the IUMMSW as well as a number of other unions thought to have Communist ties.[25] Montana's unions felt the full brunt of those attacks when the USWA decided to use allegations of communist infiltration—true or not—against Montana locals in an effort to gain membership for the USWA. Butte's copper industry locals fought tooth and nail against planned USWA raids, dividing the labor community.

Historian Laurie Mercier has argued that Cold War tensions greatly cost the area's unions, weakening possibilities for advances in the postwar years.[26] Evidence from records of the BMU, the largest of the Butte locals under IUMMSW, bears out Mercier's conclusions. Attacks by the USWA in Butte on the IUMMSW began in 1950, but were easily rebuked; the nearly six thousand members of the BMU were not particularly procommunist, but most were strong supporters of their local no matter what allegations it faced.[27]

After USWA courting, however, some longtime leaders turned away from the IUMMSW in 1953. These defectors included Bill Mason, Joe Mason, and their followers, whom the labor newspaper derided as "the Mason clique."[28] The Mason clique tried to take over the locals that year, forcing through a pro-USWA resolution during an ill-attended Christmas meeting. In response, former IUMMSW president Reid Robinson led a takeover of Butte's union hall with the backing of a large number of surprised, angry copper workers. The IUMMSW handily won a National Labor Board election to continue to represent Butte's workers in 1954.[29]

Despite this defeat, the USWA continued efforts to unsettle Butte's workers, particularly during bargaining sessions. Reports of USWA raids featured prominently in nearly every issue of the local IUMMSW paper. The steelworkers used their newspaper, the *Diggers' Truth*, and radio spots to attack the IUMMSW's leaders for being so untrustworthy and un-American that no one would partner with them in the dangerous underground. Cartoons in their newspaper pictured IUMMSW negotiators as weak and old.[30]

The IUMMSW countered in its own newspaper with appeals to local pride and allegiance to the western-based IUMMSW. As one article claimed, "Butte Built the Mine-Mill [IUMMSW]. For the record, Butte does not carry Number 1 for any other distinction, save for the fact that it was, and is the first local Union of the Mine-Mill, originally the WFM. That is a heritage that was handed down to us by our grandfathers; we have a right to continue and keep that heritage."[31] In contrast to the "pioneers of this Militant Union," the BMU labeled the Steelworkers as outsiders: "IMMIGRANTS FROM THE EAST."[32] BMU leaders argued that turning to the bigger union of "Stealworkers" actually meant forfeiting local autonomy—a transition from "freedom [to] slavery." Their newsletter's cartoons portrayed IUMMSW negotiators as large, muscular men.[33]

IUMMSW's propaganda about western pioneers proved more effective than the USWA's anticommunist arguments. Despite the defections of many locals around the United States and Canada to the USWA, Butte-area unions stayed loyal to the IUMMSW during the 1950s and early 1960s, but they expended a lot of energy doing so.

## Technological Unemployment

Despite the constant distractions of USWA raiding, Butte's union leaders began to see surface mining not simply as an attempt to split up Butte's unions, but as part of a much broader attack on other workers nationwide. Jobless Americans had fretted over the possibility of technological unemployment during the Great Depression, and believed mechanization was a major cause of massive unemployment. After World War II, those fears resurfaced. Wartime experimentation with production equipment led to a postwar obsession with new machines that could improve efficiency and reduce labor needs. Experts in engineering and management celebrated the next step in mechanization as a grand scheme that they labeled "automation."[34]

For many business owners and applied scientists, automation meant new machines that could operate for long periods without any human intervention at all. Scientists and engineers began to dream of factories that could run themselves—an image that physicists Eric W. Leaver and J. J. Brown called the possibility of "Machines without Men" in a 1946 *Fortune* article.[35] Programs that were Cold War–fueled, military-funded, and run by academics had developed programmable machine tools that, given a set of instructions, could more or less function automatically. By the mid-1950s, industrialists expected new advances in technology not just to replace human physical labor with machine strength but also eventually to replace human brain power with computers.[36]

Just as quickly as some hailed this revolution, others questioned if it meant more costs than benefits. Norbert Wiener, MIT's pioneer in cybernetics (the study of communication and controls as applied to machines), thought that increased mechanization would only mean joblessness, dislocation, and worsening work conditions for those people who could still hold down a job.[37] Some thinkers decried opposition like Wiener's as part of the era's so-called automation hysteria.[38]

Both commentators at the time and later historians have cited a set of 1955 U.S. congressional hearings on automation as proof that labor repudiated such hysteria, and instead readily accepted new technology. All eight of the invited union leaders thought automation inevitable and possibly desirable. A few labor leaders, like the CIO's Walter Reuther, argued that automation might improve the situation for production workers stuck in the drudgery of modern factory work—giving them more skilled work and possibly increased leisure time. Others worried that, if the Soviets could produce goods more efficiently than Americans, it might prove Soviet superiority. Therefore, backing automation and mechanization showed true Cold War patriotism.[39] In what has been the definitive statement on the topic, historian David Noble argues that postwar labor leaders feared being labeled Luddites at a time when Americans' growing economic prowess meant so much, so they sought to negotiate around the damage automation was bound to cause, instead of expressing outright opposition.[40]

Labor was more ambivalent about automation than such accounts suggest. David Steigerwald has recently argued that labor did not simply accept the increasing introduction of mechanized and automated technology during the postwar period.[41] Indeed, both Reuther's testimony in 1955 and additional remarks by Joseph Beirne of the Communication Workers of America expressed worry that even temporary job losses would lead to social displacement. Union leaders' testimony insisted that workers would face disaster if they were not protected from the turmoil through higher wages, shorter work weeks to spread remaining jobs, guaranteed annual wages, retraining programs, industry taxes, and displacement benefits.[42]

Labor leaders even argued with business leaders over the catch-all idea of automation. Business leaders, government officials, and engineers told Congress that automation was nothing new. To them, automation was the fruition of Taylorization and the scientific management of the workplace—beliefs that had dominated corporate thought for decades. Labor leaders instead emphasized the new features of modern automation. For them, automation meant mechanical and computerized control of production. Few industries had experimented with machines driven by other machines in the 1950s, but that fact did not seem to matter to people caught up in a panic over the sweeping idea of automation. Perhaps the definition that best reflected workers' thoughts came from James P. Mitchell, the secretary of labor. He testified, "We find that in some ways automation means precisely what each individual man on the street thinks it means, for to a great extent it is a word which produces various sorts of fears in various sorts of individuals."[43]

National commentators rightly did not label open-pit copper mining, which had existed for half a century, as part of the new trend of automation in the workplace, but Butte-area workers increasingly conflated the old idea of mechanization with the new idea of automation. Both ideas meant greater output, increased efficiencies, lower labor requirements, and often a de-skilling, or, perhaps more fittingly, a re-skilling of the workplace. In February 1956, Chicago unions hosted an Anaconda Conference for representatives of eighteen locals from around the country to plan negotiating strategy with the Anaconda Company. One major topic of discussion was the idea of automation. IUMMSW's national research director, Bernie Stern, insisted, "Although machines are being improved and output per man is being increased this way, automation, as yet, cannot replace the miner."[44]

The local labor paper in Butte—the *Mine-Mill Butte Miner*—however, did not have such a rosy assessment of the problem. Local leader George Little offered his own definition of the problem. Showing common ground with other unions like the autoworkers, he exclaimed, "If we change the word automation a little and spell it *open pit* [his emphasis] we find we have the same thing staring us in the face here in Butte."[45]

In 1956 the mechanization of open-pit mining emerged as a factor in contract negotiations between Anaconda and the unions that served the company. The destruction that the open pit caused to the landscape reflected workers' expectations of crushing blows to employee numbers. In one instance, the *Mine-Mill Butte Miner* published a photo of dirt cascading out of a massive shovel. The caption read, "Butte Hill crumbling before the onslaught of giant shovels. Keep your eye on this operation, it may mean your job."[46] Union members reported that Chester H. Steele, the vice president of Anaconda's western operations, expressed outrage at both the photo and similar statements in the *Mine-Mill Butte Miner* during one bargaining session. He reportedly asked, "Why do you print in your paper that the open pit operations will take the jobs of MEN [his emphasis]. I emphasize over and over again That It Will Not Take Away Jobs; It will Create 2500 new jobs for Butte," six hundred of which would be for miners.[47] The company claimed the addition of an open pit to an operation that also included block-caving as well as more traditional underground vein mining could mean only more employment, not less. By the end of 1956, Anaconda made vague assurances to the powerful BMU that the union would gain jurisdiction over open-pit operations in the future; BMU had also bargained a contract with its workers that included a decent pay raise.[48]

The battle over open-pit mining, however, had just begun. The price of copper dropped from about 42 cents a pound in 1956 to around 30 cents a pound in 1957.[49] Faced with a national recession and claiming that the cost of copper production was excessive, the Anaconda Company quickly took action. Chairman Roy Glover closed the Belmont and Mountain Con underground mines. An editorial in the *Mine-Mill Butte Miner* questioned the decision, complaining that Chairman Glover made "no mention of any curtailment of the monster shovels which are eating up the jobs of Butte miners." Nor had Glover laid off workers at the smelter. Editor Al Giecek argued that, as a result, it would be the surface "shovels of Meaderville" rather than the skilled miners who had long labored below Butte that "will keep the bins full."[50] The visual impact of large shovels digging into the hill remained more meaningful to worried miners than managerial assurances about mechanization.

The unions continued to criticize Anaconda's layoffs throughout 1957. Just one year before, Anaconda had claimed that the expanding Berkeley Pit signified a need for more workers and a bright future for Butte. One cynical contributor to the labor newspaper complained, "They secretly prepared for their future, and not ours, when they started gutting the Hill from the surface. Yes, everyone is saying 'Butte Is Getting Bigger and Better—At the Open Pit.'"[51] Seeing friends laid off by Anaconda's "messages of gloom," BMU members were angry that they were not sharing in Anaconda's continued profits, nor the overall consumer-driven abundance of 1950s America.[52] Anaconda had nearly 5,200 employees in 1955, but by 1959 three thousand miners and a few hundred craftspeople would be gone.[53]

Local BMU president Ernest Sjoman stepped up union attacks on Anaconda's technological changes and on automation worldwide. The concept of automation, which probably originated from late 1940s statements made by Delmar S. Harder, vice president of the Ford Motor Company, had traveled quickly to new contexts and new places.[54] In a piece Sjoman composed for the union newspaper in 1958, he critiqued a number of the latest academic works from Britain and America. First, he tackled Romanian-born British economist Paul Einzig's book, *The Economic Consequences of Automation*, which argued that, rather than downgrading labor "to the status of a servant to the machine," the process instead "must cause immense satisfaction to one's self-esteem to possess the power to set in motion and control a vast and highly involved engine." Workers could both be relieved of monotonous tasks and enjoy a great sense of power.[55] Sjoman also cited a book by British scientist Magnus Pyke, who claimed that automation would lead to "freedom from the need to do paid work [rather than to] technological unemployment." Automation would shorten working hours and lead to increased time for fishing, watching serious television, performing music, and other enriching leisure activities.[56]

Sjoman dismissed Einzig and Pyke as being out of touch with industrial workers. "Mr. Einzig has never been on a rustling line in Butte or anywhere else," Sjoman stated, or he would understand that much industrial work already required creativity. Einzig and his ilk were so blinded by numbers that they could not see the real pain suffered by decimated workforces in Butte and elsewhere around the Western world.[57]

BMU vice president Al Giecek agreed with Sjoman that workers in Europe and throughout America now faced the same threat as those in Butte. He noted the well-known case of U.S. automakers, which had astoundingly let go of 150,000 workers nationwide during 1955 and 1956. He exaggerated the number of Detroit layoffs to 250,000, but the point was clear: industrialists had placed workers under attack. Just as Walter Reuther had proposed for the United Automobile Workers Union,

Sjoman urged the BMU to fight for "shorter days and weeks with bigger pay," hopefully allowing Anaconda to avoid further layoffs.[58]

As they worked toward negotiations for the next contract, BMU leaders struggled to find solutions beyond this call for a shorter workweek. They clearly did not accept Anaconda's glowing view of open-pit mining. In one of its "chit-chat letters" sent to workers, the company continued to make the case that the economic downturn was responsible for layoffs, and not its newfound ability to replace underground workers with aboveground machines.[59]

IUMMSW's local leaders recognized that Butte was not the only mining area experiencing fears about mechanization, automation, and greater unemployment. A sharp contraction in the nonferrous workforce was happening across the American West.[60] Publications by one of Anaconda's largest rivals, the Kennecott Copper Corporation, show that workers were panicking throughout the western copper industry. Throughout the 1950s and 1960s, the editors of *Chinorama*, a Kennecott publication for employees at a New Mexican open-pit, tried to dampen workers' anger. A 1955 editorial in *Chinorama* claimed workers were simply afraid of technology that they did not understand.[61] "Automation," they explained in a later article, "is only a $2 word for change and better ways of doing your job." "Only a defeatist resents and resists change," they continued.[62]

Technological change also worried workers at Kennecott's major Utah operation, the Bingham Pit. The company's publication, *Kennescope*, tried to reassure workers that mechanization had meant only progress for the Bingham operation and the United States as a whole.[63] Throughout the 1950s and 1960s, *Kennescope* editors continually decried the unfortunate fact that "unions and many employees blame the unemployment problem on automation and mechanization," as one article put it. The editors instead argued that the "failure to modernize means certain extinction."[64]

Anaconda's paper for its workers, known as the *Headframe*, agreed with Kennecott's assertions about the importance of technological progress. During late 1950s contract debates, *Headframe* argued that an attack on one machine—as Butte's workers were doing to the pit's massive trucks and shovels—actually meant an attack on all machines, many of which were needed to produce metal from any type of mine. Citing the company's slogan, "From the Mine to the Consumer," the editors asserted that promoting mass consumption was instead the answer for suffering workers in the business of mass production.[65]

Brochures at the Berkeley Pit viewing stand also pushed the view that anyone who enjoyed mass consumption must also support the pit. They often explicitly linked tourists' and customers' lives with Anaconda's copper products. Tourists who came to wonder at the modern marvel of mass mining were also informed about the operation itself, particularly its size. Brochures emphasized people's awe at the enormous equipment, including the "mammoth trucks" that moved the material, the great cost of the operation, the natural abundance of ore that came out of the pit, and, most importantly, the "spectacular" expanse of the pit itself. This new landscape, created by nature, human labor, corporate ingenuity, and sublime technological advances, meant Butte was moving forward.[66] This positivity drifted over to some residents. Newspaper editor Rick Foote, for instance, remembers visiting the pit's viewing stand with his dad, who thought "that everything was going to be OK because of this method of mining." Residents certainly had pride in continuing to produce metals important for both war and industrial growth.[67]

Convincing suffering workers that the open pit represented progress was more difficult. In speeches to the Montana Farmers Union in 1957 and 1958, Butte native and IUMMSW district board member Ernest Salvas made common cause with farmers who similarly "work intimately with the good earth." He told them, "Some of our jobs in the metals industry are lost forever to a changing technology, just as some farm population is lost to the machines of modern agriculture." He then called for joint action against mechanization.[68]

Indeed, Butte's workers were at the forefront of a larger movement against mass production and automation, because labor's ambivalence increasingly gave way to outright attacks. After the American Federation of Labor (AFL) merged with the CIO in 1955, the combined organization allowed for a number of attacks on automation. George Meany even opened the 1963 AFL-CIO convention with a salvo against automation as "a curse to society."[69] Popular culture followed, crafting worrying portrayals of factories without workers; the standout example was Kurt Vonnegut's *Player Piano.* By the 1960s academics like Ben Seligman also began to treat the technological threat of automation in a highly critical manner.[70] In the 1960 presidential campaign, John F. Kennedy caught wind of these worries. Citing the devastation of West Virginia's coal mining region, Kennedy worried about the "dark menace [of the] steady replacement of men by machines." Once in office, Kennedy even stated that technological unemployment was "the major domestic challenge, really, of the Sixties." His Department of Labor created an Office of Industrial Automation in 1961 to study the issues, while the Manpower Development and Training Act of 1962 offered the retraining of industrial workers in industries like mining as a possible solution.[71]

Meanwhile, Butte's workers became embroiled in what would become the longest strike in the area's history up to that time. The three-year contract between the Anaconda Company and its unions expired in 1959. The IUMMSW demanded wage increases and better pensions for its workers. Seeking some way to alleviate miners' pain in an open-pit world, Butte's workforce joined 30,000 other copper workers in a nationwide strike on August 19, 1959.[72]

## Community Protests

For those who did not actually work in the industry, the major problem with open-pit mining in Butte was obvious and perhaps even more ominous: people lived on top of the copper and zinc veins. The development of open-pit mining in Butte continued an older tradition of placing industrial work in the middle of neighborhoods. Yet this particular intertwining of workplace and home space introduced the area to a number of new industrial hazards. These hazards, paired with an ever-expanding work zone, threatened to tear apart neighbors. Residents' protests thus became another front in the larger debate over open-pit mining in the Butte area.

In 1948 development work by Anaconda's Geological Department had led to the discovery of "an important tonnage of [low-grade] sulphide-zinc-lead-manganese-silver-gold ore" in the old Alice Mine. Lacking an easy way to extract zinc-sulphide ore near the surface,

Anaconda did not touch the site until 1958. That year, what geologist Edward P. Shea labeled "improvements in open-pit mining methods" led Anaconda's management to hire the F&S Contracting Company to begin extraction in Walkerville. By the time Anaconda ceased operations a few years later, workers had excavated 5.5 million cubic yards of sulfide zinc from the Alice Pit.[73]

In Walkerville the transformation of the Alice Mine introduced a number of hazards. Citizens expressed their concerns in phone calls and letters to Anaconda's Mining Engineering Department. Because there were no written restrictions on how close to a home F&S could mine, many of these complaints focused on blasting noise and falling rocks. Former resident Carol Wyatt Maunder, for instance, remembered that she "had to dodge flying rocks numerous times following a blast." Her neighbor, Gene Kuhn, "was mad as hell and called the Company up to claim that his kids almost got killed from flying rocks after a blast"; Kuhn packed up and left town. Residents Joe Panion and his wife grumbled about the severity of the blasting, claiming that it was fracturing their sewer line. John Bolton thought his basement was cracking apart. Anaconda's reply to Bolton was the same as its response to twenty-seven of the twenty-nine formal complaints it received: complete denial. According to the department's files, John Bolton's claims were "grossly exaggerated."[74]

Although phone calls drew little response from Anaconda, the Walkerville City Council reacted to the "hundreds of calls" it received.[75] On the recommendation of the small city council, belligerent mayor Jimmy Shea soon intervened. On April 10, 1958, Mayor Shea sent a letter to both F&S and Anaconda in which he aired Walkerville's grievances. One large blast had apparently thrown rocks at cars and homes; one of the stones was so big that it "would have killed someone." The company had located the pit so close to the neighborhood—it lay smack dab between the northern and southern sections of town—that north Walkerville appeared to be literally sliding into the pit. The subsidence caused by mining opened the possibility of "severe damage to water mains, gas mains, and to life and property."[76] His fears were well founded. By 1958 large cracks had begun to form in Walkerville's streets, alleyways, and yards. Eventually they also formed in houses.

WALKERVILLE MAYOR JIMMY SHEA, 1959. Shea is showing pit blast damage to the vehicle. Photo courtesy of Butte-Silver Bow Public Archives, Jimmy Shea Papers, PH296.002.004.

Anaconda blasting consultant Lewis Oriard came to Butte to observe the situation and met with Jimmy Shea. Oriard admitted he "was amazed at the large cracks" running throughout the neighborhood and conceded that blasting was likely "too heavy." Nevertheless, no significant change occurred after his visit.[77]

North Walkerville faced the brunt of Anaconda's force. The broken roads in front of its houses seemed doomed, leaving some locals unable to access their own residences. According to a telegram that Mayor Shea sent to members of the press, many "pedestrians have had to go through mud ankle deep" in order to return home. The mayor also insisted the Anaconda Company should take safety precautions on the unfenced northern rim of the pit. He reported that "children were playing on the rim of the pit and the drop off was about a hundred feet down."[78] After the company finally fenced the area, it had to keep moving back the fence—at least five times—as it spread out its operations.[79]

Walkerville officials became particularly concerned about the towering dumps, where the company had dropped overburden from the pit. Since the company placed the largest dumps on land higher than the neighborhood, runoff clogged the sewerage system with silt. Large dumps established in 1959 blocked access to the neighborhood's prized environmental amenity, its large ballpark and adjacent playground. Residents were soon selling homes to the company agent to escape what seemed to be impending disaster. Shea classified these events an industrial "invasion."[80]

Community leaders decided to take a stand. On one front, Shea tried contacting Anaconda's insurance adjuster, who said that Walkerville residents "would have to prove that damage is being done." "That," Shea insisted, "would be like proving there is a City of Denver," but the adjuster took forever to get back to the city. Walkerville's leaders also made it clear to Anaconda and F&S that they "were *not* [original emphasis] satisfied with the way you treat our citizens when they

Children Playing in Walkerville, 1959. The cracks and subsidence seen in this photo were caused by the nearby Alice Pit. Photograph by Carl Iwasaki, 1960. Copyright Carl Iwasaki estate. Photo courtesy of Butte-Silver Bow Archives, Jimmy Shea Collection, PH296.003.

complain to you about blasting."[81] Walkerville therefore put into place a blasting ordinance that asked for time restrictions, insurance against damage, and prior notification of all blasting plans. The council hoped it would give residents some protection from the noise and danger.[82]

Finally, Walkerville leaders gained a more effective point of attack. After buying the abutting property, the company set out to condemn and fence off Sibyl, Rose, and Pearl Streets. These streets had been made impassible by subsidence toward the Alice Pit. In response, Mayor Shea proclaimed that the corporate predecessor of the Anaconda Company had dedicated Walkerville's roads for public use forever in 1890.[83] In order for these streets to be discontinued and vacated, therefore, Anaconda would have to formally petition the Walkerville City Council. Anaconda gave the council its petition on December 27, 1958.[84]

Starting the next month, Shea called for public meetings to rally a wary community.[85] Notes from these intense meetings exhibit a community cracking under pressure. Some people worried about the ability of aldermen to represent residents' interests because many of them worked for the Anaconda Company. Others became upset at the people who had already sold their homes to the company. Anaconda had generally been buying the property it needed and then selling the home back to the owner for one dollar with the stipulation that the house be moved. Most residents agreed that the prices offered for property were reasonable, but minutes from the meetings suggest great tension on this issue.[86]

City leaders urged community solidarity. At an April 1959 city council meeting, Shea, after listing eight homeowners who had recently sold property and six more who were expected to sell, recounted how some residents had previously advocated that everyone should hold out or "if one sold to all sell" to keep neighbors together. Yet, these very same people had "abandoned the rest of you and your cause" because they "were the very first ones to sell." "You folks should have persuaded your neighbors not to sell," Shea argued. "*Now* [original emphasis] the Anaconda owns the ground on each side of the roads and streets they want to vacate." Once Anaconda claimed the streets, there would be little to stop it from tackling the next row of houses—without offering the city anything to pay for damages.[87]

According to the aldermen, private property owners needed to treat their property as communal, public property—subject to the will of the community. In fact, the city council's notes show that a number of residents wondered why Anaconda did not just move the whole city, as one mining company had done when Hibbing, Minnesota, faced an expanding pit decades before. Walkerville's leaders were concerned with keeping their neighbors safe and together. By the middle of 1959, residents had formed a united front as they saw advantages in practicing solidarity. There is no evidence of additional home sales until the following year.[88]

Anaconda impatiently waited for the Walkerville City Council to accept its petition. Company management wanted to follow the rich zinc vein northward and further into North Walkerville; more importantly, they also needed to expand outwards so that they could mine farther down. The already steep Alice pit, which had started at a 45-degree angle, had gained an even greater degree of slope. Mayor Shea claimed that it was almost perpendicular. Even though he was exaggerating, the slope slowed down work in the pit because large dump trucks took a long time to descend into the pit. Anaconda's leaders in Butte reported to national management that "without the needed land to expand the perimeter of the pit, the operation was unsafe," but Walkerville resisted.[89] Community leaders kept making more demands before they would accept the petition, asking Anaconda to construct a new ballpark, to repair city utilities, and to deal fairly with local residents' complaints as well as to erect fences for safety, to remain one hundred feet away from Walkerville homes, to pay for damaged streets, and to build a new road.[90]

Anaconda temporarily suspended Alice Pit operations in February 1959, claiming that the neighborhood's unreasonable demands had forced it to shut down. In local newspapers, Anaconda announced that eighty-five miners in Butte and Walkerville, eighteen workers on the Butte, Anaconda & Pacific Railway, and forty-four men at the zinc concentrator would have to be laid off.[91] An ongoing jurisdictional dispute between the Railroad Brotherhood and IUMMSW members over loading at the Berkeley Pit's yard meant that production was already experiencing hiccups.[92] Company management therefore believed that it could force Walkerville's residents to their knees by increasing the pressure from other locals who depended on mining.

Walkerville residents had few ways to publicize their side of the story. Although it would start getting out of the journalism business later that year, the Anaconda Company owned all but one of the state's major dailies in early 1959. An unspoken censorship of controversial stories meant that reporters tended to avoid most state news—especially anything critical of the Anaconda Company.[93] Mayor Shea alerted the one major daily not owned by the company, the *Great Falls Tribune,* and he contacted radio stations. People across the state became aware of the conflict.[94] In the face of increased scrutiny, Anaconda finally acquiesced to most of the neighborhood's demands. The Walkerville City Council accepted Anaconda's petition, and Alice Pit operations resumed on April 24.[95]

By the next year, however, Walkerville's leaders proclaimed that the company had broken its agreement. Summer 1959 saw record shipments—more than 277,000 tons of at least 3 percent zinc and 1.2 percent lead—but subsidence continued to increase along the pit's north boundary. Despite attempts to reduce the steepness of the slope, the ground was still crumbling at an increasing rate. Employees continued to strip the ground, but Anaconda vice president E. I. Renouard admitted that ore production would have to slow until they could safely truck all of the waste out of the pit.[96] When officials understood that the area "was going to exceed the limits anticipated in earlier plans," Anaconda tried to buy additional property in north Walkerville.[97] The divisional manager of the Montana Power Company informed the owners of ten north Walkerville households that because of subsidence in the area, it would have to terminate gas service or risk explosion. Nine sold their property. Mayor Shea said he "would rather see this [relocation] than have some one [*sic*] blown to pieces." Unlike earlier sales, these people "were actually forced from their homes whether they wanted to move or not."[98]

Residents renewed their protests. As earlier, the city council responded by using public property complaints to keep the company from infringing on residents' private property. Leaders used holdout Bill Hughes as a key figure in this battle. Because he had not yet sold his

Bill Hughes, 1960. Hughes, seen here in front of his Walkerville home, refused to sell his house to the Anaconda Company. The image shows extensive ground damage caused by the Alice Pit. Photograph by Carl Iwasaki, 1960. Copyright Carl Iwasaki estate. Photograph courtesy of Butte-Silver Bow Archives, Jimmy Shea Collection, PH296.003.

home, the fence Anaconda wanted to erect for safety would pass across his property.[99] The city council again proclaimed that Anaconda was damaging public property—including streets and sewers—despite its earlier agreement to repair all damages. "In order for the company to comply with this agreement," Shea argued, "it would be necessary for them to fill in a good part of the pit."[100]

Anaconda shut down operations in the Alice Pit again. This time, the decision came mostly because the mine's profitability had declined, although residents' continued anger had not helped. Instead of dying down, however, the conflict raged on throughout the autumn and winter of 1960.[101] Anaconda continued to move fences northward over the damaged streets and lots, claiming that it was simply trying to look out for residents' safety. Shea believed that if the company fenced off the damaged area without paying the city, though, Walkerville would never be able to afford the needed repairs. Walkerville threatened a big court case. By the autumn, Shea announced that the massive subsidence was endangering children. "I'm afraid someone will get killed," he declared.[102] Labeling Anaconda a "domineering and confiscatory" company, Shea even ordered five workmen arrested for trying to lay fence in the crumbling area. Justice of the Peace J. J. Winston dismissed the five men, but the battle lines had clearly hardened.[103]

The media came to Walkerville's rescue, starting with the *People's Voice*, the farm/labor coalition's newspaper published in Helena. Centered around this strong anticorporate publication, a loosely organized farm/labor coalition had long been the political counterbalance to the power of natural resource and railroad corporations in Montana. This tradition generally differed from that in the more conservative Rocky Mountain states contiguous to Montana.[104] The *People's Voice* described the Walkerville area as equal to "the aftermath of an earthquake. The area is ravaged by crevices, the ground is sinking, people have had to leave their homes and streets are barricaded." "Do we have government of the Anaconda Company, by the Anaconda Company, for the Anaconda Company?," Shea was quoted as asking.[105]

The national press followed. In December, *Life Magazine* sent photographer Carl Iwasaki to record the desolate landscape and the area's plight. His powerful photographs feature cracks running through homes, streets crumbling into an almost-vertical pit, and children jumping across gaping holes. Other photographs show officials standing in and almost disappearing into cracks that were at least five feet deep. Many of the images also feature an enormous mound of overburden in the background—an ominous mass towering over north Walkerville.[106]

In press releases, Anaconda provided its own message: mining was vital to Butte and halting operations risked severe economic consequences for the area. The company claimed that it had spent more than $200,000 in Walkerville "on things apart from mining." The city council replied that this money was not spent on satisfying residents' demands, and Anaconda had simply paid for some of "the heavy damage" the company had caused.[107]

Damage Caused by the Alice Pit in Walkerville, 1960. A large waste dump from the pit looms in the background. Photograph by Carl Iwasaki, 1960. Copyright Carl Iwasaki estate. Photo courtesy of Butte-Silver Bow Archives, Jimmy Shea Collection, PH296.003.

## Standing Together

Perhaps emboldened by Walkerville's success, the old Italian working-class neighborhood of Meaderville, challenged by the expansion of the Berkeley open-pit mine, began to fight Anaconda's efforts on the east side of Butte. The speed at which the Berkeley Pit was growing astounded residents. Tom Holter recalls his memories of the pit's birth: "They started the open pit in 1955. When I got out of the army in the August of '56, and I worked, my wife, well, we were at my house in McQueen, and she was supposed to take a ride. We went in her car and we drove down to the mill. I said, 'What the hell's going on here?'...What used to be a mountain there for the mines was a hole in the ground!"[108]

As Anaconda began purchasing homes in Meaderville, its citizens formed an action committee to deal with the company. The fire departments in Meaderville and Walkerville hosted meetings for residents from the threatened communities. As in Dublin Gulch, most of Meaderville's residents did not own the land—the company did. Therefore, Anaconda did not have to pay for anything but the house, which left some displaced people without enough money to buy another. Former Meaderville fire chief Angelo Petroni remembered that a number of widowed miners' wives only received $1,500 for their homes, which was not enough money to live on when "they didn't have nothing." Petroni and others therefore formed a committee and, in his words, "took on a war with the Company." Residents traveled to the state capital to ask for help, but they found it was difficult to get their voices heard by a state legislature where Anaconda had long held sway.[109]

Silver Bow's county commissioners showed some support for protestors. To make room for further expansion of the Berkeley Pit, Anaconda asked for county permission to move the access road connecting Meaderville to the Butte-to-Helena highway. The petition stated that expansion across the road "is of utmost importance to the economic welfare and industrial development of the County." This situation upset Meaderville's inhabitants; they knew that increasing the community's distance from the highway would severely damage local businesses. The area was well known for its nightlife and supper clubs, and many residents made their money from highway traffic.[110] Under pressure from Meaderville, the company withdrew its petition and instead started a new pit east of the road, which would later be joined with the Berkeley Pit. The workforce continually traveled between the two operations, taking over the road and disrupting Meaderville at large. In response, Silver Bow County ordered the arrest of two truck drivers for driving overweight vehicles across the road. A jury acquitted one of them, while the other forfeited a $50 bond.[111]

Local unions appeared quite supportive of these actions. Union workers had long connected their struggle with that of other Butte residents. In one of the *Mine-Mill Butte Miner*'s first references to open-pit mining (January 7, 1956), the labor newspaper's editor, Al Giecek, made common cause by discussing the plight of Meaderville's residents, who were the closest to Anaconda's first open-pit tests. Comparing the area to a nuclear test site, Giecek wrote, "Not long ago the residents of Meaderville had an impression of being in Yucca Flats. It seems that the boys working on the open cut are not too sure of their powder and holes. On this particular day there was a thunderous roar and windows shook, groceries and drugs left their shelves. One building had a hole bored through its roof and the rock rolling down, just missed an occupant in a chair on the outside of the building." Giecek congratulated residents on taking their complaints to F&S and Anaconda.[112] The paper's advocacy on behalf of these working-class neighborhoods should not be taken lightly—given Anaconda's control over the state's press, the labor paper represented one of the few venues to express these views.

In another article from early in Butte's open-pit days (March 31, 1956), Giecek told a sad story about Butte's past and its certain future in a piece titled "The Saga of the Sagging Hill." He first criticized residents' general acceptance of subsidence under their homes during the long history of underground mining. He described the process: "The people take this sinking and sagging as a matter of course. Their livelihood depends upon the continued operation of the mines, and certainly would not think to stop the mining over the sake of a cracking and sagging house. When discouraged, they just sell to some newcomer who is not aware of the glory hole under his home, and buy out in the 'Flats.'" Giecek then explained modern advances in mass mining, including block-caving at the Kelley Mine and the new Berkeley Pit. He wrote, "Today the city of Butte is seeing the hill not only sag from the swift removal of her belly, but now there are some of the largest electric shovels that can be bought, scooping up the surface and razing the once proud landmark."[113]

The article then connects the ongoing disappearance of old Butte residences with the plight of miners facing mechanization. "A few years ago the little hill boasted at least twelve thousand miners. Now that automation is here, she boasts a scant five thousand." Dejected, Giecek worried that no one would fight mechanization or the disappearance of a Butte that he loved: "The hill sags and disappears as three sets of mining ventures attack and disintegrate the very spot where the mighty hill once ruled."[114] Giecek kept up this fight, writing one month later that, "with the screeching and groaning of shovels; the roar of monstrous trucks; the addition of new holes, the hill is going fast."[115] The dramatic image of the open pit had become Giecek's favored symbol to discuss miners' and residents' unsure future.

Giecek need not have been so discouraged in 1956, because Butte's unions and residents continued their fight up to the 1959 battle over the next contract.[116] Unions seemed particularly keen to engage on the mechanization threat as they entered into a strike in August of 1959. Miners also saw a great deal of commonality with Walkerville residents in particular. In March, the union paper quickly reprinted an article from the *Great Falls Tribune* that first broke Walkerville's story to the rest of the region.[117] At the request of the hill town's mayor, BMU members even posted photos of Walkerville's damaged landscape—likely the ones taken for *Life Magazine*—around the union hall.[118]

Walkerville's residents similarly supported the union's fight against the company, which continued to shut down underground mines due to both the poor price of metals and the efficiency of open-pit work. During the 1959–60 strike, the Walkerville City Council angrily censored the Anaconda Company for its lack of ethics in laying off hardworking residents of the Butte area.[119] Jimmy Shea also personally worked to settle the strike through a flurry of letters—insisting on state or federal intervention.[120]

The Anaconda Company therefore faced both a recalcitrant union and an angry group of residents. The threat of mass layoffs had not worked on Walkerville. Instead, in April of 1961, the City of Walkerville launched another attack on the company. It filed a case in Silver Bow County's district court against Anaconda and F&S for damaging city property.[121] It also advocated for a resident in his own property battle with the company.[122]

## Winners and Losers

Walkerville was successful with its own case against the company. After a few years of foot dragging, the company settled out of court with generally favorable results for Walkerville; the town managed to collect $14,000 for repairs.[123] Anaconda never resumed work in the Alice Pit.

In an interview years later, Jimmy Shea lauded Walkerville as the only community to successfully fight open-pit expansion. The Anaconda Company, however, denied that Walkerville's efforts played any role in its decision to shut down the Alice Pit since the original plans for the operation were quite short term.[124] Company geologists had indicated in 1959 that sulphide zinc mining in the area was still advisable, but that the extent of deposits failed to justify expansion beyond the originally planned pit boundaries.[125] In short, the Alice Pit ran out of ore. The copper-rich Berkeley Pit, however, continued to grow, soon to eat away at Meaderville and other Butte neighborhoods, including McQueen, East Butte, the East Side, and the remains of Dublin Gulch.

Nevertheless, Walkerville's efforts should not be dismissed out of hand. Up to that time, almost no one in the state had managed to force the Anaconda Company to do anything it did not want to do. Walkerville, however, had actually managed to halt Anaconda's operations for months. Ecological sociologists have shown that members of the same community can hold widely divergent ideas about the urgency and magnitude of risks and disasters.[126] They suggest that communities facing industrial hazards and relocation tend to divide into contentious factions.[127] Walkerville neighbors instead managed to stand together. Community leaders like Jimmy Shea helped to manage definitions of risk—convincing Walkerville's residents that open-pit mining was a clear and present danger. In fact, Walkerville's residents continued to be among the loudest voices in early 1960s battles over pit expansion, despite the fact that their neighborhood played less and less of a role in Anaconda's plans.

Company employees had a more difficult time fighting the effects of open-pit mining. The 1959–60 strike has been described as a resounding success for employees by one scholar and as the "strike that broke the backs of the unions" by another.[128] Despite this disagreement, it is clear that this strike marked the last time that unions would strongly fight open-pit mechanization. Only in the 1970s did workers again question the role of the open-pit mine in Butte.

In light of later developments, the spirit shown by Butte-area residents in the 1950s and early 1960s is remarkable. According to long-held beliefs about one-industry communities, Anaconda should have easily convinced Butte's residents to lay down when it threatened their jobs. Butte's failure to do so suggests that employment was not always its first concern.[129] The initial open-pit experience was so dramatically different from underground mining for workers and community members that they instead fought back. Open-pit mining laid bare the social consequences of mechanization at the same time that it exposed working-class neighborhoods to new environmental hazards.

Being forced to confront this new form of industrial work head-on encouraged community members to protest. Even if, as historian Jeffrey Manuel puts it, "historians of labor and industry often downplayed the central role of technological change and automation in reorganizing patterns of work," Butte's residents did not.[130]

Taken as a whole, this story also challenges historians' view of a Cold War period that promoted working-class complacency and cultural conformity.[131] A counterbalance to corporate power in this mining community did exist—whether it came from local politicians, working-class residents, or union members.

## Notes

1. Many historians claim the Anaconda Company was so powerful that it faced few challenges in running both Butte and the state. See Joseph Kinsey Howard, *Montana: High, Wide, and Handsome* (New Haven, CT: Yale University Press, 1943); K. Ross Toole, *Montana: An Uncommon Land* (Norman: University of Oklahoma Press, 1959); K. Ross Toole, *Montana: A State of Extremes* (Norman: University of Oklahoma Press, 1972). Criticism of this tradition is in David M. Emmons, "The Price of Freedom: Montana in the Late and Post-Anaconda Era," *Montana: The Magazine of Western History* 44, no. 4 (Autumn 1994): 66–73; Michael Malone, "The Close of the Copper Century," *Montana: The Magazine of Western History* 35 (Spring 1985): 69–72. A moderate view is Snow, "From the Sixth Floor to the Copper Dome."

2. Bert Gaskill, "Walkerville Council Blasts Anaconda Co. for Actions By 'Financiers in New York,'" *Mont. Std.*, Feb. 4, 1960.

3. Minutes of the Regular Meeting, Oct. 12, 1960, Bound Material #230, Minutes of the BMU, 1957–1961, Western Federation of Miners/International Union of Mine, Mill, and Smelter Workers Archives (hereafter IUMMSW), Norlin Library, University of Colorado, Boulder, CO (hereafter UC).

4. Edward P. Shea, "Report on Open Pit Copper Sulphide Ore Butte District," Anaconda Copper Mining Company Geological Department, Nov. 10, 1955, p. 12, file #29801.12, Fldr., "Butte, Mine Operations," AGDC, AHC; Edward P. Shea, Memorandum no. 8, "Re: Butte District Special Development," May 20, 1955, file #29704.05, Fldr. "Progress Reports, Special Development Memos nos. 1 to 22 (Mar. 31, 1955 to Aug. 26, 1955)," AGDC, AHC; Edward P. Shea, "Memorandum no. 28, re: Butte District Special Development," Anaconda Geological Department, Oct. 6, 1955, file #29704.02, Fldr. "Progress Reports, Special Development Memos, nos. 23 to 40 (Sept. 2, 1955 to Dec. 31, 1955)," AGDC, AHC.

5. Vincent D. Perry to Clyde E. Weed, May 12, 1958, file 29802.02C, Fldr. "Butte, Mine Operations," AGDC, AHC.

6. Edward P. Shea, "Memorandum no. 19, Re: Butte District Special Development," Aug. 4, 1955, file #29704.05, Fldr. "Progress Reports, Special Development Memos nos. 1 to 22 (Mar. 31, 1955 to Aug. 26, 1955)," AGDC, AHC; Edward P. Shea,

"Memorandum no. 29, re: Butte District Special Development," Anaconda Geological Department, Oct. 17, 1955, file #29704.02, Fldr. "Progress Reports, Special Development Memos, nos. 23 to 40 (Sept. 2, 1955 to Dec. 31, 1955)," AGDC, AHC; Edward P. Shea, "Memorandum No. 83, re: Butte District Special Development," Anaconda Geological Department, Oct. 25, 1956, file #29704.04, Fldr. "Progress Reports, Special Development Memos, nos. 61 to 92," AGDC, AHC.

7. See the two sides of this debate in Noble, *Forces of Production*; David Steigerwald, "Walter Reuther, the UAW, and the Dilemmas of Automation," *Labor History* 51, no. 3 (Aug. 2010): 429–53.

8. Finn, *Tracing the Veins*; Swibold, *Copper Chorus.*

9. On geographies of labor, see Peck, "The Nature of Labor."

10. Marcosson, *Anaconda*, 275. See also John Hutl, "The Greater Butte Project," *Engineering and Mining Journal* 152, No. 10 (Oct. 1951): 72–81.

11. Ed Johnson, "A. C. M. Company's Development Project Among Top 1947 Stories," *Mont. Std.*, Jan. 1, 1948. See also "Greater Butte Mining Project Launched with Ground Breaking for Kelley Shaft," *Mont. Std.*, May 9, 1948; "McGlone Outlines Extensive Mining Plans," *Great Falls Tribune*, Dec. 24, 1947; Chester H. Steele, "How the Greater Butte Project Was Born: Development, Sampling and Block Caving Tests Were Foundation of Great Plan," *Mining Congress Journal* (Mar. 1949): 32–39; "From A to Z with Anaconda," *Investor's Reader* 1, no. 27 (June 27, 1956): 1–13.

12. "'Richest Hill' Subsiding," *Great Falls Tribune*, Oct. 15, 1953.

13. Corbett interview, by Mercier; Hoy interview; Hileman interview; Matteuchi interview; McHugh, "The Gulch and I"; Edwin Dobb, "Viewpoint: Location, Occupation, Juxtaposition, Interpenetration: Notes on an Erotics of the Mining City," *Buildings and Landscapes* 17, no. 1 (Spring 2010): 1–12.

14. On silicosis and block-caving, see Rosner and Markowitz, *Deadly Dust*, 204; On block-caving and labor, see Hovis and Mouat, "Miners, Engineers." On open-pit mines during the 1950s, see Thomas R. Navin, *Copper Mining and Management* (Tucson: University of Arizona Press, 1978), 39–41.

15. Joe Mason, "Our Union Today. . . Its Program and Objectives," *Miners Voice*, June 13, 1953, Fldr. 5, Box 341 of *Miners Voice* (Butte, MT), IUMMSW, UC.

16. Joint Negotiations Committee, Montana ACM Locals, "Report to Meeting of the ACM Council, Denver, Colorado," Aug. 10, 1953, Fldr. "MMSW Wage Policy and Bargaining, Anaconda Negotiations, 1953," Box 3, International Union of Mine, Mill, and Smelter Workers/United Steelworkers of America Collection (hereafter USWA), Penn State Historical Collections and Labor Archives, University Park, Pennsylvania (hereafter PSHCLA).

17. Edward P. Shea, "Report on Open Pit Copper Sulphide Ore Butte District," Anaconda Copper Mining Company Geological Department, Nov. 10, 1955, p. 12, file #29801.12, Fldr. "Butte, Mine Operations," AGDC, AHC.

18. Robert "Bob" Schrank, international rep. of IUMMSW, to Bernard "Bernie" W. Stern, research director, Denver, Apr. 7, 1955, Fldr. "Robert Schrank," Box 168, IUMMSW, UC.

19. Bernard W. Stern, to Robert Schrenk, Apr. 29, 1955, Fldr. "Robert Schrank," Box 168, IUMMSW, UC.

20. Oct. 5, 1955 entry, Bound Material #229, Minutes of the BMU, 1955–1956, IUMMSW, UC.

21. BMU to Members, June 26, 1956, Fldr. 27, "Raids USW-Montana Mine-Mill," Box 258, Microfilm LA1148, USWA, PSHCLA.

22. For the quote on splitting the labor movement, see Nov. 2, 1955 entry. For the quote on Company threats, see Oct. 26, 1955; both in Bound Material #229, Minutes of the BMU, 1955–1956, IUMMSW, UC.

23. "Look Ahead Miners!" *Diggers' Truth* (Butte, MT) I, no. 11 (Aug. 25, 1956): 1, Fldr. 29, "Raids USW-Montana-Steelworkers Publications," Box 258, Microfilm LA1148, USWA, PSHCLA.

24. Finn, *Tracing the Veins*; Mihelich, "The Richest Hill on Earth."

25. Harvey A. Levenstein, *Communism, Anticommunism, and the CIO* (Westport, CT: Greenwood, 1981), 330–40; F. S. O'Brien, "The 'Communist-Dominated' Unions in the United States Since 1950," *Labor History* 9 (Spring 1968): 184–209.

26. Mercier, *Anaconda*, 144–80.

27. Ibid., 144–45, 150; Mason, "Our Union Today," IUMMSW, UC.

28. "Hysteria Grips Steel Raiders as Court Orders Butte Vote," *MMBM*, Feb. 1954, copy in Fldr. 5, Box 341, IUMMSW, UC.

29. Art Moralez, "Rank and File Talks," *MMBM*, Dec. 11, 1954; Mercier, *Anaconda*, 150–53.

30. "Cannibals Are Here Again," Mar. 3, 1956, *MMBM*; Earl Gregory, "Butte Miners Union Makes History," Feb. 4, 1956, *MMBM*; Al Giecek, "Around the Hill," "This Is What Steel-Minded Butte Union Leaders Said in 1949," Mar. 3, 1956, *MMBM*; Mercier, *Anaconda*, 163.

31. "An Open Statement to Attorney General Brownell, the FBI, and the President's Republican Party," Feb. 4, 1956, *MMBM*. See also Al Giecek, "Around the Hill with Al Giecek," Aug. 11, 1956, *MMBM*.

32. First quote is from "Stick with Your Union," *MMBM*, Mar. 7, 1959. The second is from Earl Gregory, "Brother Member of Butte Miners Union No. 1 Think It Over," *MMBM*, Mar. 7, 1959. See also Barney Rask, "Act Now," Sept. 29, 1956, *MMBM*; "All Around the Hill with Al Giecek," May 18, 1957, *MMBM*; "Birthday of the Miner's Union," June 13, 1957, *MMBM*; Tom Breslin, interview by Arthur S. Weinberg, July 15, 1968, United Steelworkers of America Oral History Collection, PSHCLA.

33. First quote is "Stealworkers Make Gains, and How," June 9, 1956, *MMBM*. Second quote is Barney Rask, "Act Now," Sept. 29, 1956, *MMBM*. See also "Raiders and Union Busters," Mar. 7, 1959. Blaine Beakey, "Drippings from the Faucet," Apr. 11, 1959, *MMBM*; Mercier, *Anaconda*, 165.

34. Amy Sue Bix, *Inventing Ourselves Out of Jobs? Americas's Debate Over Technological Unemployment, 1929–1981* (Baltimore, MD: Johns Hopkins University Press, 2000).

35. Eric W. Leaver and J. J. Brown, "Machines without Men," *Fortune* (Nov. 1946), 192–204.

36. Noble, *Forces of Production*, 1–231.

37. Norbert Wiener, *Cybernetics: or Control and Communication in the Animal and the*

*Machine* (New York: John Wiley & Sons, 1948), 37–38; Norbert Wiener, *The Human Use of Human Beings: Cybernetics and Society* (Boston: Houghton Mifflin, 1950), 186–89.

38. George Terborgh, *The Automation Hysteria* (New York: Norton, 1966).

39. *Automation and Technological Change*, Hearings Before the Subcommittee on Economic Stabilization of the Joint Committee on the Economic Report, Congress of the United States (Washington, DC: U.S. Government Printing Office, 1955), 101–2, 596–98, 615–17. See also Bix, *Inventing Ourselves*, 245–49, 256; A. J. Hayes, "Labor's Aims in Adjusting to the New Technology," Feb. 1959, in U.S. Bureau of Labor Statistics, *Impact of Automation: A Collection of Twenty Articles About Technological Change, From the Monthly Labor Review*, Bulletin No. 1287 (Nov. 1960): 28–31.

40. Noble, *Forces of Production,* 248–49.

41. Steigerwald, "Walter Reuther," 429–53.

42. *Automation and Technological Change*, 103, 107–10, 120, 125–26, 455–60; James R. Bright, *Automation and Management* (Boston: Harvard University Press, 1958), 199.

43. *Automation and Technological Progress*, 262–63.

44. George Little, "The Anaconda Conference and Other Things," *Mine-Mill Butte Miner*, Feb. 18, 1956.

45. Ibid.

46. *Mine-Mill Butte Miner* (hereafter *MMBM*), June 9, 1956.

47. Al Giecek, "Around the Hill," *MMBM*, June 30, 1956. On the numbers of miners required, see Ernest Sjoman, "Automation and Technological Progress of Industry," *MMBM*, Feb. 8, 1958.

48. Mar. 7, 1956 entry, Bound Material #229, Minutes of the BMU, 1955–1956, IUMMSW, UC; Contract Agreement between the Anaconda Company and the International Union of Mine, Mill, and Smelter Workers, 1956, roll #1, Index #1 "Anaconda Co., LU1 to Zaretti Brothers," Microfilm Roll LA2225PSU, USWA, PSHCLA; Mercier, *Anaconda*, 173; Finn, *Tracing the Veins,* 54–55.

49. USGS, data series 140, "Historical Statistics for Minerals and Material Commodities in the United States," minerals.usgs.gov/ds/2005/140/#data.

50. Al Giecek, "Editorial," *MMBM*, Apr. 13, 1957.

51. "Birthday of the Miner's Union," *MMBM*, June 13, 1957.

52. Al Giecek, "Around the Hill," *MMBM*, June 13, 1957.

53. Sjoman, "Automation and Technological Progress."

54. On the origins of the word *automation,* see "Business," *Time*, Mar. 19, 1956.

55. Paul Einzig, *The Economic Consequences of Automation* (New York: W. W. Norton, 1956), quotes on 193, 233.

56. Magnus Pyke, *Automation: Its Purpose and Future* (New York: Philosophical Library, 1957), 176–91; quote is on 179.

57. Sjoman, "Automation and Technological Progress."

58. Al Giecek, "Around the Hill," *MMBM*, Feb. 8, 1958. On automakers' figures, see Steigerwald, "Walter Reuther," 436. See also Ernest Sjoman, "A Brief Look into the Past on Miner's Union Eightieth Anniversary," *MMBM*, June 15, 1958.

59. Al Giecek, "Around the Hill," *MMBM*, Feb. 8, 1958. See also "The Pressure Is On: Let Us Talk It Over," *MMBM*, Oct. 26, 1957.

60. "The Bargaining Program for '59," *MMBM*, Apr. 11, 1959.

61. "Who's Afraid of Automation?," *Chinorama* (Hurley, NM: Kennecott Copper Corporation, Chino Mines Division), July 1955, 1, J. Cloyd Miller Library, University of Western New Mexico.

62. "Some Talk About Horses, Cyclone Classifiers, Automation, and Jersey Bulls," *Chinorama*, July 1955, 2, 7.

63. "Half Century of Progress," *Kennescope* (Salt Lake City, UT: Kennecott Copper Corporation, Utah Copper Division), May 1956, 1, in J. Willard Marriott Library Special Collections, University of Utah (hereafter UU).

64. First quote is "Better Methods Mean Progress and Security," Sept./Oct. 1962, 12–13, *Kennescope*, UU. Second is F. C. Green, "Modernization Means Job Security," Apr. 1959, 4, *Kennescope*, UU. See also "The Plodding Diplodocus," May–June 1962, *Kennescope*, UU.

65. *The Headframe* (Butte, MT: Anaconda Company's Western Operations, Communications Department), Jan. 1959, MHS.

66. On brochures at the Berkeley Pit, see LeCain, "'See America the Bountiful,'" 5–17. The quotes are from "Berkeley Open Pit Draws Thousands of Tourists," *Anaconda Company Trailsman*, Aug. 15, 1956, 1; "Berkeley Pit One of Montana's Most Spectacular Sights," *Anaconda Company Trailsman*, Apr. 1, 1960, 1. On the technological sublime, see David E. Nye, *The American Technological Sublime* (Boston: MIT, 1996).

67. Foote interview.

68. First quote is Ernest Salvas, "Address to Annual Convention—Montana Farmers Union, Great Falls, MT," Oct. 23, 1957, Fldr. "Salvas, Ernest, 1958 to," Box 168, IUMMSW, UC. The second is Ernest Salvas, "Address to Annual Convention—Montana Farmers' Union, Great Falls, Montana," Oct. 22, 1958, Fldr. 14 "Montana Locals—General," Box 289, IUMMSW, UC.

69. Damon Stetson, "Meany Declares Automation Evil," *New York Times*, Nov. 11, 1963.

70. Kurt Vonnegut, *Player Piano* (Originally New York: Dell, 1952, 1980). Ben B. Seligman, *Most Notorious Victory: Man in an Age of Automation* (New York: Macmillan, 1966).

71. First quote is Kennedy quoted in Bix, *Inventing Ourselves*, 256. Second quote is in "Transcript of President's News Conference on Foreign and Domestic Affairs," *New York Times*, Feb. 15, 1962.

72. Mercier, *Anaconda*, 174–75; Finn, *Tracing the Veins*, 54–55.

73. Quotes in Edward P. Shea to R. S. Newlin, C. H. Steele, V. D. Perry, and E. I. Renouard, "Alice-Rainbow Vein Diamond Drilling Progress Report," Jan. 2, 1959, file #29705.04, Fldr. "Butte Mine Operations 1937–1979, Silver Bow County," AGDC, AHC. See also Charles C. Goddard Jr. to R. S. Newlin, C. H. Steele, V. D. Perry, and E. I. Renouard, "Alice-Rainbow Vein Surface Diamond Drilling Final Report," Mar. 10, 1959, file #29705.05, Fldr. "Butte 1937–1979, Mine Operations, Silver Bow County," AGDC, AHC; Kearney, *Butte Voices*, 234.

74. First quote in Kearney, *Butte Voices*, 234. Second in R. D. Piper to E. I. Renouard, Fldr. "North Walkerville," Box 5B, Dave Piper Records, MC388, BSBPA. Third in Dave Piper to P. J. Berget, "re: inspection of a one and one-half story, six room, frame

dwelling at 1620 North Main St.," Fldr. "Main St. North/St. Lawrence School," Box 3C, MC388, BSBPA.

75. "Anaconda Blames Walkerville for Need to Close Alice Pit, Shea Denies Claim," *Great Falls Tribune* (Great Falls, MT), Feb. 27, 1959.

76. James R. Shea to F&S Contracting Company, Apr. 10, 1958, file 3, Fldr. 1, Box 1, James Shea Papers, Personal Collection 104 (hereafter PC104), BSBPA.

77. Jimmy Shea, handwritten note, Dec. 1958, file 2, Fldr. 1, Box 1, PC104, BSBPA.

78. James R. Shea to Bill Zadick, editor of *Great Falls Tribune*, telegram of Feb. 26, 1959, file 3, Fldr. 1, Box 1, PC104, BSBPA. See also "Anaconda Blames Walkerville," *Great Falls Tribune.*

79. Undated Walkerville city meeting notes, file 2, Fldr. 1, Box 1, PC104, BSBPA.

80. James R. Shea to Bill Zadick, Feb. 26, 1959, file 3, Fldr. 1, Box 1, PC104, BSBPA. See also "Shea Claims Company withholding Facts Not the City," news release, file 3, Fldr. 1, Box 1, PC104, BSBPA.

81. "Areas of Disagreement," Mar. 2, 1959, file 2, Fldr. 1, Box 1, PC104, BSBPA.

82. Walkerville City Council Meeting Minutes, Apr. 20, 1959, file 2, Fldr. 1, Box 1, PC104, BSBPA.

83. "Alice Pit Shutdown Is Caused by Petition Rejection," *Mont. Std.*, Feb. 26, 1959; Proposed Court Case, The City of Walkerville v. The Anaconda Corporation, no date, pp. 3, 5, 8, 10, file 3, Fldr. 1, Box 1, PC104, BSBPA.

84. Council Resolution Draft, Dec. 27, 1958, file 3, Fldr. 1, Box 1, PC104, BSBPA.

85. "Attention Walkerville Property Owners," *Butte Daily Post*, Jan. 8, 1959, file 1, Fldr. 1, Box 1, PC104, BSBPA.

86. "Areas of Disagreement"; "Alice Pit Shutdown"; "Anaconda's 'Block-Caving' Irks"; Frank Quinn, "Court Action May Be Necessary to Settle Issue Between Anaconda Co. and Walkerville," *Mont. Std.*, Oct. 25, 1960.

87. Walkerville City Council Minutes, Apr. 20, 1959, file 2, Fldr. 1, Box 1, PC104, BSBPA.

88. Ibid.; "Areas of Disagreement."

89. Quote in Undated Walkerville city meeting notes, file 2, Fldr. 1, Box 1, PC104, BSBPA. See also Kearney, *Butte Voices*, 235–36; Charles C. Goddard Jr., "Memorandum No. 206 re: Montana Operations," Mar. 5, 1959, File 30001.07 "Butte, Mine Operations," Fldr. "1959–Butte District Reports, Memos #198 to #226," AGDC, AHC.

90. C. N. Steele, Anaconda Company Vice President, to Mayor and City Council of Walkerville, Mar. 6, 1959; C. N. Steele, Anaconda Company Vice President, to Mayor and City Council of Walkerville, Apr. 13, 1959, both in file 3, Fldr. 1, Box 1, PC104, BSBPA.

91. "Alice Pit Shutdown." On negotiations with Walkerville, see Charles C. Goddard Jr., "Memorandum No. 208 re: Butte District," Mar. 19, 1959 and Charles C. Goddard Jr., "Memorandum No. 213 re: Butte District," Apr. 23, 1959, both in file 30001.07 "Butte, Mine Operations," Fldr. "1959- Butte District Reports, Memos #198 to #226," AGDC, AHC.

92. Charles C. Goddard Jr., "Memorandum No. 206 re: Montana Operations," Mar. 5, 1959, file 30001.07 "Butte, Mine Operations," Fldr. "1959- Butte District Reports, Memos #198 to #226," AGDC, AHC; "The Railroad Strike," *MMBM*, Mar. 7, 1959.

93. Swibold, *Copper Chorus*, 285–310.

94. "Anaconda Blames Walkerville," *Great Falls Tribune*.

95. Proposed Court Case, PC104, BSBPA; Anaconda Company to James R. Shea, Apr. 20, 1959, file 3, Fldr. 1, Box 1, PC104, BSBPA; Charles C. Goddard Jr., "Memorandum No. 214 re: Butte District," Apr. 30, 1959, file 30001.07, Fldr. "Butte, Mine Operations 1959- Butte District Reports, Memos #198 to #226," AGDC, AHC.

96. E. I. Renouard, "Memorandum No. 226, Butte District," Aug. 12, 1959, file 30001.07 "Butte, Mine Operations," Fldr. "1959- Butte District Reports, Memos #198 to #226," AGDC, AHC.

97. "Anaconda Co. Replies to Hill City Mayor, Claims Facts Withheld," *Mont. Std.*, Oct. 3, 1960.

98. Undated Walkerville City Council Meeting notes, file 2, Fldr. 1, Box 1, PC104, BSBPA.

99. "Walkerville Councilmen Deny Company's Request in Pit Dispute," *Mont. Std.*, Oct. 6, 1960. See also James R. Shea, Mayor of Walkerville, Press Release, Jan. 9, 1961, in file 3, Fldr. 1, Box 1, PC104, BSBPA.

100. "Walkerville," Oct. 1960, in file 3, Fldr. 1, Box 1, PC104, BSBPA.

101. Quinn, "Court Action May Be Necessary."

102. "Walkerville Mayor Blames Pit for 'Serious Situation,'" *Mont. Std.*, Oct. 2, 1960.

103. James Shea Press Release, Dec. 5, 1960, file 3, Fldr. 1, Box 1, PC104, BSBPA. Arrests and dismissal in "Controversy Over Pit Stepped Up, Then Become Quiet," *Mont. Std.*, Oct. 8, 1960.

104. On the farmer-labor coalition, see Michael P. Malone and Dianne G. Dougherty, "Montana's Political Culture: A Century of Evolution," *Montana: The Magazine of Western History* 31, No. 1 (Winter 1981): 44–58; William C. Pratt, "Rural Radicalism on the Northern Plains, 1912–1950," *Montana: The Magazine of Western History* 42, no. 1 (Winter 1992): 42–55. On Butte labor's support of the "only state-wide independent newspaper," see John Fonk, Secretary-Treasurer of Montana Mine-Mill Legislative Committee, "Report of Mine-Mill Legislative Committee Meeting, Eagle's Hall, Helena, Mont., Feb. 19, 1961," District Office #1, Fldr. "Salvas, Ernest, 1958 to," Box 168, IUMMSW, UC.

105. "Anaconda's 'Block-Caving' Irks."

106. "Walkerville, Montana—A Mining Town's Undermined Streets," *Life Magazine*, Dec. 12, 1960, 39–40. Additional Iwasaki photographs from this session in Butte are in PH296.003, PC104, BSBPA.

107. "Shea Claims Company Withholding Facts Not the City," PC104, BSBPA.

108. Holter interview, by Leech.

109. Quotes are in Petroni and Petroni interview. See also Holter interview, by Leech; Matteuchi interview; Tom Holter, interview by Mary Feriter, Apr. 12, 2005, Butte, MT, transcript, OHC, BSBPA; June Merzlak, interview by Ray F. Calkins, July 16, 1980, Butte, MT, transcript, OHC, BSBPA; Joseph Samuel Leeper, "The Changing Urban Landscape of Butte, Montana" (PhD diss., University of Oregon, Eugene, 1974), 85–87; Kearney, *Butte Voices*, 191, 220–21.

110. Petition quoted in Lewis T. Poole, "Company to Expand Pit, Asks Change in Road," *Mont. Std.*, Mar. 3, 1960.

111. "Controversy Over Pit Stepped Up," *Mont. Std.*, Oct. 8, 1960; Maxwell, "Hill City Group," *Mont. Std.*, Jan. 23, 1961.

112. Al Giecek, "Around the Hill," *MMBM*, Jan. 7, 1956.

113. Al Giecek, "Around the Hill," *MMBM*, Mar. 31, 1956.

114. Ibid.

115. Al Giecek, "Around the Hill," *MMBM*, Apr. 21, 1956.

116. On the miners' union's continued anger at open-pit mining, see *MMBM*, Fldr. 5, Box 341, IUMMSW, UC: Mar. 31, 1956, June 9, 1956, June 30, 1956, Apr. 13, 1957, June 13, 1957, July 6, 1957, Feb. 8, 1958.

117. "Anaconda Blames Walkerville," *MMBM*, Mar. 7, 1959, in Box 341, Fldr. 5, IUMMSW, UC.

118. Minutes of the Regular Meeting, Oct. 12, 1960, Bound Material #230, Minutes of the BMU, 1957–1961, IUMMSW, UC.

119. Gaskill, "Walkerville Council Blasts Anaconda Co."

120. "Solving Our Problems: We Find Government Can't Settle Strike," *Mont. Std.*, Nov. 15, 1959.

121. *City of Walkerville v. Anaconda Co.*, District Court of Second Judicial District, MT, Apr. 18, 1961. See a copy in Box 1, Fldr. "Walkerville Alice Pit Correspondence 1954–1960," PC104, BSBPA.

122. "Walkerville Council to Intercede in Hill City Property Problem," *Mont. Std.*, Nov. 3, 1960; E. I. Renouard to James R. Shea, Nov. 15, 1960, file 3, Fldr. 1, Box 1, PC104, BSBPA.

123. *City of Walkerville v. Anaconda Co.*, Plaintiff's Complaint and E. I. Renouard to James R. Shea, Nov. 15, 1960, file 3, Fldr. 1, Box 1, PC104, BSBPA; Check copy, Dec. 26, 1962, Fldr. "Correspondence Legal Papers 1960s," Box 1, PC104, BSBPA.

124. Kearney, *Butte Voices*, 235–36; *Mont. Std.*, Oct. 25, 1960.

125. Charles Goddard Jr., "Alice-Rainbow Vein Surface Diamond Drilling Final Report," Mar. 10, 1959, file 29705.04, Fldr. "Butte Miner Operations, 1937–1979." See also Shea, "Alice-Rainbow Vein Diamond Drilling Progress Report," Jan. 2, 1959, file 29705.04, Fldr. "Butte Mine Operations 1937–1979"; and Charles C. Goddard Jr., "Memorandum No. 203 re: Montana Operations," Feb. 10, 1959, file 30001.07 "Butte, Mine Operations," Fldr. "1959-Butte District Reports, Memos 198 to 226," AGDC, AHC.

126. Stephen Kroll-Smith and Stephen R. Couch, "What Is a Disaster? An Ecological-Symbolic Approach to Resolving the Definitional Debate," *International Journal of Mass Emergencies and Disasters* 9 (1991): 355–66; Eric Tucker, ed. *Working Disasters: The Politics of Recognition and Response* (Amityville, NY: Baywood Publishing, 2006).

127. Thomas E. Shriver and Dennis K. Kennedy, "Contested Environmental Hazards and Community Conflict Over Relocation," *Rural Sociology* 70, no. 4 (2005): 491–513; Stephen R. Couch, "Environmental Contamination, Community Transformation, and the Centralia Mine Fire," in *The Long Road to Recovery: Community Responses to Industrial Disaster*, ed. James K. Mitchell (New York: United Nations University Press, 1996), 60–85; Stella M. Capek, "The Environmental Justice Frame: A Conceptual Discussion and an Application," *Social Problems* 40, no. 1 (1993): 5–24.

128. Mercier, *Anaconda*, 175; Finn, *Tracing the Veins*, 56.

129. On the conflict, or lack thereof, between the environment and jobs for workers, see Scott Dewey, "Working for the Environment: Organized Labor and the Origins of Environmentalism in the United States," *Environmental History* 3, no. 1 (Jan. 1998): 45–63; Andrews, *Killing for Coal;* Montrie, *To Save the Land and People.*

130. Manuel, *Taconite Dreams*, 219.

131. Those who claim that this period represents conformity and complacency include Noble, *Forces of Production;* Reinhardt, "Drowned Towns"; Whitfield, *The Culture of the Cold War.* On the other side of this debate stands Liesl Miller Orenic, *On the Ground: Labor Struggle in the American Airline Industry* (Urbana: University of Illinois Press, 2009); Steigerwald, "Walter Reuther."

CHAPTER 4

# WORK

In 1985 former Anaconda Company employee Al Beavis reflected on his working life in the mines of Butte, Montana. As general mine foreman at the recently closed Berkeley Pit, Beavis was narrating an oral history for writer Teresa Jordan. When asked how difficult it had been for him to cross the line from labor to management, Beavis redirected the question. The "hardest thing" in his thirty-six years of working for the Anaconda Company was not becoming a boss—it was "going from underground to open pit." He stated that he "knew everything there was to know about underground, but you had to forget that, you were going into open pit property."[1] His reaction was not unusual among workers. By 1959 the Anaconda Mining Company was well on its way to focusing on open-pit mining, a switch that would be made permanent in 1974, when the underground mines closed for good. Those Anaconda employees who kept their jobs found their work descriptions in the pit to be altered radically and their labor placed under increased scrutiny.

What had changed was more than just organization of the shop floor, as labor historians typically put it. Open-pit mining instead reshaped the working landscape, termed "workscape."[2] In an open-pit mine, machines, human labor, and natural processes became integrated in an envirotechnical system, or, more simply, a workscape.[3] Workscape does particularly well to describe the dynamics of a mine, which is neither a static stretch of land nor a simple workplace, but is instead both a natural landscape and a factory that expands with its work. As Timothy LeCain puts it, this kind of factory had one main goal: "The cheap and efficient disassembly of the environment in order to extract the tiny percentage of copper that the society of the day considered valuable."[4]

In the underground workscape, miners used a complex set of skills to locate and extract high-grade ore from veins underneath the ground. Workers in an open pit instead gathered the most material, whether waste or ore, as quickly as possible. Because ore was more broadly disseminated in a large open-pit mine than in a typical underground operation, workers needed to dig up much more of it in order to make a profit. Separating the ore from waste, which had once been part of the underground miner's responsibility, became a process increasingly performed aboveground, primarily by machines and chemical metallurgic processes.[5]

The switch to an open-pit workscape does not simply lead into a typical mass production tale about the de-skilling of workers. Anaconda's management favored open-pit mining because it made mineral extraction more efficient and because it greatly increased the safety of the miners. Company workers accepted these benefits, but they were more likely to see the shift to open-pit mining as creating a mostly detrimental work life, primarily because it took labor whose distance from management, union bosses, and other community members was its most defining feature, and instead starkly put it out in the open. Workers certainly fretted over the decreasing number of employees needed and the more repetitive nature of work, but they were perhaps more worried about the lack of camaraderie, the removal of a sense of masculine independence, and their loss of status in the community, all of which they saw as a consequence of performing industrial work in the open. Butte's unions struggled mightily against those consequences, but they had no answer to the company's truthful claims that overseas copper had marginalized Butte, nor could they help it when continuing layoffs and internal divisions sapped their strength. Over two decades, the open-pit workscape and its social consequences became firmly established in Butte.

## Bigger and Faster

The Anaconda Company kicked the Berkeley Pit into full-scale ore production in the late 1950s. By 1959 local and national publications featured the fully operational Berkeley Pit. The *Mining Congress Journal* and a Bureau of Mines circular provided detailed but staid discussions

The Anaconda Company Trailsman

VOL. 4. No. 1. BUTTE, MONTANA, APRIL 1, 1960 THE RICHEST HILL ON EARTH

The Anaconda Company's

BERKELEY PIT

ORE BINS

ORE STOCKPILE

BELT CONVEYOR

CRUSHER

THE PIT IS NAMED FOR THE BERKELEY MINE, ONCE A FAMOUS LODE-MINING PRODUCER IN THE MEADERVILLE AREA.

The reddish-brown material is worthless waste and averages 250 feet thick. A total of 200 million tons of waste will be removed.

The grey material beneath the waste is ore and averages 200 feet thick.

Five large electric shovels are available. The huge shovels use 6 cubic yard buckets and have a capacity of 8 to 9 tons.

Berkeley Pit One of Montana's Most Spectacular Sights

*Anaconda Company Trailsman* Cover, 1960. This promotional newsletter, placed at the Berkeley Pit's viewing stand, gave visitors a better idea of how the pit operated. Its authors often worked to convince readers of the pit's technological and natural sublimity. *Anaconda Company Trailsman* 4, no. 1 (April 1, 1960): 1. Image courtesy of the Montana Historical Society.

about the growing pit. The *Anaconda Company Trailsman*, however, gave a much more exuberant celebration of the achievement. Made available at the Berkeley Pit viewing stand, the publication described the massive scale of the Berkeley Pit as well as the immense power of the equipment and invited tourists to visit the observation deck at noon, when workers usually set off blasts. Anaconda also promoted its operations in a series of television commercials and in employee newsletters. All the media made it evident that open-pit mining was a sublime technological marvel. Increasingly larger machinery helped to extract material, and more-efficient systems began to process it.[6]

Open-pit mining started with the Geological Department. Well-known for its geological work, Anaconda served as a virtual graduate school for geologists. To encourage continual planning, the Geological Department operated under a Ton-for-Ton Program, begun during underground mining's heyday. For every ton of ore that the company mined, the Geological Department had to identify another ton that could be mined in the future. Each ton added to the known reserves.[7] A small cadre of engineers and geologists constantly evaluated the ore body, its economic potential, and the stability of the rocks making up the Berkeley Pit's walls. Geologists used churn or diamond drilling machines to drill large-diameter holes. Rock picked up by drills helped geologists to delimit the ore body. In the late 1950s geologists sampled ore every two hundred feet, then sent data to the Metallurgical Research Department and the assay office. At the assay office, employees determined the percentage of copper, silver, and other minerals in each sample.[8]

Based on those data, engineers would make production forecasts on a long-term, annual, semiannual, monthly, and even daily basis. They would then stake the ground with flags—red for rock containing ore, white for waste.[9] Open-pit planners had varying degrees of success in rationalizing the inorganic world of rocks. No one was completely certain about what lay underground before the miners dug it up. Samples could only tell part of the story and pit walls were sometimes liable to crumble.[10]

The actual process of open-pit mining became relatively simple. Berkeley Pit workers initially spent much of their time removing the 200 to 250 feet of reddish-brown, oxidized overburden that they deemed worthless. Below it lay a gray area that held the ore.[11] Copper atoms bond with other elements, and most of Butte's copper ore came in the form of chalcocite, a dark gray mineral with one atom of sulfur and two of copper; enargite, a mineral containing four sulfur, one arsenic, and three copper atoms; or bornite and chalcopyrite, which contained copper, iron, and sulfur encased in a quartz porphyry mix.[12] The bonding meant that large quantities of ore had to be gathered in order to get enough valuable minerals to make mining worthwhile. It also meant that digging and processing ore exposed many harmful materials, including arsenic and sulfur.

Anaconda's paper played up the "investment of many million dollars" that went into its massive equipment. Although the confined spaces of traditional underground mining constrained the possible size of equipment, the large open space of the pit made big machines possible. To keep equipment functioning well, the company required

Early Work at the Berkeley Pit. Photograph by Cleveland S. Bishop. Butte–Silver Bow Public Archives, Postle-Bishop Photograph Collection, PH406.04.

a large Mechanical Department. Maintenance work was often done on the go, with trucks carrying gasoline, lubricants, and repair equipment going to work sites to maintain shovels, drills, and tractors.[13]

For other maintenance work, the Berkeley Pit operation included a number of large steel structures northeast of Meaderville. The main shop building housed fourteen service bays for enormous rigs, shovels, and trucks, as well as a few offices and a tool room. Right next to that shop sat a large warehouse, a paint shop/high-pressure washing room, a fueling area, and a house with oil pumps. Next door, a number of employees worked in the main pit offices, where the pit superintendent, his clerks, the foreman, shift bosses, engineers, timekeeper, and powder foreman held court. The upper floor of this same building was a change house, where crews could take a shower and remove their grimy, sweaty clothes. For the sake of security, a large wire fence surrounded the whole area, which was well lit at night, since the Berkeley Pit typically operated around the clock through three eight-hour shifts.[14]

Anaconda's pit-mining operation started with drilling. During its early years, the Berkeley Pit employed four rotary drill rigs that used sharp, rotating drill bits to penetrate earth. An operator and a helper/sampler ran the drills, gathering geological samples along the way. They usually drilled seven holes, seven feet into the ground, twenty feet from each other.[15]

Next, blasting crews—comprising a truck operator, three powder men, and a helper—loaded either 160 pounds of explosive into holes drilled in ore, or four hundred to eight hundred pounds of explosive drilled in waste rock (damaging the latter was not a concern). Anaconda's explosive of choice was ammonium nitrate saturated with a more combustible element. In 1959 crews mixed one gallon of diesel fuel into each eighty-pound bag of ammonium nitrate. A blasting crew would hook lines between all of the holes for each blast and then trigger it.[16]

Ore loading began after the blasts broke up the rock. In 1959 Anaconda used five large electric shovels, which had a capacity of eight to nine tons, and one smaller diesel shovel. Each shovel crew consisted of two men: an operator and an oiler. For the sake of efficiency, mammoth Euclid and Mack trucks, whose capacity was at about thirty-six tons, would pull up on both sides of the shovel at the same time. Then the shovel operator would load broken-up ore into trucks—eight thousand to ten thousand tons per shovel during each eight-hour shift. In support of the shovel operation, ten bulldozers pushed broken rock into piles for loading and cleaned up whatever did not make it into trucks the first time.[17]

Once the trucks were loaded, their drivers took over. Approximately a hundred drivers a shift drove material up the benches, which were flat areas carved into the oval-shaped pit. They spiraled, counterclockwise, around, and then out, of the pit. Already a two-mile journey in 1959, the drive out of the pit grew longer in the ensuing years. The length was a logistical problem, because trucks in 1959 moved, on average, fourteen miles an hour. The truck driver headed either to a waste dump, if the load had less than 1 percent ore; to a leach dump, if it was more likely to have 1 to 4 percent ore; or to the crusher, if it carried ore in excess of 4 percent.[18]

Waste dumps were primarily located north of the pit, in the valley of Silver Bow Creek, though the company used other areas as pit development continued. Using several dump sites simultaneously was necessary to avoid traffic congestion. A truck therefore might drive only a few thousand feet, or it might drive several miles. Dumpmen directed the traffic within each site. Dumps grew ever taller as truck drivers drove up their slopes to deposit their loads at the top. The amount of waste was enormous. Anaconda estimated that its stripping ratio of waste to ore was 1.47 to 1 in 1959, and that ratio became much less favorable in subsequent years.[19]

Rock with minor ore-bearing potential went to the leach dump at the extreme northeast side of the pit. There drivers dumped material on an 1,800-square-foot asphalt pad. A sprinkling system then dripped a weakly acidic solution on top. The watery solution percolated through the pile, dissolving metals and collecting them at the bottom. The metal-bearing liquid, along with water pumped daily from the underground mines, was then piped to the precipitating plant, located to the east of the Berkeley Pit. Within the plant, sweeper operators made sure that copper was extracted. To accomplish that task, Anaconda added shredded tin cans into the solution. The cans contained high levels of iron and that reactive metal exchanged ions with the copper solution. Copper deposited on the iron, from which it was easier to extract.[20]

Rock containing more than 4 percent ore went to one of three dumping sites next to a crusher on the southeast side of the pit. Management reserved one dumping site for material that included timber, left over from old underground workings, which were now being excavated in the new open pit. Debris could be picked out from the rock here. All the ore-bearing rock then was dumped into a hopper, which sat above the crusher. The hopper fed into a vibrating screen that segregated small material from large rocks, propelling the larger ones into the crusher's gaping jaws. The crusher, which the company paper assured readers was "one of the largest crushers of its kind," was undoubtedly big: it weighed 250 tons and measured 64 feet by 84 feet.[21]

By 1959 transporting crushed ore involved highly mechanized, even automated, processes. The crushed ore, now at under nine inches, first passed downward into a surge bin below the ground, where foreign material could be extracted from the rock flow. A conveyer belt that was forty-eight inches wide then moved the ore, at a speed of four hundred feet per minute, up through a six hundred-foot-long concrete tunnel, and above the ground. The belt dropped material into 5,000-ton-capacity ore bins or into a 35,000-ton-capacity stockpile bin. Continuing to use gravity to their benefit, workers released ore from the bins straight into 75-ton-capacity railroad cars.[22] Employees of Anaconda's Butte, Anaconda & Pacific Railway then took over. They trekked loaded cars to the town of Anaconda, about twenty-five miles west, where the crushed ore was concentrated and smelted. By 1962 this system produced about 215,000 tons of ore each week.[23]

Like the rest of the extraction system, the ore processing system in the town of Anaconda relied on large machinery. Workers delivered pit ore from train cars to a grinding area, where ball and rod mills crushed it into even smaller bits. A ball mill involved a rotating cylinder that was horizontal or slightly inclined. Partly filled with steel balls, it used friction to grind material into tiny bits. Instead of a cylinder, rod mills used a rotating drum with steel rods inside to smash and grind ore into fine particles.

The next step of the process, concentration, demanded that particles be particularly finely ground. Flotation operators mixed ground ore with chemicals in a vat. During the froth flotation process, they "frothed" the vats with compressed air and skimmed a concentrate of 27–30 percent copper from the top.[24]

The smelter was the next stop. There the application of heat liquified the concentrate, dissolving the bonds between copper and the other atoms—including sulfur, arsenic, and other less valuable minerals—in it. Copper and other valuable minerals would sink, and the remaining impurities would rise to the top, where they were removed. The solid, blackened waste, called slag, would be deposited outside the smelter. The more concentrated substance, containing copper, iron, sulfur, and precious metals, then went to a converter, where blowing air burned off the sulfur and oxidized the iron. Sulfur dioxide and arsenic oxide flew into the air as gaseous waste. Although engineers tried to control this air pollution, atmospheric waste still resulted in a number of dead zones surrounding the smelter, where vegetation and crops died and animals sickened. Processing the lower-grade ore that open-pit mining made possible meant more waste. Dust, gas, and smoke were thus increasing problems for smelter workers as well as nearby residents. Finally, smelter workers poured the molten ore into a form. At that point, the product was known as an anode and was about 98 percent copper.[25]

Smelter workers shipped the anodes hundreds of miles to Great Falls, in north-central Montana, where complex metallurgical processes continued. There, Anaconda workers placed the anode into a solution containing sulfuric acid and a bar of pure copper known as the cathode. When workers shot electricity into the solution bath, oxidation occurred at the anode. The impure copper anode released its metal ions, which traveled across the solution to the cathode. At the cathode, the opposite reaction occurred, joining copper ions. The anode was thus robbed of its mass, while the cathode increased in size, adding additional pure copper metal to itself. More impurities dropped to the bottom of the solution during this transformation, and the final product of more than 99 percent pure copper became marketable.[26]

Great Falls shipped the pure copper to other sites, often Anaconda-owned facilities, which turned the metal into wire and other useful products. Anaconda owned plants that used copper in Great Falls; Kenosha, Wisconsin; Buffalo, New York; and Paramount, California. Anaconda was thus an important, vertically integrated business, bringing copper "from the mine to the consumer," as its advertising campaign phrased it.[27] The large Montana operations were interconnected with an even more massive industrial-consumer system. Anaconda was a modern global corporation, and sites like Butte fueled its existence.[28] Open-pit mining simply made the mining that fed into this system more efficient, although, as Timothy LeCain points out, it was an efficiency only made possible by the open-pit regime's reliance on cheap carbon-based energy and the population's acceptance of massive amounts of waste as part of mining.[29]

The processes described above remained fairly constant during the heyday of open-pit mining in the 1960s and 1970s. The scale of the operations increased, however, as Anaconda sought greater efficiency. Technology became bigger and bigger. The focus on size began with explosives. By 1967 crews were even using more explosives per blast than in the past; charges reached 320 pounds in ore. By the early 1970s the typical number of holes per blast in the pit had also risen—from a handful to a hundred or more on occasion.[30]

Operations to load and move ore similarly increased in scale. In 1959 the equipment involved forty-three trucks with thirty-six-ton capacity, six electric shovels with a six-yard capacity, four mounted rotary drills, and ten bulldozers. By 1966 Anaconda was running ninety-five trucks, ranging in capacity from sixty-five to one hundred tons, called "ukes" colloquially; eleven electric shovels, and nine drills. It also now hosted busses to transport workers from worksite to worksite and an increasingly "large fleet of auxiliary vehicles" that included powder trucks, sprinkler and fuel trucks, and bulldozers. By 1973 workers drove 135 large trucks, ranging from a hundred to two hundred tons in capacity and with tires that were twelve feet high. They also operated fifteen shovels, nine of which now had a fifteen-yard capacity; twenty-one bulldozers; thirteen drills; ten loaders; seven graders; four powder trucks; a fire truck; an ambulance; and many smaller vehicles.[31] Anaconda was truly following a well-worn industry attitude when it came to open pits: "bigger is better."[32] Workers were soon moving more than 250,000 tons of material every day.[33]

The Berkeley Pit likely encouraged the use of big haul trucks in other open-pit operations, including the largest in the United States, Kennecott's Bingham Canyon Mine in Utah. Using trucks instead of the typical railroad tracks to move ore and waste out of pits made surface operations more flexible, fluid, and efficient. Every time the pit needed to be expanded, workers did not have to move railroad track. In each instance when waste needed to go to a different dump site, trucks were better able than trains to change course. Flexibility equaled efficiency in an open pit because it was difficult to predict exactly how solid the ground would be or how much ore could be uncovered.[34]

A bigger operation dug out an exponentially bigger pit. In 1966 the pit was more than a mile wide and 600 feet deep. By 1967 it was almost 7,000 feet, or 1.3 miles, east to west—the length of 19.5 football fields—and 3,500 feet, north to south. By 1968 it was 4,500 feet north to south. By 1973 its north to south growth had already reached more than a mile, or 5,600 feet, wide. Just driving into and out of the Berkeley Pit in 1973 took some doing, because there were twenty-five miles of haulage roads within the pit that year. By 1974 the pit had grown 1,300 feet deep. Just one year after that, the depth of the pit would surpass the height of the Empire State Building at 1,500 feet. Anaconda was soon rightfully bragging that the Berkeley Pit was "the largest all-truck operated copper mine in the U.S."[35]

The scale of the pit amazed workers. A 1973 article by a *Seattle Times* journalist reported employees' descriptions of the Berkeley's strange allure; "'It's awesome,' said one man, a Pit employee. 'You get to the bottom and look 1,500 feet up across the terraces. If the sun is setting or the stars and moon are out in the clear air, it's like watching scenes from *2001* [the film].'" Another employee compared the pit to the pyramids of Egypt: "Both are big and brown and monuments to a hell of a lot of human work," he joked.[36]

Beyond the pit itself, the company also sought bigger equipment and faster transport. Anaconda bragged that it "remains a leader in this trend" too.[37] Management had deemed that it would be cheaper to do more mineral processing in Butte. By 1964, rather than ship tons of crushed, ore-bearing material to the town of Anaconda, the company moved it on a conveyor system to an 80,000-ton stockpile, or directly into storage bins. From those locations, conveyor belts then moved it, at a rate of about 4,500 tons per hour, straight into the brand-new Clyde E. Weed Concentrator in Butte. The concentrator included one main plant, thirteen additional buildings, a twenty-six-mile-long pipeline to bring fresh water from Anaconda, and more pipeline to transport waste up to the Yankee Doodle Tailings pond, an earthen dam north of the pit. The plant employed the first large-scale autogenous pebble mills, which grinded rock on rock. Due to these processing changes, the smelter in the town of Anaconda could receive smaller amounts of concentrates rather than trainloads of crushed rock, hence cutting transportation costs.[38]

As the unions feared, one result of mechanization and automation was job loss. The increasingly sophisticated, large-scale technologies in the copper industry led to what Timothy LeCain terms "an astounding increase in labor productivity." Equipment operated by the pull of a lever accomplished the work once done by dozens, or hundreds, of men. As LeCain puts it, hydrocarbon energy had replaced human labor.[39]

## Marginalization and Union Struggles

In the 1950s Butte's unions had some success fighting open-pit mining, but by the early 1960s overseas competition and the declining number of required mine workers had irrevocably damaged the unions' bargaining position. During the very long 1959–60 strike, Butte's unions found themselves in a particularly bad place. A recession coupled with Anaconda's mechanical improvements meant the loss of many workers from the employment rolls.[40] By June 1959, just four years after the start of open-pit mining, Anaconda's Butte workforce had dropped from 4,586 to 2,159 employees. Meanwhile, production numbers from that same period had jumped—from 7,960 to 8,430 tons of recoverable copper. A tipping point had been reached. In 1957, 48 percent of Butte's copper ore came from open-pit mines. By 1959 the number had risen: more than 63 percent came from open pits.[41]

During the 1959 negotiations, Anaconda wanted to guarantee that nothing in the new contract would be "construed as preventing the use of improved methods or mechanical equipment."[42] Butte's unions were initially determined to resist the company's "right to manage" stand, but, after holding their ground for months, the unions dropped outright opposition. They insisted instead on better communication about impending changes and some input into them, as well as both company attention to those who lost their jobs and better conditions, benefits, and pay for those who remained at the workplace.[43] Bargaining reached an impasse, and a strike that lasted more than six months began on August 19, 1959.

The strike forced Butte's unions to relax their resistance to open-pit mining. The company pointed out that Butte's high-cost operations were at a disadvantage in competing against Kennecott's low-cost, completely open-pit operation in Utah.[44] Overseas mines were even more of a challenge than domestic ones. Indeed, some of those overseas operations were Anaconda's own. Having built an empire in the state of Montana during the first half of the century, Anaconda looked abroad in the following decades. By 1959 Anaconda's two open-pit mines in Chile, Chuquicamata and El Salvador, were together producing 350,000 tons of copper, five times the 70,000 tons produced in Butte. Output from the company's Mexican mines was also growing.[45] With high productivity and low labor costs available elsewhere, Anaconda could afford to be less concerned about Butte than at any previous time.

The union countered that Butte was central, not marginal, to Anaconda's future. The company had recently made large investments there, and the area had massive ore reserves. Since the start of open-pit mining, Butte workers had more than doubled their hourly production, while the company's labor cost per pound of mined copper had been cut in half.[46] With company profits rising rapidly, Anaconda could not claim poverty when faced with requests for better benefits. The research department of the international IUMMSW argued, "The tremendous labor savings accruing to the companies in this industry as a result of rising productivity afford plenty of room for granting the substantial wage increases included in this year's demand."[47]

Butte's union leaders were caught in a bind. The rationale for touting the productivity of their workers rested on acceptance of the job losses that open-pit mining entailed. The union hoped to win a small raise for the workers who remained but could not help those who had already lost jobs. The less-productive underground mines were in special danger; attempts to raise wages there could prompt the company to close them, perhaps for good. As Anaconda's lead geologist wrote, "Too bad for the deep mines, because continued labor increases are simply going to put them out of business no matter how many deep shafts are sunk."[48]

With so many members laid off due to the decrease in underground work, IUMMSW locals tried to make job security, in some form, a major bargaining issue. Butte's locals particularly worried that Anaconda was using layoffs as a way to get rid of older workers, who had already "given the best years of their working life." They were convinced that the company's new, "fantastically strict physical examinations" were a typical Anaconda policy: to "junk the older models... men, or machines...and get a new one." The company was replacing bodies, broken down by the tough underground environment, with shiny new equipment.[49] Locals fought for, and eventually won, limitations on company physical examinations during firing, hiring, and rehiring decisions. More importantly, Butte's locals won the institution of seniority, based on the length of employee service, and central hiring for the first time.[50]

After the intervention of federal mediators, the strike finally ended on February 15, 1960. Anaconda, however, never reopened the Anselmo or Emma zinc mines. It hired fewer employees back and continued to shrink the underground workforce, thus dampening the power of the biggest, most radical local: the BMU.[51]

Anaconda used Butte's competitive disadvantages as an issue to hold down wages during the rest of the 1960s. The company's slogan, "Working hard to keep Montana metals competitive," highlighted Butte's parlous position during bargaining in 1962.[52] A full-page ad placed by Anaconda in the local newspaper warned miners: "The vein mines of Butte are in a precarious position costwise and under the pressure of increased costs might have to be closed and we would turn to the Berkeley Pit for all of our production. Should that ever happen most of the Butte crafts and other jobs would be lost."[53]

The number of Anaconda Company workers in the Butte region fell from 6,493 in 1954 to only 2,494 in 1960. It improved to 3,702 by 1964, but the outlook was not encouraging. Unions had made a few gains in benefits and pay over the past two contracts, but payroll costs were falling relative to the value of output. In 1954 payrolls made up 51 percent of the total value of mineral production, but by 1964, they took up only 29 percent.[54] As one 1964 bargaining document made clear, "Automation, larger capacity machines, engineering improvements, [and] technological developments" meant that Butte workers faced the same problem as mining operations elsewhere.[55] Seeing job security as the "overriding problem at the bargaining table," the IUMMSW made it a major bargaining issue again that year.[56] The company, however, continued to claim that it had "extended itself farther than the economic conditions of Montana operations presently justify." Workers gained only incrementally, continuing to receive lower wages than those paid by similar copper companies.[57]

## The End of Block-Caving

Anaconda's continued shift to surface mining during the 1960s ended the temporary recovery in employee numbers. The situation was especially bad at the Kelley Mine, Butte's sole low-grade underground mine. When introduced in 1947, underground block-caving held the promise of keeping Butte's mines running for decades. By the late 1950s, however, Anaconda geologists began to question its suitability. As geologist Vin Perry put it, "Why use higher-cost underground methods when the present Anaconda Plant can be filled to capacity with low cost Pit ore?"[58]

Late in 1962 Anaconda executives decided to close the Kelley Mine. In addition to reducing the need for workers, shutting the Kelley would enhance the Berkeley Pit. The two projects were not only competing for the same ore reserves, but block-caving in the Kelley also threatened slope stability on the northwest edge of the pit, an area that was already experiencing subsidence. Coping with the problem had already forced engineers to reduce the general slope of the pit from 45 to 38 degrees in affected areas.[59] One form of mass mining, the open pit, had replaced another, underground block-caving. The irony was not lost on Berkeley Pit superintendent John Killoy, who had served as the Kelley superintendent in 1952. As the employee newsletter put it, Killoy was now "tearing asunder that which he has wrought."[60] Discontinuing the Kelley allowed the Berkeley Pit to include more of the Kelley workings, expanding the pit's boundaries farther west into Butte.[61]

By the mid-1960s the company had begun to use the large, 35 feet by 7 feet, concrete-encased Kelley shaft, left over from the Kelley Mine operation, as a way to hoist ore from the three vein mines still in operation. The company did some block-caving for zinc in the Badger Mine during the 1960s, but otherwise Anaconda's once-extensive underground had narrowed to the Mountain Con, Steward, and Leonard Mines.[62]

## Surface Safety

Even as layoffs continued, Anaconda felt little need to apologize for open-pit mining's effects. On the contrary, management was proud of the important contribution to increased safety that open-pit mining made. In the early twentieth century, underground mining was the most dangerous industrial occupation in the United States. The deadly conditions in Butte's underground workscape were matched only by mines in the Transvaal Province of South Africa. Butte's miners had long experienced an uncertain, risky workscape that included frequent accidents, the possibility of underground fires, and machinery malfunctions; as a group they suffered thousands of cases of silicosis, a debilitating disease caused by breathing mine dust.[63]

Bringing work to the surface quickly eliminated the greatest of mining's dangers. It was simply easier to see oncoming problems in the brightly lit surface, and the open pit provided a larger space in which to maneuver dangerous equipment. Anaconda's 1956 pamphlet, titled "Anaconda Safety Rules," contained sixty-four pages on underground safety, but only sixteen on open-pit dangers.[64] Miners recognized how much safer the surface was. After getting acidic water in his eyes and being almost killed by a falling rock underground, miner William Dudding decided to switch to the open pit. Another miner related that his wife convinced him to transfer to the open pit because of its safety.[65]

Operations in the Berkeley Pit still had their dangers. The repetitive nature of haul-truck driving or shovel operation could mean falling asleep, especially for drivers on midnight shifts. Malfunctioning explosives could also mean danger. Employees at the Berkeley Pit were digging up so many old underground mine tunnels that they ran into additional safety issues. In the area of the old Pennsylvania Mine, a blasting crew uncovered old workings that were still on fire from years before. The movement of people within the pit could also cause problems. Backing up wide trucks in crowded areas entailed risks, as did the weaving of pickup trucks, driven by managers, between the massive equipment. In August 1964 twenty-six employees suffered injuries when a bus taking them to different work spots nosed into a trench.[66]

Other dangers related to the fact that this workscape was outdoors. Some loved this feature of the Berkeley Pit operation, but others complained about their exposure to the weather. Depending on the mine, temperatures below ground could be cold or hot, regardless of conditions at the surface.[67] During Rocky Mountain winters, however, the weather at the surface was always freezing. The average January temperature was 15.1 degrees Fahrenheit. After working in the pit during the winter graveyard shift, miner Dan Aguilar spent his postwork mornings shivering next to a heater at home. The experience made him consider going back underground. Since Butte stands at 5,500 feet, right next to the Continental Divide, freezing temperatures could also happen at other times during the year.[68] Weather conditions could be treacherous in the pit: fog, snow, ice, and mud caused many a runaway truck.[69] To alleviate the problem, engineers never let the roads out of the pit exceed a seven-degree grade. The warehouse area also included an electric heater island, where drivers could plug in trucks that would otherwise be unable to start during particularly cold weather.[70]

Pit workers bickered about the quality of the air. Truck driver Bill Long said that a common Berkeley Pit joke went, "If you can't see it, smell it, or taste it, it ain't real air."[71] The deadly, silica-rich dust that plagued underground mines was absent, but exhaust fumes from diesel machinery and flying dirt from blasting and digging filled the open pit's air, even if such hazards were somewhat dispersed. Despite efforts by graders to maintain the roads in and out of the pit, large trucks continued to kick up dirt. Even the general manager of Montana operations had to admit that a dusty haze often hung over the pit, but the company's studies of diesel exhaust did not uncover a solution.[72]

Anaconda's safety engineers improved safety procedures to lessen workscape problems. Systems monitored areas liable to subside due to weak ground. Water from a sprinkler truck was routinely on hand in order to keep dust down during both drilling and driving. Safety cones, radio alerts, and a loud alarm signaled each impending blast in the pit. Protective footwear, safety glasses, hardhats, and gloves became mandatory.[73]

Union and management leaders were safety conscious and the open-pit workscape allowed them greater oversight than they had in dark spaces underground. By 1970 Butte's unions had asked for an ambulance to be on call, the availability of heat during the winter, better blasting regulations, and improved clothing. Like the underground, the open pit soon gained artificial light so as to fight back against nature's work-halting, unsafe darkness. All parties agreed that the space around offices and warehouses needed to be well lit, as pit operations typically ran all night. Workers also had to mount floodlights on shovels and drills and post reflectors on haulage roads, while placing continually flashing lights on both supervisors' trucks and service vehicles. Propane lanterns lit the dump areas, although a federal mine inspector in 1970 claimed that they were still lit inadequately. Floodlighting extended to the entire pit in the early 1970s. Nevertheless, power failures remained frequent, and electrical shock continued to be a potential danger because trucks and tractors could trample power lines that ran along the ground.[74]

Achieving full cooperation with safety guidelines was difficult. The federal inspector noted in 1970 that some employees failed to wear the required hardhats and glasses. He also saw some carelessness with explosives and electricity.[75] A lack of proper communication could prove dangerous, even if everyone was following protocol. In one case, a huge tank that workers mistakenly believed to be empty tilted when the fuel inside shifted. A worker on top of that tank fell forty-three feet. In 1969, thirteen disabling injuries occurred in the Berkeley Pit, a number that was only seven fewer than the twenty in Butte's underground mines that year.[76]

Despite some serious accidents, the radically different organization of the surface workscape ensured that open-pit labor was demonstrably safer than underground work. The surface environment simply offered fewer major hazards, and many of those could be easily resolved through the increased oversight that the pit allowed. By the mid-1970s, most injuries in the pit were due to "people falling off ladders on the trucks" or similar minor incidents and were not the result of such underground dangers as subterranean explosions or falling rocks.[77] Butte historian James D. Harrington's research in the Butte–Silver Bow Clerk and Recorder's Office estimated that 2,350 mining-related fatalities occurred in the Butte area from 1889 to 1974. Of that number, only five happened in the Berkeley Pit; the rest happened underground. After underground mining ceased, only two additional deaths were the result of mining operations before Anaconda went out of business in the early 1980s. Moreover, none of these figures take into account the thousands of chronic and deadly illnesses, such as those from silicosis, most of which resulted from careers underground, not from work in the pit.[78] Trends in Butte stood true for the rest of the metal-mining industry. A 1970s analysis showed that the chance of an accident underground was three times as great and the severity twice as great as in a pit.[79]

An anecdote captures how safe the open-pit workscape was. After spending his life involved in underground mining, Butte resident John Thatcher thought that he would never recommend the dangerous mining profession to anyone he cared about. He was willing, however, to let his son work in an open-pit mine because the environment was so much safer.[80] The open-pit workscape enabled safety by offering fewer unseen dangers, greater oversight, and more predictability.

## The Monotony of Supervised Labor

The surface might have meant increased safety and productivity, but workers who experienced both kinds of mining found the pit lacking. Underground miners often worked on contract; the company paid them according to the distance of tunnel they dug or the amount of ore material they removed. Writer Teresa Jordan, who interviewed many former Anaconda employees, discovered that the thrilling danger of underground work and the satisfaction miners gained from working on contract—in which "the harder you worked the more you were paid"—vanished in pit labor.[81] Workers experienced in both environments lamented that the freedom, camaraderie, prestige, and macho work culture of mining similarly disappeared in the pit setting. The causal factor for these problems seemed obvious to workers: industrial work once hidden from view was now done in the open—and workers saw that as to the detriment of their work culture.

The safety of the open-pit workscape, coupled with the repetitive tasks it required, led to boredom. Job descriptions for employees in the Berkeley Pit, concentrator, and precipitation plant emphasized the performance of mostly routine or semi-routine tasks.[82] As miner Al Beavis explained, "Boredom is real bad on trucks. And on drills, anything in the pit.... It gets awful monotonous." Dan Aguilar agreed that even the length of a shift felt longer in the pit. Due to the difficulty of setting up equipment in tight underground spaces, "By the time you got anything straightened out and got anything done, it was time to go home. Down in the pit, they got you to the job right away, and it seemed like a long 8 hours, 12 hours."[83]

Miners especially missed the independence that distance and darkness in the underground had given them. Because contract miners were paid according to the amount of work they accomplished, the company had little need to supervise. One supervisor put it this way: "You either put out or you didn't get paid."[84] For mine workers like Victor Giuseppe Segna, this system meant that "you're your own boss in the mines."[85] Due to the great distance between teams of workers, even employees who were not working on contract could go for hours without seeing a boss. Miner George Tarrant told a story about his onetime partner who would turn off his lamp and snooze in the pitch black whenever the boss left. Just before the boss came back four hours later, the partner would awaken, say, "We'd better make a showing," and put in a few minutes of work.[86] The nature of the payment system made such cases rare, but their existence highlights miners' freedom from supervision.

The open-pit workscape removed this independence. Anaconda tried a piecework system in the open pit similar to what it used underground, but the system was apparently not functional because, in an open pit, "the machinery puts out, not the man." The greater prevalence of machines in the pit also meant that an important ratio changed—the ratio of workers to managers. The company needed fewer workers to operate machines, but it needed more managers to oversee the work.[87] The expansive open space in the pit, described by an environmental engineer as "blue sky outlined with rock," also affected workers' independence. Instead of seldom seeing a boss during his eight-hour shift, Dan Aguilar "felt like the whole world was watching me" in the Berkeley Pit. Shift bosses in the pit saw everything.[88]

By the mid-1960s, the company had built a new control tower in the pit, furthering management's surveillance capability. Equipped with radio and closed-circuit television, the tower allowed engineers to more closely manage all drilling, blasting, loading, dumping, crushing, and conveying. As one Bureau of Mines publication reported, "Close engineering control of all mining activity is necessary to develop the pit according to plan and to maintain ore production at the desired grade and tonnage level. Ore volume and grade are forecast and checked daily, bench elevations and ramp development are closely watched, and bench expansions are projected and controlled."[89] Anaconda posted a couple of dispatchers in the tower, ensuring that each truck drove to the correct shovel and then to the correct dump in a timely fashion.[90] The tower served a purpose somewhat like eighteenth-century social theorist Jeremy Bentham's panopticon, a place from which everyone could be watched and, through the threat of possible observation, controlled.[91] Engineers similarly managed each step in the concentrating process from a central control room; their closed-circuit television cameras could scan eight separate areas.[92] Unlike for efficiency and safety, little evidence suggests that easier surveillance was an important reason for Anaconda's transition to the pit, as it was for asbestos mine operators in Canada, but Anaconda did come to embrace its new advantage.[93]

## Workers Driven Apart

Beyond decreased control over their own labor, former underground miners seemed saddened by open-pit work's lack of togetherness, which was another product of the pit's expansive spaces. One Anaconda employee recalled that when you are in an underground mine, that's a "definite brotherhood . . . because your life is in so much danger and you have to be able to count on those guys that are around you."[94] Due to safety concerns, almost all underground employees worked with partners—usually the same partner every day, sometimes for many years. People underground depended on each other—the motorman, the parts runners, the contract miners, the shift bosses—and they would talk to each other all the time, both about work and about life outside work. Miner Dennis Murphy proclaimed that this camaraderie cut across ethnic boundaries, while fellow miner Floyd Bossard believed camaraderie below the surface translated aboveground as well, into the community of Butte at large.[95]

The work environment in the open pit encouraged division, not cooperation. Driver Jane Bartscher said that open-pit machine operators "didn't have to communicate with others if they didn't want to."[96] In the pit, things were impersonal. As another former Anaconda worker put it, "The shovel operators went to work and worked all day by themselves. The truck drivers got in their truck and they drove all day by themselves."[97] Instead of working with partners, people in the open pit tended to have extended interactions with others only in the lunchroom. For parts runner Dennis Murphy, lunch underground meant sitting down with a few coworkers in a well-ventilated spot where people could vent their troubles and engage in "tomfoolery, playing." In the pit, people were still friendly, but the real togetherness was gone.[98]

There were additional repercussions outside the workplace. After an underground shift, for instance, Al Beavis would hang out with his partner, but this same socialization did not occur after a Berkeley Pit shift.[99] Each worker was involved with more people in the pit than in the underground, but on a more superficial level. For miner Dan Aguilar, new employees who had never worked underground were partly responsible for the open pit's "looser companionship. . . . You had your real close friends, but nothing like underground."[100]

Anaconda's move to open-pit mining also meant a change in the power structure between Butte's unions, further dividing workers. Starting in the 1930s, Butte had functioned with one powerful union—the BMU—and a few small crafts unions. The first big change to this structure happened with the introduction of open-pit mining. A representation election in 1956 led to the certification of the International Union of Operating Engineers as the representative of many open-pit workers. Leaders of the BMU became quite concerned about this development, because it meant the loss of what had traditionally been their complete jurisdiction.[101]

As open-pit mining grew in importance from the 1950s onward, the crafts unions also grew in power, while the BMU went into decline. Describing this change, Anaconda Company labor manager Ray Tilman stated, "As you went away from the underground mining and you went up to the shop—you had electricians, machinists, mechanics... [—] the tradition of the shops carried over to the pit and the concentrator and that really started a lot of the problems, all the unions started choosing up jobs."[102] Instead of a workscape where underground miners ran the show, the open pit meant that electricians and teamsters gained. By the 1960s the open-pit workscape involved thirteen unions with increasingly equal power, even if the IUMMSW, mostly in the guise of the BMU, still housed a larger proportion of the employees.[103]

Industrial unions, like the USWA, struggled in the new workscape. The USWA's District 38 director in California, Charles J. Smith, explained to the international union's officers in 1967 that the nature of open-pit mining made it difficult to organize industrial-type unions. Unlike the sinking of underground shafts, which mining companies usually carried out by themselves, preparatory work for open-pit projects was typically carried out by construction firms, which owned the necessary giant trucks, cranes, and shovels. Their employees were more likely to be members of construction craft unions, such as the Teamsters and Operating Engineers Unions, rather than of industrial unions, like the IUMMSW or the USWA. The preference of the National Labor Relations Board for craft unions meant further divisions in the organized workforce for each open pit.[104] Instead of one industrial union representing all of the workers at a mine site, the Teamsters, Operating Engineers, and other craft unions acted on their own and sometimes had little communication with the IUMMSW or the Steelworkers. Nor did these various unions share the same approach to bargaining. As historian Michael Malone has remarked, the Teamsters and Operating Engineers, both of whom represented many workers at the new open pits, "leaned much less to the left" than industrial unions and appeared less willing to challenge metal mining companies.[105]

Mechanic Dana Miller complained that these changes hurt workscape camaraderie in Butte. When one union went on strike, everyone could be thrown off work, which could lead to bad feelings between different unions' members.[106] In 1962 unions affiliated with the IUMMSW settled before their contract expired, but many of these workers still sat idle because the Teamsters Union, annoyed by a subcontracting dispute, struck on its own, shutting down everything at the Berkeley Pit. Anaconda tried to set the unions against each other by blaming the Teamsters for affecting everyone else. The company acquired a court order directing the Teamsters to resume production, but the Montana State Supreme Court blocked it. The work stoppage ended only after the Brotherhood of Railroad Trainmen and the Brotherhood of Locomotive Firemen and Enginemen intervened with the Teamsters.[107]

Wildcat strikes, unauthorized by the unions' leaders and led by rank-and-file members, became more frequent, partly because the membership bristled under the increased supervision of union leaders, all of which was made more possible by the surveillance-friendly workscape. Teamsters, mine-mill workers, and other tradesmen staged a number of wildcat strikes and walkouts during the early 1960s. During the 1962 and 1963 construction on the Weed Concentrator, for example, seven short walkouts occurred due to jurisdictional disputes and complaints about supervisors.[108]

The biggest unions, including the IUMMSW, with which many Butte workers were affiliated, and the USWA, with which Anaconda's smelter workers were affiliated, soon realized that they needed cooperative arrangements for organizing and bargaining.[109] Exhaustion bred the new spirit of togetherness. The old battle, in which the Steelworkers made often-unsubstantiated accusations of communist leanings against IUMMSW, had not helped workers at the bargaining table. Such tactics had led the smelter union to defect to the USWA in 1962. As a result, the steelworkers gained a foothold in Butte when the concentrator and its USWA workers moved there from the town of Anaconda two years later.[110]

A weakened international IUMMSW fully merged with the USWA on July 1, 1967. Although Butte locals had stuck strongly with the IUMMSW, the international IUMMSW had suffered many defections to the Steelworkers in both Canada and elsewhere in the United States.[111] A. C. Skinner, the IUMMSW president, said that a merger was necessary to face a now internationally organized copper industry. Worker relations were no longer handled by local mine superintendents and managers, and so, argued Skinner, unions needed to follow suit. One reporter memorialized the merger as the end of the nonferrous industry's "Wild West."[112] Unionists hoped the merger would tip the balance of power back to their side by rejecting the ad hoc, loosely organized bargaining sessions of the past. The USWA therefore convinced other industrial and craft unions to negotiate as an industrywide coalition.[113]

Following the merger, Butte's copper unions joined their fellows around the nation to strike all the major copper producers. At issue was the workers' desire to have negotiations focused at the company level rather than at each mine. As the *Wall Street Journal* put it, "In the past, management wielded a big negotiating stick over a scattered array of separate unions and kept labor costs well in check through sub-par wages."[114] Companywide bargaining would mean common expiration dates for contracts, uniform wages for equal work, and "simultaneous settlements for all properties of any company." The companywide approach would limit management's ability to play unions against each other and would facilitate the removal of inequities among operations in different places.[115]

Efforts by western copper corporations to resist companywide bargaining prolonged the strike, which dragged on for eight months. The corporate view was that job descriptions and pay scales should vary by locality because each place had workers doing different jobs in divergent situations. Anaconda's management especially worried that companywide bargaining would limit its flexibility in dealing with marginal properties, such as those in Montana.[116]

Bernard Rask, a Steelworkers' representative from Butte, accused R. Lewis Brown, Anaconda's director of labor relations, of rehashing the same argument. Since the 1950s, he stated, the company had claimed that Butte operations were so marginal that "if the Union couldn't settle for a cheaper settlement than was made in the rest of the industry, that there was a good possibility that some of the underground operations would have to shut down." By using this excuse,

Rask complained, Anaconda had not made a financial settlement with Butte workers equal to the rest of the industry since 1956.[117] The sides were set—Anaconda refused to accept a unified front by its unions, while the unions were sick of the company lording its overseas operations over workers.

The great length of the 1967–68 strike exacerbated problems. The stoppage became not only the longest strike in Montana's history, but also the longest industrywide strike in American history. *U.S. News and World Report* described local reactions as mixed. Some thought unions needed to take a stand. A Butte minister, on the other hand, saw the strike as outside of local concerns, claiming, "The miners didn't want to go out in the first place."[118] As during previous strikes, miners took whatever jobs they could find, while the number of people on unemployment shot up. Women had to hold families together and scrounge for food during a particularly harsh winter. Meanwhile, many Butte residents left for California, Oregon, and Washington, where jobs were more plentiful. The environment suffered as well. With the workers on strike, the water treatment plant used in processing copper closed. As untreated effluent seeped into the Clark Fork River, its water turned red, until the State Fish and Game Department won a restraining order against the company and union.[119]

The eventual settlement, brokered by the White House, did not provide for companywide bargaining, but the separate agreements reached in Montana generally contained the same provisions for vacations, health care, and other benefits.[120] The appearance of more unity belied increasing problems for many locals, including those in the Butte area. Anaconda reopened its underground mines in 1968 on a reduced scale. It did not reopen the Leonard Mine, and it curtailed operations at the Mountain Con and Steward Mines, the two still-operating vein mines left. Many Butte workers were thus thrown into unemployment as Anaconda laid off 1,300, reducing the number of people able to enjoy the benefits that unions had won.[121]

Despite attempts at unity, divisions among the unions continued to be a day-to-day problem because the open pit enabled continual surveillance by union stewards. Underground miners were renaissance

Berkeley Pit Trucks. The hardhats' different colors differentiate between types of jobs in the pit. Photo courtesy of Butte-Silver Bow Public Archives, Clinton Peck Collection, PH308.07.

men, used to performing all the necessary skills to get the job done. Miner Dan Aguilar recalls, "[Underground] we done all our own work. We were our own bolt cutters. They more or less looked over it because it was hard, like on night shift to get an electrician or a welder." People would simply help out where needed, reducing bottlenecks. "Down in the pit," however, "they pressed that jurisdiction lines a lot more than they did in the mines."[122]

Surface mines enabled Anaconda's management and union leaders to observe workers, the colors of whose hardhats depended on their roles, doing tasks outside the jurisdiction of their union. Sticking to the rules sometimes interrupted the flow of work. Because only bosses could drive pick-up trucks and only operating engineers could deal with shovels, for example, moving a light from one shovel to the next took two people—a boss and an operating engineer. In this jurisdictional hodgepodge, employees within each single union also had different job classifications. Shovel operators, drill operators, and bulldozer operators were all members of the Operating Engineers Union, but someone who had seniority on a shovel did not have his seniority carry over to a drill or a bulldozer. Employees had to stick to one task.[123]

According to welder Ron Garbarino, the company's continuing reduction in its number of employees added to this jurisdiction-related inefficiency. Fewer workers meant longer waits for the completion of tasks. The time to move electrical lines, for example, increased as the number of available electricians fell. Other employees shared his belief that the company contributed to slowdowns and that it exaggerated the impact of jurisdictional disputes. One productivity solution offered by the company, for instance, was simply to put more trucks and shovels in the pit; doing so, however, was as likely to increase congestion, and hence slow down work, as to raise the amount of transported ore.[124] Former contract miners were frustrated by pit conditions. As one employee put it, "A lot dreaded going to work because they couldn't do what they knew they could do, contribute, produce a day's work."[125] Management generally supported this segregation of jobs, refusing to let employees train for new tasks. In 1971, though, frustrated company leaders decided to try a different approach.

During contract negotiations that year, Anaconda proposed more flexibility in allowing workers to cross traditional craft lines while performing work incidental to their jobs. Crafts union members feared the company proposal would entail three setbacks for them: the loss of jobs because crew numbers could be cut, the removal of limits on how hard or fast a worker had to toil, and the acceptance of arbitrary controls over what had been deemed typical local practices. After an exhausting eleven-week strike, most crafts gave up some of their privileges in return for improved health and welfare benefits.[126] Even after bargaining ended, though, the unions remained very wary of company demands to cross jurisdictional lines.[127]

During the 1970s strikes revealed further community divisions. Wildcat strikes, in particular, suggest how fractured union power had become. One in 1973 and another in 1975 threw unsuspecting employees, who were not previously aware of others' grievances, out of work for a few weeks.[128] Every three years, when contracts expired, bigger, all-union strikes typically took place. Those strikes after

MINERS UNION DAY, BUTTE, MONTANA. Photo courtesy of Butte-Silver Bow Public Archives, Clinton Peck Collection, PH308.23.

1968 often grew out of broader issues for the international coalition of unions, not local problems, so dissatisfaction grew among Butte workers who resented what was to them an unnecessary loss of paychecks during each stoppage. When the Nonferrous Industry Conference, which included international union leaders, called a strike in 1977, many of BMU's members objected. A group of unhappy miners' wives even formed a group called WHY to question union leadership about the strike.[129]

## KINGS NO MORE

Tensions between former underground miners and workers hired specifically for surface operations also lessened labor camaraderie in the Berkeley Pit. Longtime Butte miners viewed underground mining as difficult, rewarding, and prestigious work. Partly because so many were paid by what they produced, "Miners knew how to work," recalls Tom Holter, and they performed this labor in a dark, dangerous "real mine."[130] The contract miner stood at the top of the workscape hierarchy. As one Butte resident puts it, "Before the Pit, the contract miners were it!"[131] They were kings at work, at the bars, and around town. Bragging rights went to those who made the top of the contract board by digging out more ore than the other guys.[132]

In the open pit, miners lost their place in the pecking order. As miner Dan Aguilar recalls, "I felt—how can I put it—little in the pit. You had your electricians, you had your boilermakers, your pipefitters, and they were the kings down there. Where I went, I was a laborer, I felt like nothing." He had to deal with employees who had never been underground, employees who he claimed were "more belligerent, argumentative... spoiled." New surface employees, in his opinion, simply were not used to hard work.[133] The value that miners had placed on work in the dark tunnels below seemed to disappear in the light above. As another former employee put it, "Miners, they prided themselves on being miners. I never heard a truck driver ever say he was proud of being a truck driver."[134]

The tensions between older miners and new employees shows that Butte not only experienced some de-skilling but also a partial change in labor regime. Some underground miners transferred to open-pit mining, but many stuck with underground work. Dan Aguilar stated that many of his fellow Hispanic workers "were strictly underground miners" and left Butte with the advent of the open pit.[135] When the first U.S. open-pit copper mine began in turn-of-the-century Bingham, Utah, southeastern European and Asian immigrants arrived to join in work that now required less training.[136] In Butte, a similar process occurred on a smaller scale, but the replacements for skilled miners in the Pit were usually locals, including some women.

Industrial underground mining has long been described by scholars as exhibiting an "exaggerated masculinity."[137] That was particularly true for the heroes of this subterranean world: contract miners. Anaconda employee Jim Killoy explained that contract mining "turned most people on. Moving rock, making good wages—a lot of it had to do with the macho thing."[138] Women had once been heavily involved in mine labor around the world, but as mining became more industrialized, capitalized, centralized, and globalized, employers and communities normalized female exclusion from the workforce.[139]

In the United States, women had worked in nonunion and private mines but definitely not in larger mining camps. Even taking women into underground mines was taboo in late nineteenth- and early twentieth-century mining camps. Butte miners never fully practiced this common prejudice. Tales existed both about women who worked underground while disguised as men and about the company sometimes allowing wives to enter the tunnels.[140] Women, however, seldom openly worked underground, especially because the prevailing belief was, as one miner put it, that mining "would have been too hard of work for a woman."[141]

This masculine world would begin to change in the late 1960s and early 1970s. The first woman hired to work in the Berkeley Pit was Ann Nettles, who, starting in 1969, served as secretary to the pit's general superintendent, Pablo Balentine. The local newspaper featured her as the "First Lady of the Pit," celebrating the pride that this "attractive, petite matron" had in her work. She even refused to wash the part of the windshield with her pit clearance sticker. The article suggests that Nettles fit in with the Berkeley's "he-men," and it contained a couple of pictures of her wearing a hardhat, but the reporter spent almost as much time asking Nettles about hobbies, children, and the candy dish she put on her desk as about her work.[142]

With mining companies facing legal pressure, more women followed. A 1967 amendment to the Civil Rights Act of 1964 added a prohibition against discrimination by federal contractors and subcontractors on the basis of sex. The amendment forced the mining industry, one of the last fully male bastions, to open up. As a series of discrimination lawsuits went forward in the United States, in the 1970s mining companies began to hire women for jobs beyond the secretarial pool.[143]

Few Anaconda employees welcomed the possibility of women entering their all-male workspace. According to truck driver Bill Long, one of his more macho colleagues prophesized, "You ain't never gonna see a woman driving truck in this pit." Just two weeks later, in April 1972, the first women began driving haul trucks.[144] In 1974 female pit truck driver Jane Bartscher joined a workforce that now included a few women who operated graders, others who worked as operating engineers, and still others at the waste dumps. She felt like an intruder who had to perform better than the men. She experienced little outward harassment, although "sometimes someone would throw rocks at her windows." Women also had to wait quite a while until Anaconda provided them with a separate bathroom, and that facility was an inferior one with lights that always flickered.[145]

Despite inroads, very few women ever worked at the pit in Butte. Men who later left the area to find new jobs were surprised by the prevalence of women at other western workplaces.[146] Surface coal mining on the western plains, for instance, was so new in the 1970s that it did not have a long history of male domination; hence, women were able to negotiate a lasting role in that industry.[147] North American hard-rock mines, on the other hand, experienced financial problems just as women became more established in the late 1970s and early 1980s, causing many women to either leave the industry or to be laid off, as they typically held little seniority.[148] Although challenged, male working culture continued to predominate in Butte until the end of the Berkeley Pit. When the Berkeley Pit closed down, one employee recalled, some Butte men resented the fact that more jobs seemed to be open in the service industry, which they considered women's jobs, than for men who were used to the "macho mindset" of industrial mining.[149]

## Devaluing the Underground

Not all of the changes in work life that male miners found unsettling were due to open-pit operations. During the same era, the underground experienced changes to its workforce. Anaconda's recruitment of coal miners from the South during the 1950s and 1960s troubled the close-knit community of underground workers. Longtime Butte residents looked at these new miners, whom they called "Okies," as a different breed, especially because former coal miners did not understand hard-rock operations, which sometimes meant increased accidents or near-accidents.[150] Likewise, Butte's underground increasingly featured more machines and fewer men, but also hosted increasing inefficiencies, purportedly due to employers or unions, depending on whom a person asked.[151]

The changes underground, however, did not match in drama or impact those generated by the open pit. For those affected, the difference between the underground world and the pit was stark. For them, the social products of the underground workscape—masculine independence, camaraderie, and prestige—had all but disappeared. Men's status might have been challenged by the introduction of women, but the reorganized pit workscape was far more important to their complaints and fears.

Given workers' worries, it is surprising that union leaders failed to perceive the impending demise of Butte's underground. During the late 1960s and early 1970s, the issue of wage parity across the nonferrous, metal-mining industry had become particularly important to Butte's locals. Wage rates for Anaconda workers trailed by more than twenty cents per hour those of their peers at American Smelting and Refining Company, Phelps Dodge, and Kennecott—the other employers who made up the "Big Four" of the U.S. copper industry. Reeling from overseas business setbacks, Anaconda came to the conclusion that it had to come up to the standards of Kennecott in order to stay in business. An agreement between the company and its unions in the 1971 contract led to a job evaluation program that promoted wage equity but also led to a number of layoffs.[152]

For the unions, agreement to the job evaluation program was tantamount to acceptance of open-pit mining as the norm. The program, which took place between 1972 and 1974, reevaluated each job on the qualifications, training, skills, effort, and intelligence needed to perform it as well as on the physical environment and dangers in which it was carried out. Reclassification established a new numbered list of pay grades. During this process, both the company and unions came to agree that machine-operating positions deserved as much or more pay as dangerous underground work. According to the new scale, open-pit jobs often held by members of the Operating Engineers Union, who ran complex equipment, scored best on the factors relevant to high grade and pay, while jobs held by underground workers lagged behind, despite their exposure to trying physical conditions.

Comparing pay grades between 1959, when open-pit operations became important to Butte, and 1972–1974 makes evident the lessening status of underground miners. In 1959 contract underground miners, shaftmen, station tenders, and engineers were at the top pay grade (number 7) in the contract. Other underground jobs, including motorman, swamper, and vent man, were at levels six and five. Loaders at the Berkeley Pit were only at grade three.[153] In 1972 the union and company agreed that shovel operators should sit five pay grade levels above, mechanics two levels above, and Berkeley Pit production truck drivers at the same level as the once-mighty underground miners.[154]

Underground mining's heyday was over. Many of the good veins near the surface had already played out; it proved increasingly difficult to move men and equipment into smaller remnant veins. The percentage of Butte's ore production from traditional vein mining had precipitously dropped from 100 percent of the total in 1940, to 54 percent in 1955, to 32 percent in 1961, and to 14 percent in 1970. Skilled underground employees similarly dropped in number. Anaconda's total workforce in Butte fell from 6,006 employees in 1950 to around 2,000 employees in 1974.[155] Mass mining had reshaped the workscape, completely changing what was expected of and experienced by workers. As the open-pit factory expanded farther into Butte, mining would soon challenge other community members, creating new industrial hazards and altering the city's shape.

## Notes

1. Beavis interview.
2. Andrews, *Killing for Coal*, 125.
3. On envirotech, see Reuss and Cutcliffe, *Illusory Boundary*; Sara B. Pritchard, "An Envirotechnical Disaster: Nature, Technology, and Politics at Fukushima," *Environmental History* 17, no. 2 (Apr. 2012): 219–43.
4. LeCain, *Mass Destruction*, 132.
5. Hovis and Mouat, "Miners, Engineers."
6. "Butte, One of World's Greatest Mining Areas," *Anaconda Company Trailsman*, July 1973, 3, VF "ACM: General Mining," BSBPA; "Berkeley Open Pit Draws Thousands of Tourists"; "Berkeley Pit One of Montana's Most Spectacular Sights," 1; "Employees Shown on Television," *The Anacondan*, Apr. 1965, 4, MHS.
7. E. I. Renouard, "Let's Talk It Over," *The Anacondan*, Dec. 1964, 3, MHS.
8. E. I. Renouard, "Berkeley Pit Operation," *Mining Congress Journal* 45, no. 6 (June 1959): 51–54, 60.
9. McWilliams, "Mining Methods and Costs," 15; Gardiner interview.
10. On mining and difficulties rationalizing the natural world, see Andrew Isenberg, *Mining California: An Ecological History* (New York: Hill and Wang, 2005), 23–52.

11. "Berkeley Pit One of Montana's Most Spectacular Sights," 1.

12. John B. Hutl, "Berkeley Pit Crushing and Conveying Plant," *Engineering and Mining Journal* 158, no. 12 (Dec. 1957): 102–24; LeCain, *Mass Destruction*, 66.

13. Quote is in "Berkeley Pit One of Montana's Most Spectacular Sights," 1. See also Monninger testimony, U.S. Senate Subcommittee on Minerals, Materials, and Fuels of the Committee on Interior and Insular Affairs, *Closure of Underground Copper Mines: Hearing to Review the Anaconda Company's Plans to Close Its Underground Copper Mines in Butte, Montana*, 93rd Cong., 2nd sess., Nov. 23, 1974, 30; McWilliams, "Mining Methods and Costs," 29.

14. "Berkeley Pit Is Fascinating Sight," *Anaconda Company Trailsman*, May 1, 1962, 4, VF "Anaconda Company Trailsman," MHS; McWilliams, "Mining Methods and Costs," 29–32; Renouard, "Berkeley Pit Operation," 52–53; T. J. Shepich, "Health and Safety Inspection: Berkeley Pit, The Anaconda Company, Butte, Silver Bow County, Montana" (Metal and Nonmetal Mine Safety: Rocky Mountain District, Bureau of Mines, Department of the Interior, 1970), Fldr. "Anaconda Co. 1967, 1968, 1970 Negotiations," Box AX/B40/HCLA/03098, Wage Department, USWA, PSHCLA, 1.

15. "Anaconda Maps Greatest Expansion," *Mining World* (Nov. 1956): 56–61; Hutl, "Berkeley Pit Crushing," 108; Shepich, "Health and Safety Inspection," 3; "Berkeley Pit One of Montana's Most Spectacular Sights," 1; Renouard, "Berkeley Pit Operation," 60; McWilliams, "Mining Methods and Costs," 25.

16. "Berkeley Pit One of Montana's Most Spectacular Sights," 1; Hutl, "Berkeley Pit Crushing," 109–10; Renouard, "Berkeley Pit Operation," 60; McWilliams, "Mining Methods and Costs," 24–25.

17. "Berkeley Pit One of Montana's Most Spectacular Sights," 1; McWilliams, "Mining Methods and Costs," 27–28; Renouard, "Berkeley Pit Operation," 60.

18. Bill Long, *Pit Driver: Reminiscences of a Berkeley Pit Haul-Truck Driver* (La Habra, CA: Self-published, 2007), 63–66; Gardiner interview; Jane Bartscher, interview by Teresa Jordan, Nov. 22, 1985, Butte, MT, partial transcript and notes, OHC, BSBPA; "Montana's Copper Land," *Anaconda Company Trailsman*, May 1966, 1–2, VF "Anaconda Company Trailsman," MHS; Hutl, "Berkeley Pit Crushing," 108; McWilliams, "Mining Methods and Costs," 27.

19. "Job Descriptions," Fldr. "Anaconda Co.—Benchmark Proposals," AX/B40/HCLA/03098, USWA, PSHCLA; Shepich, "Health and Safety Inspection," 3; Renouard, "Berkeley Pit Operation," 52, 60; Hutl, "Berkeley Pit Crushing," 128; McWilliams, "Mining Methods and Costs," 15.

20. "Copper Leaching in Butte," *The Anacondan*, Mar. 1964, 1–2, MHS; "Leaching and Precipitation," *Anaconda Company Trailsman*, June 1968, 3, VF "Anaconda Company Trailsman," MHS; "Tin Can Mine," *Anaconda Company Trailsman*, July 1973, 3, VF "ACM: General Mining," BSBPA; "Berkeley Pit Is Fascinating Sight," 4; McWilliams, "Mining Methods and Costs," 15.

21. "Montana's Copper Land," 1966, 1–2. See also Al Darr, "Anaconda Co. Wrestling Hauling Costs," *Mont. Std.*, Nov. 30, 1973; Renouard, "Berkeley Pit Operation," 54; "Berkeley Pit One of Montana's Most Spectacular Sights," 1; Shepich, "Health and Safety Inspection," 2.

22. Hutl, "Berkeley Pit Crushing," 102–24; McWilliams, "Mining Methods and Costs," 32–41; "Berkeley Pit One of Montana's Most Spectacular Sights," 1.

23. "Berkeley Pit Is Fascinating Sight," 1.

24. McWilliams, "Mining Methods and Costs," 44; "Job Descriptions"; Trux Fisher in Fisher and Corcoran interview; Quivik, "Smoke and Tailings," 383–92.

25. Discussion of dead zones from LeCain, *Mass Destruction*, 66–69. More on the process in Mercier, *Anaconda*, 158; Quivik, "Smoke and Tailings," 383–92.

26. Quivik, "Smoke and Tailings"; "Berkeley Pit One of Montana's Most Spectacular Sights," 1.

27. LeCain, *Mass Destruction*, 188–206; "Berkeley Pit One of Montana's Most Spectacular Sights," 1.

28. Christopher Schmitz, "The Rise of Big Business in the World Copper Industry 1870–1930," *Economic History Review* 39, no. 3 (Aug. 1986): 392–410; Alfred D. Chandler. *The Visible Hand: The Managerial Revolution in American Business* (Cambridge, MA; Harvard University Press, 1977), 362.

29. LeCain, *Mass Destruction*, 129–37, 206–8.

30. "Montana's Copper Land," *Anaconda Company Trailsman*, June 1967, 1–2, VF "Anaconda Company Trailsman," MHS; "Butte One of World's Greatest Mining Areas," July 1973, 3.

31. Quote is "Butte One of World's Greatest Mining Areas," July 1973, 3. See also "Montana's Copper Land," 1966, 1–2. See also Ed Bonner, "Mining History of Butte," Sept. 1962, page 11, Fldr. 11, Box 132, MC169, MHS; Braach interview; Michael J. Sullivan Sr., interview by Michael J. Sullivan Jr., Oct. 18, 1999, Butte, MT, tapes and transcript, OHC, BSBPA; Bossard interview.

32. Enyatollah Bozorgebrahimi, "The Evaluation of Haulage Truck Size Effects on Open Pit Mining" (PhD diss., University of British Columbia, Vancouver, 2004), 1.

33. "Montana's Copper Land," *Anaconda Company Trailsman*, June 1968, 1–2, VF "Anaconda Company Trailsman," MHS.

34. "Trucks Make Marginal Open-Pit Mining Pay"; Navin, *Copper Mining and Management*, 39–41; Haycraft, *Yellow Steel*, 30–31. See also in *Kennescope* (Bingham, UT), J. Willard Marriott Library, Special Collections, UU: "Green Light Means Grow," Mar.–Apr. 1963, 3–6; "The Big Switch Begins at Mine," Nov.–Dec. 1963, 4–5; "They Out-Man the Abominable Snow: More Than a Match for the Mountains," Jan.–Feb. 1964, 6–7.

35. Quote is "Butte One of World's Greatest Mining Areas," July 1973, 3. Numbers in "Montana's Copper Land," 1966, 1–2; "Montana's Copper Land," 1967, 1–2; "Montana's Copper Land," 1968, 1–2; "Butte One of World's Greatest Mining Areas," *Anaconda Company Trailsman*, Aug. 1974, 2, VF "Anaconda Company Trailsman," MHS; "Into the Pit," *Time*, Sept. 22, 1975; Bonner, "Mining History of Butte."

36. Johansen, "Butte, MT: Sitting Atop a Fortune."

37. The Anaconda Company, Montana Mining Division Open Pits, File #29906.01A, Fldr. "Butte, Mine Operations, 1961–1985," AGDC, AHC.

38. Trux Fisher in Fisher and Corcoran interview; "World's Most Modern Concentrator," *Anaconda Company Trailsman*, June 1965, 2, VF "Anaconda Company Trailsman," MHS; "The Concentrator," *Anaconda Company Trailsman*, July 1973, 2, VF "ACM: General Mining," BSBPA; "Welcome to the Weed Concentrator," file 29906.01A, Fldr. "Butte, 1961–1985, Mine Operations," AGDC, AHC.

39. LeCain, *Mass Destruction*, 207.

40. John Clark to BMU members, May 8, 1959, Fldr. 27, "Raids USW–Montana Mine-Mill," Box 258, Microfilm LA1148, USWA, PSHCLA.

41. "Background Information on the Anaconda Strike," Butte, MT, Feb. 4, 1960, 10, Fldr. "MMSW Wage Policy and Bargaining, Anaconda 1959," Box Paterno/GST/Y05.01, Wage Policy and Bargaining, USWA, PSHCLA.

42. "Anaconda Company Proposals Are Given Miners," *Mont. Std.*, Oct. 17, 1959. See also "Let's Get Back to Work," Anaconda Company Advertisement, *Mont. Std.*, Oct. 19, 1959; "The Company's Right to Manage... A Statement on the Copper Strike Offer," Anaconda Company Advertisement, *Mont. Std.*, Oct. 21, 1959; "List of Company Proposals," Oct. 16, 1959, Fldr. "MMSW Wage Policy and Bargaining, Anaconda, 1959," Box 3, Paterno/GST/Y05.01, USWA, PSHCLA.

43. "Industry Officials Say Right to Manage Has Become Big Issue in Copper Strike," *Mont. Std.*, Oct. 11, 1959; "Man Against the Machine: But Strike Won't Settle the Issue," *Mont. Std.*, Oct. 20, 1959; "Report from Mine-Mill Joint Negotiating Committee," Sept. 10, 1959, Fldr. "MMSW Wage Policy and Bargaining, 1959, Anaconda," Box 3, Paterno/GST/Y05.01, USWA, PSHCLA.

44. "Company Claims Union Asks Package of 41.84," *Mont. Std.*, Nov. 26, 1959.

45. Finn, *Tracing the Veins*, 57; Mine-Mill Research Department, "The Anaconda Company: Comments on Financial Analysis," May 1959, Fldr. "Wage Policy and Bargaining, 1959, Research Department Company Financial Analysis, Paterno/GST/Y05.02, USWA, PSHCLA; "What's Best for Butte and Montana," Anaconda Company Advertisement, *Mont. Std.*, Oct. 11, 1959.

46. "Background Information on the Anaconda Strike," USWA, PSHCLA, 1; "Where It Hurts," *Mine-Mill Research Report* 2, no. 8 (June 1959): 1, Fldr. 10 "MMSW Wage Policy and Bargaining, Anaconda 1959," Box 1, Paterno/GST/Y05.01, USWA, PSHCLA.

47. "Productivity: A Key Issue," *Mine-Mill Research Report* 2, no. 7 (May 1959): 1, Fldr. 10 "MMSW Wage Policy and Bargaining, Anaconda 1959," Box 1, Paterno/GST/Y05.01, USWA, PSHCLA. See also *MMBM*, June 13, 1959, and Feb. 10, 1960, in Fldr. 5, Box 341, IUMMSW, UC; National Anaconda Conference Minutes, May 22–23, 1959, Chicago, IL, Fldr. "MMSW Wage Policy and Bargaining, 1959, Anaconda," Box 3, Paterno/GST/Y05.01, USWA, PSHCLA; "Report from Mine-Mill Joint Negotiating Committee," Aug. 14, 1959, Fldr. "MMSW Wage Policy and Bargaining, 1959, Anaconda," Box 3, Paterno/GST/Y05.01, USWA, PSHCLA.

48. Vincent D. Perry to Reno H. Sales, Aug. 19, 1959, File 310701, Fldr. "Reno Sales Correspondence," AGDC, AHC.

49. First quote is "The Mine-Mill Story or 'Are Montana Workers Worth Less?' Butte KXLF-TV, Helena KXLJ-TV, Great Falls KFBB-TV," Sept. 21, 1959, Fldr. "MMSW Wage Policy and Bargaining, 1959, Anaconda," Box 3, Paterno/GST/Y05.01, USWA, PSHCLA. Second and third quotes are "Mine-Mill Reports: Butte and Helena TV Show, Re-Broadcast Over Great Falls Radio," Oct. 19, 1959, Fldr. "MMSW Wage Policy and Bargaining, 1959, Anaconda," Box 3, Paterno/GST/Y05.01, USWA, PSHCLA.

50. "Copper Industry Strike Ends in Victory," Fldr. "MMSW Wage Policy and Bargaining 1960–1961 Settlement (Chain)," Box Paterno/GST/Y05.02, USWA, PSHCLA; Ernest Salvas to John Clark, Irving Dichter, Nathan Witt, roll #1, Index #1, "Anaconda Co., LU1 to Zaretti Brothers," Microfilm Roll LA2225, USWA, PSHCLA; "Union Asserts Wages Hiked 15 to 35 Cents," *Mont. Std.*, Feb. 12, 1960.

51. *Wage Chronology: The Anaconda Co. (Montana Mining Division) and the Steelworkers, 1941–77,* Bureau of Labor Statistics, Bulletin 1953 (Washington, DC: GPO, 1977), 6; Frank Quinn, "Ratification Ends Copper Strike: Workers Are Returning to Jobs," *Mont. Std.*, Feb. 16, 1960; "Labor's Share in Mining Income," *Mine-Mill Research Report*, no. I (Mar. 1965) in Fldr. 10 "Mine Mill Research Report 1957–1966," Box Paterno/GST/Y04.17, USWA, PSHCLA; "1,283 Workers May Find Jobs in Butte," *Mont. Std.*, Feb. 13, 1960.

52. For the above tactic and responses to it, see "Talking About Jobs," Anaconda Advertisement, *Mont. Std.*, July 6, 1962; Frank Quinn, "Butte Has Threat of Strike," *Mont. Std.*, July 14, 1962; "An Announcement to Our Employees and the General Public," *Mont. Std.*, Aug. 25, 1962.

53. "Talking About Jobs."

54. "Labor's Share in Mining Income," *Mine-Mill Research Report* 1 (Mar. 1965): 2, Fldr. 10 "Mine Mill Research Report 1957–1966," Box 1, Paterno/GST/Y04.17, USWA, PSHCLA.

55. "The Answers About Mine-Mill's 1964 Demands," 1964, Fldr. "MMSW Wage Policy and Bargaining Material Issued 1964," Box 4, Paterno/GST/Y05.02, USWA, PSHCLA.

56. Al C. Skinner, Asbury Howard, and Irving Dichter to Local Unions, Oct. 3, 1963, Fldr. "MMSW Wage Policy and Bargaining Material Issued 1964," Box 4, Paterno/GST/Y05.02, USWA, PSHCLA. For on-job security as defining issue during 1964 Mine-Mill bargaining, see also *Mine-Mill News* Bulletin, Jan. 22, 1964, Fldr. "MMSW Wage Policy and Bargaining, National Wage Policy Conference, Tucson, AZ," Box 4, Paterno/GST/Y05.02, USWA, PSHCLA; "Labour's Bargaining Demands in 1964," *Mining* (London) 262, no. 6701 (Jan. 24, 1964): 1–2, Box 4, Paterno/GST/Y05.02, USWA, PSHCLA; "Mine-Mill's Major 1964 Bargaining Demands," n.d., Fldr. "MMSW Wage Policy and Bargaining Material Issued 1964," Box 4, Paterno/GST/Y05.02, USWA, PSHCLA"; "Mining Management to Fight Union Demands," *Metal Mining and Processing* (Apr. 1964): 39–41, 52.

57. "Negotiations '64 . . . A Report to Our Employees and the Public," Anaconda Company Advertisement, *Mont. Std.*, July 21, 1964. On workers' small gains, see these articles in *Mont. Std.*: "Teamsters Offered New Pact by Company, Will Vote Today," July 30, 1964; Frank Quinn, "Union Must Ratify It: Miners, Company Agree on Contract," Aug. 19, 1964; Frank Quinn, "Mine-Mill Approves New Contract 5 to 1," Aug. 26, 1964; Frank Quinn, "Butte Operating Engineers Ratify New Labor Contract," 4, 1964; Frank Quinn, "Company, Craft Unions Agree," Oct. 2, 1964.

58. Vincent D. Perry to Clyde E. Weed, May 12, 1958, File 29802.02C, Fldr. "Butte, Mine Operations," AGDC, AHC.

59. On the Kelley's problems, see E. I. Renouard to R. S. Newlin, Nov. 20, 1961, File 29802, Fldr. "Kelley Mine Operations, Butte Mines," AGDC, AHC; Charles C. Goddard Jr., Anaconda Geological Department, "Results Obtained in Recent Deep Pit Development and Recommendations for Future Work," Memorandum, June 29, 1961, File 29801.16, Fldr. "Butte, Mine Operations," AGDC, AHC; Edward P. Shea to Martin K. Hannifan, Feb. 28, 1962, File 29801.14, Fldr. "Butte, Mine Operations," Document 14, AGDC, AHC; Miller, "Production History."

60. "Pit Mining Takes Superintendent Back to Old Stand," *The Anacondan*, Sept. 1965, 8, MHS.

61. Edward P. Shea to Martin K. Hannifan, Feb. 28, 1962, File 29801.14, Fldr. "Butte, Mine Operations," Document 14, AHC; "Berkeley Pit," Jan. 1, 1965, Map, File 120807.02, AGDC, AHC.

62. "Digging Ore... a Mile Below Surface"; "The Underground," 3.

63. Emmons, *Butte Irish*, 148–52; Shovers, "Perils of Working"; Brian Lee Shovers, "Miners, Managers, and Machines: Industrial Accidents and Occupational Disease in the Butte Underground, 1880–1920" (master's thesis, Montana State University, Bozeman, 1987), 63–84.

64. Anaconda Company Bureau of Safety, *Anaconda Safety Rules*, "Six Safety Booklets: 1942–1946, 1953–1956," BSBPA.

65. William Dudding, interview by Logan Dudding, Nov. 20, 2005, Butte, MT, tapes and transcript, OHC, BSBPA; Aguilar interview. See also Petritz interview.

66. Bartscher interview; Tom Holleran, interview by Taylor Flotre, Nov. 12, 2005, transcript, OHC, BSBPA; Sullivan Sr. interview; McCarthy interview, by Leech; Gardiner interview; Shepich, "Health and Safety Inspection," 4; Frank Quinn, "26 Berkeley Pit Workers Hurt," *Mont. Std.*, Aug. 18, 1964.

67. Holter interview, by Leech; Braach interview.

68. Montana Department of State Lands, *Preliminary Environmental Review for a Proposed Expansion of Anaconda Company's Berkeley Mining Complex* (Helena, MT: Sept. 20, 1976), miscellaneous Box 2, Butte Model Cities Collection (hereafter BMCC), BSBPA; Aguilar interview.

69. See also Bartscher interview; Sullivan Sr. interview.

70. Renouard, "Berkeley Pit Operation," 52–53; McWilliams, "Mining Methods and Costs," 29–32; Shepich, "Health and Safety Inspection," 1.

71. Long, *Pit Driver*, 6.

72. Jenks, "An Interview."

73. Shepich, "Health and Safety Inspection," 2–8.

74. On inspections and unions, see Shepich, "Health and Safety Inspection," 1–8; "Butte Local Issues," July 20, 1971, Fldr. "Anaconda Co.—1971 Negotiations," Box AX/B40/HCLA/03098, USWA Wage Department, PSHCLA. On lighting, see McWilliams, "Mining Methods and Costs," 29–32; Renouard, "Berkeley Pit Operation," 52–53; "Berkeley Pit Put in Better Light," *Mont. Std.*, Mar. 18, 1973.

75. Shepich, "Health and Safety Inspection," 6–8.

76. Holter interview, by Leech; James D. Harrington, "Mining Related Fatalities," May 5, 2007, BSBPA.

77. Bartscher interview.

78. Harrington, "Mining Related Fatalities."

79. Navin, *Copper Mining and Management*, 35.

80. John Thatcher, interview by Lucas Thatcher, Apr. 5, 2005, Butte, Montana, transcript, OHC, BSBPA.

81. Jordan's comments in Bruce Chamberlain and Carol Chamberlain, interview by Teresa Jordan, June 23, 1986, Elko, NV, partial transcript and notes, OHC, BSBPA.

82. "Job Descriptions."

83. Beavis interview; Aguilar interview. Similar comments in Jim Killoy, interview by Teresa Jordan, Apr. 10, 1986, Butte, MT, notes, OHC, BSBPA.

84. Chamberlain and Chamberlain interview.

85. Segna and Zanchi interview. See also Corbett interview, by Mercier; Forsell interview.

86. Tarrant interview.

87. Chamberlain and Chamberlain interview.

88. First quote is Tom Malloy, interview by Teresa Jordan, May 21, 1985, Butte, MT, partial transcript and notes, OHC, BSBPA. Second is Aguilar interview. See also Chamberlain and Chamberlain interview.

89. McWilliams, "Mining Methods and Costs," 15.

90. "Montana's Copper Land," 1966, 1–2; Gardiner interview.

91. Benthem's panopticon is most famously discussed in part III of Foucault, *Discipline and Punish.*

92. "World's Most Modern Concentrator," *Anaconda Company Trailsman,* 1967, MHS; "The Concentrator," 2.

93. Fritz Cirkel, *Asbestos: Its Occurrence, Exploitation, and Uses* (Ottawa: Mines Branch, Department of the Interior, 1905), 30. See also Van Horssen, *A Town Called Asbestos,* 20.

94. Al Hart and Tim Little, interview by Teresa Jordan, Apr. 17, 1985, Butte, Montana, cassette tapes and notes, OHC, BSBPA.

95. Murphy interview; Bossard interview. See also Hart and Little interview; Ayres interview; Watters interview.

96. Bartscher interview.

97. McCarthy interview, by Leech.

98. Murphy interview.

99. Beavis interview.

100. Aguilar interview.

101. BMU to members, June 26, 1956, Fldr. 27, "Raids USW–Montana Mine-Mill," Box 258, Microfilm LA1148, USWA, PSHCLA; U.S. Department of Labor, *Wage Chronology,* 4; "Around the Hill with Al Giecek," *MMBM* 4, No. 11, Aug. 24, 1957, IUMMSW, UC.

102. Ray Tilman and Julie Tilman, interview by Teresa Jordan, Jan. 21, 1986, Butte, MT, cassette tapes and notes, OHC, BSBPA. See also Jim Ballard and Cheryl Ballard, interview by Teresa Jordan, June 10, 1986, Wright, WY, partial transcript and notes, OHC, BSBPA.

103. W. D. Williams to F. D. Mitchell, June 29, 1972, Box AX/B40/HCLA/03098, United Steelworkers Wage Department, USWA, PSHCLA.

104. Charles J. Smith to I. W. Abel, Joseph P. Molony, and Walter J. Burke, 1967, Fldr. 3 "Merger MMSW Strike Material 1967," Box 71, "USWA Treasurer's Office," Paterno/GST/U/06.05, USWA, PSHCLA.

105. Michael P. Malone, "The Collapse of Metal Mining: An Historical Epitaph," *Pacific Historical Review* 55, no. 3 (Aug. 1986): 455–64, quote on 460.

106. Dana Miller, interview by Teresa Jordan, June 8, 1986, Butte, Montana, partial transcript and notes, OHC, BSBPA.

107. Articles in *Mont. Std.*: Frank Quinn, "Strike Closes Pit But Mines Operate," July 17, 1962; Frank Quinn, "Mine-Mill and Company Agree," July 27, 1962; "Pickets Will Stay, Say Teamsters," July 28, 1962; "Court Bridles Teamsters, Rail Workers," Aug. 2,

1962; Frank Quinn, "Work Resumed in Butte Mines: Teamsters, Crafts, Anaconda Co. Meet Next Week," Aug. 3, 1962; Frank Quinn, "Picket Line Halts BA&P Engine," Aug. 15, 1962; Arthur Hutchinson, "Supreme Court Issues Order in Butte Strike," Aug. 15, 1962; Frank Quinn, "BA&P Is Target of Dynamiters," Aug. 16, 1962; Lewis T. Poole, "Hearing of BA&P's Action Against Teamsters Recessed," Aug. 17, 1962; "Teamsters Declare They, Company Still 'Far Apart,'" Aug. 24, 1962.

108. Renouard, "Let's Talk It Over," 3.

109. "Joint Organizational Pacts Between United Steelworkers of America, the International Union of Operating Engineers and the International Brotherhood of Teamsters, Chauffers and Warehousemen: Union Representation in the Non-Ferrous Open-Pit Mining Industry," 1967, Fldr. 2 "Merger MMSW Strike Material 1967," Box 71, "USWA Secretary-Treasurer's Office," Paterno/GST/U/06.05, USWA.

110. George Kalafatch to Irving Dichter, "Letter to Membership," Mar. 16, 1964, in Fldr. "George Kalafatich," Box #107, "International reps reports, ca. 1963–64, International Union General Alph. Files, A–L, 1936–41 (Pres.)," IUMMSW, CU; Mercier, *Anaconda*; "Steelworkers Retain Bargaining Job in Anaconda Voting," *Mont. Std.*, July 2, 1964.

111. Mercier, *Anaconda*, 182–84; "Steel Workers, Mill Union, Plan to Merge," *Colorado Gazette Telegraph* (Colorado Springs), Jan. 19, 1967.

112. Frank Pitman, "Turbulent Wild West Chapter Ending," *Rocky Mountain News* (Denver, CO), Jan. 29, 1967. See also "Roughneck Mine Mill Re-enters AFL-CIO," *Loveland Colorado Reporter*, Feb. 1, 1967.

113. "Asarco Chief Assails USW Strike Goal," *Journal of Commerce*, Oct. 26, 1967, Fldr. 5 "Merger MMSW Strike Material 1967," Box 71, USWA secretary-treasurer's office, Paterno/GST/U.06.05, USWA, PSHCLA; Victor Riesel, "Inside Labor: Final Conflict," Fldr. 5 "Merger MMSW Strike Material 1967," Box 71, USWA secretary-treasurer's office, Paterno/GST/U.06.05, USWA, PSHCLA; U.S. Department of Labor, *Wage Chronology*, 2–8.

114. Philip Hawkins, "Copper's Cobwebs: Current Strike May Shake Off Industry's Old Sales Notions," *Wall Street Journal*, Feb. 2, 1968.

115. "Striking Locals Reaffirm Determination to Win Uniformity," *Nonferrous Strike* Bulletin, Mar. 1, 1968, Pittsburgh, PA, Fldr. "Copper Nonferrous," Box 2730, AX/B40/HCLA/02730, Bernard Greenberg Papers, PSHCLA. See also Joseph Molony in "What the Copper Strike Is All About," in "Seventh Constitutional Convention of the American Federation of Labor and Congress of Industrial Organizations, 1967," Dec. 11, 1967, Proceedings, Bal Harbour, Florida, 5–6, Fldr. "Copper Nonferrous," Box 2730, AX/B40/HCLA/02730, Bernard Greenberg Papers, PSHCLA; National Nonferrous Strike Coordinating Committee, "Copper Panel Report," Fldr. 8, "Merger MMSW Non Ferrous Strike Material 1968," Box 71, "USWA Secretary-Treasurer's Office," Paterno/GST/U.06.05, USWA, PSHCLA; United Steelworkers of America, "Facts About the Copper Strike," 1967, Fldr. 3 "Strike Material 1967," Box 71, "USWA Secretary-Treasurer's Office," Paterno/GST/U/06.05, USWA, PSHCLA; United Steelworkers of America, "Brief Union Summary of the 1967 Copper Strike," Fldr. 4 "Merger MMSW Strike Material 1967," Box 71, "USWA Secretary-Treasurer's Office," Paterno/GST/U.06.05, USWA, PSHCLA.

116. Testimony of John Will and R. Lewis Brown, "Transcript of Proceedings of the Special Panel Appointed to Assist in Settlement of the Copper Dispute," Feb. 1, 1968 (CSA Reporting Co., Washington, DC, 1968), 292–93, 312–13, Fldr. 15 "1968 Negotiations: Transcript of Copper Dispute (Special Panel)," Box 76, Paterno/GST/UO6.10, USWA secretary-treasurer's records, USWA, PSHCLA.

117. Testimony of Bernard Rask, "Transcript of Proceedings of the Special Panel Appointed to Assist in Settlement of the Copper Dispute," USWA, PSHCLA, 389–90, 392; quote on 390.

118. "The Toll of a 5-Month Strike: Who Is Hurt and How Much," *U.S. News and World Report*, Dec. 11, 1967, 94–96, quote on 95.

119. "The Toll of a 5-Month Strike"; United Steelworkers of America, "This Fight Must Be Won: With the Help of Our Women It Will Be Won!" 1967, Fldr. "Copper, Non Ferrous," Box 2730, AX/B40/HCLA/02730, Bernard Greenberg Papers, PSHCLA; Finn, *Tracing the Veins*, 149–58; "Strikes: River of Poison," *Newsweek*, Aug. 28, 1967, 70.

120. Maclovio Barraza, interview by Alice M. Hoffman, Dec. 10, 1969, Denver, CO, transcript, USWAOHC, PSHCLA; Vern Curtis, interview by Greg Giebel, Dec. 11, 1969, Denver, CO, transcript, USWAOHC, PSHCLA; U.S. Department of Labor, *Wage Chronology*, 7–9.

121. Joseph P. Molony to Senator Mike Mansfield, Jan. 13, 1968, Fldr. 9 "Merger MMSW Nonferrous Strike Material 1968," Box 71, "USWA Secretary-Treasurer's Office," Paterno/GSt/U.06.05, USWA, PSHCLA; John B. Place testimony, *Closure of Underground Copper Mines*, 7. The Leonard re-opened briefly during the 1970s. See "Anaconda Co. Problems Aren't Insurmountable," *Mont. Std.*, July 1, 1971.

122. Aguilar interview. See also Jim Sullivan and Dolores Sullivan, interview by Teresa Jordan, June 9, 1986, Wright, WY, partial transcript and notes, OHC, BSBPA; Chamberlain and Chamberlain interview; Kovacich interview.

123. Gardiner interview; Erin Lapham, interview by Teresa Jordan, Mar. 7, 1986, Butte, MT, partial transcript and notes, OHC, BSBPA; Shepich, "Health and Safety Inspection," 4.

124. Ron Garbarino, interview by Teresa Jordan, Apr. 7, 1986, Butte, MT, partial transcript and notes, OHC, BSBPA. See also Gardiner interview.

125. Ray Tilman in Tilman and Tilman interview.

126. Articles in *Mont. Std.*: Rick Foote, "Craftsmen Still Hope for Talks," Aug. 6, 1971; "Butte Electrical Workers Accept Anaconda Offer," Aug. 7, 1971; "Copper Talks: No Comment," Aug. 9, 1971; "Crafts Again Reject State Copper Offer," Aug. 12, 1971; "Anaconda Sends Letter to Crafts," Aug. 13, 1971; William J. Clark, "Butte Machinists Vote No," Aug. 18, 1971; "Strike Talk to Resume," Aug. 23, 1971; "Montana's Copper Talks Are Going to Washington," Sept. 8, 1971; William J. Clark, "Anaconda Copper Pact Worked Out: Ratification Vote Is Next Week," Sept. 18, 1971; William Clark, "Unions Vote Overwhelmingly to End State Copper Strike," Sept. 23, 1971.

127. Tuesday, Oct. 26, 1971, Minutes of Anaconda Job Evaluation Meetings, Fldr. "Anaconda Co. General Correspondence 1971," Box AX/B40/HCLA/03098, Wage Department, USWA, PSHCLA.

128. "Wildcat Strike Closes Anaconda," Sept. 14, 1973, *Mont. Std.*; "Court Order Ends Wildcat Strike," Sept. 15, 1973, *Mont. Std.*

129. *Mont. Std.*, July 14, 1977, July 21, 1977.

130. Holter interview, by Leech.

131. Kovacich interview.

132. Gardiner interview.

133. Aguilar interview.

134. McCarthy interview, by Leech.

135. Aguilar interview. See also Holter interview, by Leech.

136. Gunther Peck, "Padrones and Protest: 'Old' Radicals and 'New' Immigrants in Bingham, Utah, 1905–1912," *Western Historical Quarterly* 24, no. 2 (1993): 157–78.

137. "Introduction," Gier and Mercier, *Mining Women*, 5.

138. Killoy interview. On the stereotypes about mining men in Butte during this period, see Finn, *Tracing the Veins*, 110–16.

139. Kuntala Lahiri-Dutt and Martha Macintyre, eds., *Women Miners in Developing Countries: Pit Women and Others* (Burlington, VT: Ashgate, 2006); Mercier, "Bordering on Equality."

140. Wayland Hand, "The Folklore, Customs, and Traditions of the Butte Miner," *California Folklore Quarterly* 5 (Jan. 1946): 19–20; Watters interview; Basso, *Meet Joe Copper*, 148.

141. Ray Burns, interview by Brad Richins, Nov. 23, 2005, Missoula, MT, transcript, OHC, BSBPA.

142. Betty Ann Raymond, "A Light Touch on a Heavy Operation: Mrs. Nettles, First Lady of the Pit," *Mont. Std.*, Dec. 28, 1969.

143. Carletta H. Savage, "Women Coal Miners: Another Chapter in Central Appalachia's Struggle Against Hegemony" (master's thesis, West Virginia University, Morgantown, 1999), 15–33; "Women Coal Miners," *Mont. Std.*, Dec. 23, 1973; "Ladies Keeping Big Trucks on the Road," *Mont. Std.*, Dec. 2, 1973.

144. Long, *Pit Driver*, 35–36.

145. Bartscher interview. See similar stories in Katherine G. Aiken, *Idaho's Bunker Hill: The Rise and Fall of A Great Mining Company, 1885–1981* (Norman: University of Oklahoma Press, 2005), 168–204; Kristen R. Yount, "Women and Men Coal Miners: Coping with Gender Integration Underground" (PhD diss., University of Colorado, Boulder, 1986), 326.

146. Sullivan and Sullivan interview; Ballard and Ballard interview.

147. Jessica Smith Rolston, *Mining Coal and Undermining Gender: Rhythms of Work and Family in the American West* (New Brunswick, NJ: Rutgers University Press, 2014).

148. Mercier, "Bordering on Equality," 40–44; Clement, *Hardrock Mining*, 89–93.

149. Tilman and Tilman interview.

150. Enid W. Stearn, "Coal Miners Move West, Help Fill Underground Ranks," *Metal Mining and Processing* (Jan. 1965): 29–31. See also Helen Nicholls, interview by Teresa Jordan, Apr. 23, 1985, Butte, MT, partial transcript and notes, OHC, BSBPA; Al Hooper, interview by Teresa Jordan, Apr. 11, 1985, Butte, MT, partial transcript and notes, OHC, BSBPA; Carden interview; Dick Skates, interview by Teresa Jordan, Feb. 5, 1986, Butte, MT, partial transcript and notes, OHC, BSBPA.

151. Watters interview; Hooper interview, by Jordan; Kovacich interview.

152. Al Skinner, interview by Helmut Golatz, Mar. 3, 1975, Tucson, AZ, transcript, USWAOHC, PSHCLA; Oct. 28, 1971, Minutes of Anaconda Job Evaluation Meetings, Fldr. "Anaconda Co. General Correspondence 1971," Box AX/B40/HCLA/03098, Wage

Department, USWA, PSHCLA; The Anaconda Company and the United Steelworkers of America, "Brief in Support of Pay Increases Challenged by the Pay Board," Aug. 14, 1972 Pay Board Case 07979, Fldr. "Anaconda Co.—General Correspondence, 1972–74," Box AX/B40/HCLA/03098, Wage Department, USWA, PSHCLA; George Hagglund to Thomas Striegel, Oct. 17, 1973, Fldr. "Anaconda Co.—General Correspondence, 1972–74," Box AX/B40/HCLA/03098, Wage Department, USWA, PSHCLA.

153. "Proposed Changes: Butte," Fldr. "MMSW Wage Policy and Bargaining, 1959, Anaconda," Box 3, Paterno/GST/Y05.01, USWA, PSHCLA.

154. "Job Descriptions."

155. On remnant veins, see Frank Monninger testimony, *Closure of Underground Copper Mines*, 30. On mine production numbers, see Hannifan, "Merits of Open Pit Vs. Underground Mining in Butte"; Frank Monninger testimony and John B. Place testimony, *Closure of Underground Copper Mines*, 43, 15. For employee numbers, see Marshall Kaplan, Sheldon P. Gans, and Howard M. Kahn, *Butte Community Renewal Program Report* (Kaplan, Gans and Kahn Report to the Butte Model Cities Agency, Dec. 20, 1972), 1–3, BMCC, BSBPA; "Governor Says 14% of Butte's Jobs on Line," *Mont. Std.*, Nov. 26, 1974.

# PART III

# FEEDING THE FACTORY, 1955–75

CHAPTER 5

# HAZARDS

BEFORE OPEN-PIT MINING BEGAN, residents of Butte had lived with underground mining and its attendant hazards for three-quarters of a century. Locals debated the underground's risks—particularly when it came to ground subsiding into underlying mines. Widespread settling of the neighborhood south of the business district caused sewers to shift, water and gas lines to break, and homes to creak and crack. From 1940 to about 1960, the Anaconda Company received subsidence complaints on more than 1,500 properties. Settling about 45 percent of these claims was expensive.[1]

When the Anaconda Company switched the majority of its Butte operations from underground to aboveground mining, it understandably sought to better manage its liability, but the dispersed and intrusive nature of risks in open-pit mining instead created as many financial risks to the company as underground mining had caused. The problem was particularly acute for the company because its biggest open-pit mine, the Berkeley Pit, sat on the edge of the city. As the pit expanded, the boundaries of the operation gobbled up outlying neighborhoods. Living space increasingly became workspace. For residents, the open pit therefore became more than a new way to mine—it was a new way to live.

The company received many complaints about damage from its new operations. Even families with company employees—including members of Berkeley Pit blasting crews—complained.[2] Letters and phone notes from complainants, and engineers' responses to them, reveal the intensity of the sensory experience for those who lived near the Berkeley Pit as they coped with new sights and smells, vibrations and sounds. Like the power plants, chemical plants, and dams that Joy Parr describes in *Sensing Changes*, the massive undertaking of open-pit mining transformed people's sensory experience with and understanding of their environment.[3] Like that book and other recent studies, the story of open-pit blasting in Butte shows a divide between public and expert definitions of risk.[4] Indeed, residents often sensed danger and damage that company engineers could not find with their scientific instruments. Engineers quickly recognized how important lay perception was. They soon learned to manage the company's risks with three steps—by refining the process through which they responded to complaints, by regulating residents' sensory experience of industrial work, and by spending more money, buying property held by worried residents.

The open-pit mine thus completely shifted the risk culture in Butte. Although an open pit increased the amount of land facing mining's effects, it did not always increase industrial hazards. Mine subsidence and poisoned water had long intruded Butte's residential neighborhoods. The open-pit workscape also clearly lessened risks for miners, who no longer had to bear the brunt of mining's impact in the form of severe injuries and death. The open pit, however, literally took the roof off mining. Hazards and risks from blasting, digging, and hauling became so widely dispersed into the environment that more people experienced mining's sights and sounds without facing, or being able to prove, immediate damage from them. A more universal distribution of risks beyond miners and the edges of poor neighborhoods makes Butte a good example of what theorist Ulrich Beck calls a "risk society," where vague, common risks spread fear, despite the fact that most people in Butte were actually safer with the open-pit regime than they had been with the underground.[5] Although the company was therefore able to pay less for actual damages, it likely paid as much or more for perceived risks, particularly through the purchase of additional property.

## The Wild West of Hazard Complaints

The steps for processing ore were the same for rock taken from both underground mines and open pits. The switch to the latter, however, increased the volume of material extracted, and so the company wanted to increase the efficiency of its operation. Rather than ship all the material twenty-six miles to the town of Anaconda, as it had in the past, in 1964 the company built a new concentrator—the Weed Concentrator—next to the Berkeley Pit. After the rock had been crushed into very fine powder and water had been added, the material became known as slurry. The slurry was then transported into an enormous area for the purpose of froth flotation. Flotation began to separate ore from the rest of the material. There were 1,296 flotation cells, which were essentially big vats, ranging in size from 150 to 325 feet in diameter. The reservoir for these cells held 8 million gallons of water: 3 million fresh, and 5 million reclaimed. During froth flotation, a slurry of crushed rock with copper was dropped into the vats, along with chemicals. Agitating the vat coated the copper molecules with chemicals that made them repel water, attach to air bubbles, and rise to the surface. Workers skimmed the concentrated ore, which was now at 27 to 30 percent copper, off the top of the vat and shipped it to the town of Anaconda, where the company smelter used intense heat to concentrate the ore to between 98 and 99 percent copper.[6]

Copper-free refuse, known as tailings, sank to the bottom of the vat in staggering amounts. The manager of the concentrator said that for every one hundred tons of ore-bearing rock put into the concentrator, less than one ton became concentrated ore. Official figures show that in 1965 only 50,000 tons became final copper concentrates, with the rest (950,000 tons) simply discarded as tailings. Workers pumped the often-acidic tailings, which could be environmentally hazardous, to areas where the materials could be more safely contained. The company eventually used the Yankee Doodle Tailings pond, a massive earthen dam north of the Berkeley Pit, as its tailings holding area.[7]

Locating the concentrator in Butte created headaches for the company. One man who lived just south of the concentrator complained bitterly about the dust and odors coming from it and the Berkeley Pit. He even inserted a leaf covered in dust in his letter to the company. His wife had a respiratory problem, which her physicians at the Mayo Clinic told her could easily be exacerbated by Anaconda's operations.[8] Similar complaints came through Anaconda's phone lines.[9]

The process of pumping wet tailings up to the pond also caused complaints. In 1964 and in 1972–73, problems with the pumps meant issues with the line transporting tailings through the McQueen neighborhood. The result was slime running into the streets. Both events happened during the winter, which left the neighborhood with icy, dangerous conditions before work crews could clean up the mess.[10] In summers, if the ponds dried out, more problems could occur. Because tailings contained finely ground rock, dust storms would be the result. Company personnel therefore had to keep the ponds wet.[11]

In the end, the company received few complaints about processing work in Butte. While the pit had expanded the amount of land used for tailings dams, few residents worried at first about the massive environmental dead zone sitting above them.[12] Most of the complaints that streamed into company headquarters instead had to do with its new open-pit extraction work. This work repeatedly encroached on nearby neighborhoods. Company engineer Nelson Jones pointed out that every time operations moved into a new area, "bordering residents would register complaints about our operations."[13]

Dust was a constant complaint. Trucks moving rocks and materials through the neighborhoods inevitably kicked up dirt. One resident of a McQueen house cursed the company for the condition of his siding, as it was "chalked, yellowed, and covered with years of accumulated dust." Residents also became concerned about the diesel fumes and clatter that the "heavy and noisy equipment" made.[14] To lessen the amount of hauling, Anaconda typically dumped overburden, which is unprocessed rock with no ore in it, as close as it could to the pit. As a result, mounds of waste rock grew right next to residential areas. The dumps themselves were not only eyesores but also attracted children at play, which created another source of worry.[15] In total, the Anaconda Company received miscellaneous complaints about floods, tailings, trucks, dust, and fencing (or the lack thereof around mining areas) on 248 properties, sometimes getting many complaints for each location.

Noise and damage from blasting were the most common complaints. From the start of open-pit mining until Anaconda concluded its Butte operations in the early 1980s, the company received blasting-related complaints for 493 properties, with many residents complaining multiple times. Some residents accepted that noise and hazards were, to some extent, part and parcel of being members of a mining community. For example, two ladies came into the Legal Department in March of 1973 and were "not seriously upset," even though one was losing tenants due to the encroaching operations.[16]

Others were more unsettled by blasting, even if they did not seek any money or a larger response from the company. In one case, the new manager of the Finlen Hotel, an institution on the east side of the CBD, wrote to Anaconda: "I have heard old stories about the loud welcome often extended in the 'Wild West' but I hardly expected to be so greeted upon my arrival in Butte. While relaxing in my Apartment here in the hotel at about 4:40 PM, December 27th, I was startled by a blast which, as a matter of fact, put a crack in the window glass next to me. I hope the powder charge isn't always as large."[17]

No amusement and much stronger anger were obvious in other complaints. Living right next to a noisy industrial workplace affected not only the buildings but also people's mental health. As two residents proclaimed about the "company harassment" of blasting, "our homestead jumps, quivers, and quakes not less than two times daily, adding then to nervous anxiety to each of us."[18] During one short-lived open-pit project, the Syndicate Pit in northwest Butte, Mrs. Flynn claimed that she was unable to sell the home she owned adjacent to the operation. Blasts happened within 150 feet of her property. She cleaned up fragments that she alleged came from some of the blasts, and recorded new cracks in the brick veneer and granite-block foundation.

Once Anaconda had abandoned the Syndicate Pit, Mrs. Flynn finally sold her home to the Ledling family. Mr. Ledling's experience as a boss at an underground copper mine, the Mountain Con, turned out, however, not to have prepared him or his wife for life in the vicinity of open-pit operations. When mining resumed in the area, they were "practically knocked off our feet." Blasting kept their six-year-old up at night. Worried about her gas pipes and her heat, Mrs. Ledling had finally "had it" and composed an angry letter to the company. She closed with a line that "even the Anaconda Co. has no right to interfere with the physical well-being of people. I'm worried, losing sleep, and seeing my home fall apart." Anaconda's response was initially to try to calm

McQueen Neighborhood. This photograph shows just how much Berkeley Pit work encroached on Butte neighborhoods. Photo courtesy of Butte-Silver Bow Public Archives, Clinton Peck Collection, PH308.20.

her, but seven years later, the Ledlings were complaining again. This time, Mrs. Ledling alleged that noon blasts at the Berkeley Pit were increasing the already extensive damage.[19]

Residents often perceived these hazards as dangerous as well as obnoxious. This perception initially appeared to be beneficial for the company, which needed people to move if it was going to expand the pit. In 1974 geographer Joseph Leeper encountered locals who explained to him that the industrial use of nearby property encouraged residents to sell. As the "slag heaps, noise, and fouled air" that came with open-pit mining invaded each neighborhood, homeowners would sell, "hoping to get a little bit better deal than their neighbors."[20] In interviews, many residents mention industrial hazards as the major factor that convinced them to sell. For some, the smoke and earthshaking booms caused by blasting were the catalyst for moving; for others, it was the noise from massive trucks or dust from the dumps.[21] The same situation held true in other open-pit communities. In Asbestos, Quebec, open-pit blasting in the 1960s convinced many people to relocate, making that mine seem "an ever more dangerous and invasive presence."[22]

Expanding operations in Butte similarly spread the associated problems deeper into the community. A 1973 blast that shot rocks throughout the East Butte area generated multiple complaints. The civil engineer who handled most of them reported on the "general feeling" of the residents: "You will not buy us out. Are you trying to blast us out?"[23]

Expansion of the pit therefore had an impact like the blockbusting in Eastern cities, but the people of Butte were quickly selling their homes in anticipation that industrial hazards, rather than the arrival of minority groups, would soon cause their property values to fall.[24] Anaconda's Land Department did not have to work hard to convince locals of dangers and discomforts that they could see and feel. As one letter to the company proclaimed, "The inexorable crunch of your machinery has rendered and continues to render—not only our peace of mind intolerable but our property values depreciating." Others claimed that they were denied Federal Housing Authority–insured loans if the house was located near pit operations.[25] Anaconda's Engineering and Legal Departments, which handled damage claims, would often immediately refer complaints to the Land Department, in the belief that a home or business purchase was imminent. Homes that were not quickly purchased sometimes "would end up on a solitary island surrounded by pit operations," as geographer Joseph Leeper put it.[26]

## Liability and Property Damage

The dynamics of expansion were not completely positive for Anaconda. As the pit grew outward, toward heavily populated areas, complaints about blasting unsurprisingly increased. Residents sometimes organized to resist. In the early 1970s, for example, people in the Columbia Gardens neighborhood met to complain about dust pollution and the erection of hazardous fencing. They debated "a possible confrontation with the Anaconda Co.," contemplating whether they should raise a joint legal fee.[27]

Open-pit operations therefore became a new liability issue. Anaconda feared it would end up paying for at least as many settlements as underground mining operations had generated. In 1960 Dave Piper, a Butte engineer who later headed the department, reported, "Plans were formulated to examine the cause of complaints, the extent of the damage, if any, caused by vibrations, the noise level and its effect on the complainers, the types of blasting and other factors that caused anxiety among nearby home owners."[28]

Anaconda's handling of its potential liability for blasting in the short-lived Syndicate Pit revealed the company's legal reasoning. The company received a number of "complaints of damage and disturbance" during the pit's operation through 1963. C. J. Hansen, author of a memorandum on this issue, explained that, although courts elsewhere had begun to adopt a broad view about blasting damage—considering depreciated values, annoyance, or mental distress—Montana's remained focused on physical property damage. Anaconda was therefore quite lucky. In regards to property damage, two conflicting doctrines came into play. One held that an owner "should be allowed the fullest possible use of his property," including the right to develop land with explosives. Under that doctrine, "there is no liability for damages unless negligence is proved against him." Unfortunately for Anaconda, Montana tended toward a second view, which posited absolute liability. Under that doctrine, there would be absolute liability for blasting damage, even if no negligence was involved.[29]

According to Anaconda's assessment of this second doctrine, the company was most vulnerable when flying debris from blasting caused damage. Even if no negligence existed, the blaster was liable if debris showed up on another's property.[30] Anaconda thus always settled complaints about flying debris. Denying responsibility was impossible when rocks flew through roofs, into car windows, or—in a few unfortunate situations—directly at people.[31]

Despite Anaconda's worries about "vexed and annoyed" residents, early attempts to stop flying debris failed.[32] When the company was blasting near the southwest end of the Berkeley Pit, it posted workers to stop traffic in anticipation of imminent blasts. In one situation, however, workers prematurely left their posts and flying rock went through the front window of a passing car. Luckily for the company, the injured driver did not pursue punitive damages, which would have entailed costs far beyond the driver's medical expenses and car repairs.[33]

Damage from flying debris, though uncommon, was evident, but the negative effects of ground vibrations and air concussions were more dispersed and hence harder to prove. According to Dave Piper, most of the energy from an explosive was expended in shattering rock, but the excess was either "transmitted through the earth in the form of waves" or it escaped through the air as "sound or concussion."[34] Ground vibrations seemed most likely to cause significant damage and were the basis for most complaints about blasting. Company engineers therefore focused early research on ground movement. They bought a seismograph in 1959, ran periodic blasting tests in 1960, and began to coordinate efforts between the Berkeley Pit Mining Department, the Mining Engineering Department, and the Legal Department.[35]

Efforts to deal with ground vibrations became more systematic in 1968, when Lewis L. Oriard returned to Butte as Anaconda's blasting consultant. Oriard was the chief geophysicist at Woodward-Clyde & Associates in Oakland. At the time, he was consulting on drilling, blasting, and dredging techniques by the Panama Canal Company, but Oriard also had previous experience in Butte. In the 1950s he had consulted with both the F&S Contracting Company and Anaconda during early Alice Pit and Berkeley Pit operations.[36]

Oriard provided Anaconda's engineers with new seismometers and airwave detectors, telling them to keep better track of each blast. Anaconda began to take permanent seismic records of every blast it fired in the Berkeley Pit. Residents, who did not have such equipment, had long asked company personnel to come to their houses and to sit with them during a blast. Starting in the late 1960s, engineers would arrive at their houses with a seismograph to measure the blast's severity, then show residents the record.[37]

Top Anaconda executives in New York agreed that technical data were the best defense against "possible damage suits and/or injunctions."[38] As Oriard explained, "There is a continuing need to prove that the ground vibrations are kept below potentially harmful levels. It seems that when this proof is lacking, many speculations arise about the severity of the vibrations, and the unwarranted damage claims tend to increase substantially in both number and value."[39]

Anaconda's engineers scoured the U.S. Bureau of Mines' recommendations and looked through the research in leading engineering journals.[40] By 1962 the Bureau of Mines had settled on peak particle velocity as the major motion descriptor to use when measuring vibration. The figure of 2.0 inches per second became the standard for the construction industry and government agencies; any ground waves moving above this line would signal caution first, then blasting damage. Into the 1970s, Oriard continued to claim that only above 2.0 inches per second would caution for structures be warranted. Only in 1980, as Anaconda was closing up shop, did the standard lower significantly, to between 0.5 and 2.0 inches per second peak particle velocity as safe, depending on a variety of additional factors.[41]

Both Oriard and Anaconda's Mining Engineering Department were satisfied with the results of blasting tests. A typical Berkeley Pit blast caused movement at 0.6 inches per second. Sometimes blasts came closer to the 2.0 inches per second point, with a measurement of 1.7 inches/second at structures closer to the pit. Oriard dismissed such results, stating that vibration intensity decreases very rapidly after the first thousand feet, even if the vibration itself carried on (and thus could be perceived) for great distances.[42]

Besides distance from the blast, a few other factors determined the possibility of damage. The composition of the ground affected the blast's intensity. Deep alluvium led to lower frequency, longer duration, and more damage; shallow overburden generated higher frequency, shorter duration, and less harm. Vibrations could vary with the size of explosive charges and they could even cause cave-ins when they occurred near the site of old underground mines.[43] Unsurprisingly, the amount of damage from blasting also reflected the workmanship and age of a house. Older houses already showing wear and tear were more susceptible to damage.[44] None of these factors, however, proved as important as particle velocity, at least in the technical literature of the time.

Anaconda relied on technology to solve the problems of surface movement. A typical explosion ignited two hundred to four hundred pounds of ammonium nitrate in a set of blasting holes. Anaconda had reduced itself to detonating only nine holes at a time so as to limit ground vibration.[45] New interval delay blasting techniques enabled the company to set off holes in a series rather than simultaneously. This system meant that the company could blast two holes at a time, blowing up many more holes—up to 150 or 200—without increasing the surface movement. For most blasts, a crew would first drill holes into the rock in a pattern. It would load the holes with ammonium nitrate, mixed with some sort of fuel (like coal dust, oil, or other sensitizing agents). To ensure detonation, it would then prime the holes with detonating cord and often other fast-exploding materials, like nitramite, gelatin dynamite, or a puck, which was a high-explosive detonator that looked like a hockey puck. Next, the crew would connect the holes together with the detonating cord, which essentially operated as a fuse. The crew would then attach delay connectors to the detonating cord as a way to time the blast. Finally, it would initiate the blast, using either an electric blasting machine or a more manual system of a nonelectric detonating cord. (Anaconda used both types of initiators at different times.)[46]

Research from other locations showed that this delay technique lessened the amount of ground vibration, and the company had success experimenting with a number of methods to slow down the detonation sequence. Engineers also kept improving the monitoring system, taking more care near heavily populated areas and examining the reasons for unusually strong blasts.[47]

Because solving the vibration damage problem was relatively easy, the company did not suffer from many successful claims. After a spike in the late 1960s, the seismograph readings of pit blasting became "generally lower and in the safe area indicating that an excellent job is being done by the blasting people," in Nelson Jones's words.[48] By 1972 Anaconda engineer James P. Connell bragged to the Butte Association of Life Underwriters that "the intensity of blasts now made in the pit are only one half that allowed in safety standards set by the American Society of Civil Engineers."[49] Vibration damage was often not obvious, and technology gave the power of expertise to the company. Anaconda's lawyers often dismissed claims with the statement, "We respectfully wish to advise you that the denial of your claim by The Anaconda Company is based strictly on an analysis of the scientific and engineering information available."[50]

Technical data could dissuade even the most disaffected claimants. In 1970 John Corr, owner of the Race Track Bar, which sat one mile southeast of the Berkeley Pit's limits, "claimed structural damage in the amount of $12,500, and had witnesses to testify to the extreme vibrations in his place of business." His representation was the Holland brothers, "who have been anti-Company generally." Corr was one of the first Butte residents to bring a lawsuit against the company for structural damage from blasting. His case was so worrisome to Anaconda that it became the deciding factor in hiring Lewis Oriard on a more permanent basis.[51]

Nevertheless, Anaconda prevailed. As the manager of Montana's mining operations, Martin Hannifan, reported, "Thanks to the excellent scientific data recorded on each blast, and Mr. Oriard's skill in explaining the meaning of such data, the case was dropped before it came to court." Anaconda had worried that this case would be a test that, if successful, would lead to multiple lawsuits all around the Berkeley Pit, but such was not the case. As Hannifan wrote, "This first test case will undoubtedly discourage others who have been registering complaints and threatening lawsuits."[52]

In addition to an increased focus on technical variables, Anaconda also strengthened the one-by-one claims system it had developed in dealing with complaints about subsidence from underground mining. The process involved detailed investigations. While the Mining Engineering Department had an engineer inspect the premises, the Legal Department contacted the Abstract Department for a written report on the title chain. Engineers also began a file for each piece of property, recording any conversations with the claimant (or, in a small number of cases, with the client's attorney), the structure's distance from the pit, related observations about the building, and even the level of noise in each area. The file also recorded whether the affected property was in an area targeted for expansion and property purchase. Investigators examined claims, looking for patterns of relationship among "friends, relatives, neighborhood groups." Finally, Engineering sent copies of all correspondence and documents to Legal. Legal then followed up with a letter to the claimant, explaining the probable cause of the alleged damage, often including statements about Anaconda's general monitoring procedures.[53] Throughout this process, Anaconda tried not to set precedents for insurance payments. The company therefore refused to deal with insurance companies, and instead dealt directly with the claimants.[54]

Beyond the technical defense, Anaconda had other ways to say no. Sometimes it simply referred complaints to F&S Contracting Company, which handled early blasting in the Alice and Berkeley Pits.[55] Likewise, the company faced no liability on property it owned or leased, and thus complainants who were renting from Anaconda's Kenwood Realty were out of luck. Much of the area into which the Berkeley Pit was expanding fell into that category.[56] Finally, Anaconda pointed to possible alternative causes of fallen ceilings and cracked walls, including slammed doors, rapid changes in humidity, and deteriorated plaster.[57]

Anaconda sometimes accepted responsibility for damage. In 1972 it paid small amounts for repairs of windows that broke in Butte's business district after a Berkeley Pit blast. Even then, the company commented that the affected windows had been "very poorly supported."[58] Residents believed there was "an indirect admission" of company guilt when blasting crews occasionally changed their practices but, in almost all cases, letters from the Legal Department to homeowners vehemently denied causing any damage.[59]

Anaconda's engineers were largely right; it was uncommon for open-air blasting, because it spread vibrations so far afield, to cause serious damage.[60] Partly because of that fact, and because residents' senses could be "disproven" by Anaconda's experts, the difference in the numbers of complaints Anaconda received for underground mining's effects and what it received for open-pit mining is stark. It received complaints on more than 1,500 properties due to underground subsidence, settling 45 percent of them, many for large amounts of money, but Anaconda only settled 7.5 percent of the nearly five hundred blasting complaints it received from 1955 to 1983.

However, these numbers fail to tell the full story of the financial problems Anaconda faced, which had far more to do with perceived risks than actual hazards. Concepts of liability were subject to changing attitudes. Blasting consultant Lewis Oriard was concerned that Anaconda would receive a suit that alleged "nuisance, invasion of privacy, or environmental pollution," situations which "leave more room for subjective opinion" than previous cases alleging damage.[61]

Management hence faced another, bigger, problem: increasingly savvy residents. Some residents merely wanted acknowledgement of their suffering or help making repairs.[62] Others, however, were aware that they were in an "authorized acquisition area" and kept track of the encroaching hazards to bargain for a better price for their property.[63] Some frequently filed complaints not with Mining Engineering but with the Land Department, which handled home purchases. Instead of asking for money to help with repairs, they proffered purchase prices for which they would be willing to sell. Others never registered formal complaints at all—instead threatening various company personnel that they would, and that they'd bring their own lawyers.[64] According to the engineers, even residents without plausible claims to damages often used impending hazards as a bargaining chip. Paying more for property that Anaconda wanted was not part of the company's plan, but the dispersed nature of open-pit risks made that situation probable. Nor did Anaconda desire the increasingly common prospect of buying property it did not want in areas in which it had no major plans to mine.[65]

## The Human Response Problem

Beyond financial considerations, Anaconda's management began to see open-pit hazards as a big public relations problem—a problem that perhaps even threatened the future of its Butte operations. In the 1960s Anaconda took a cue from the New York Trap Rock Corporation, whose limestone and dolomite quarries had come under fierce attack from nearby urban areas. Its quarries were such a "blasting nuisance" that it "was on the verge of being put out of business." Besides seeking technical solutions, New York Trap Rock successfully introduced a public relations campaign, showing the complainants its data, visiting with them about what actually caused damage, and, most importantly, trying overall to make residents feel at ease.[66]

Like New York Trap Rock, Anaconda came to see educating the public as key to improving community relations. By 1968 the new head of public relations for the firm, Julian Hayes, expressed the need for "better public contact," making sure that everyone who dealt with the public was conversant in blasting methods and standards. Orders came down from Anaconda's main New York office to conduct "an extensive public relations program."[67] Following New York's directive, Butte officials engaged in more one-on-one visits with complainants. In 1973, after a company lawyer visited with two "cordial" ladies, he still suggested that members of the Mining Engineering Department visit their homes. Doing so "might have more public relations value than anything else."[68]

Engineers also became more sensitive to people who could cause trouble beyond their damage claim. In 1972 grocer Bob Lee called about blasting. He alleged blasting "is shaking up his store quite a bit, causing the front door to pop open and generally making him feel as if it were unsafe to be inside the premises." Although angry, Lee did not claim any actual damage and so, if the company worried only about liability, it might have disregarded the call. A company lawyer, however, suggested the matter be treated as urgent. Lee was a city councilman, he noted, "and, I understand an opponent of the Company." Because Lee was "someone who can give us considerable trouble if he is not handled properly," engineers made his case a priority.[69]

Anaconda not only handled individual cases with public relations care, but also developed a broader public relations approach. Describing an altercation with two irate people, Anaconda lawyer J. J. Picarelli explained that many complaints he handled involved "no legal problems or claims." Instead, they were broader complaints about open-pit mining's intrusion into each new area—in this case, the Columbia Gardens. He suggested that additional "communication be made to the residents of Columbia Gardens with respect to the facts of the Continental-East operation with which they are going to have to live, e.g., 24 hour operations."[70] Lewis Oriard agreed that engineers should "try to prepare the public for blasting operations before they begin."[71]

Anaconda's teams were soon following this advice. By the 1970s, other experts on the subject agreed that educating the public before the start of blasting in each area greatly improved public relations.[72]

During the 1960s and 1970s, Anaconda's public relations efforts came to focus on the sensory experience of blasting, and not the damage from it. When the company's tests suggested that damage occurred very irregularly but the complaints continued, engineers, lawyers, and work crews rightly concluded that blasting complaints had more to do with perception than damage. Pit blasting spread vibrations out across such an expansive area that many could sense its effects. Anaconda therefore took steps to manipulate people's sensory experience and hence lower its financial, legal, and political risks. The company sought to show laypeople that what their bodies sensed was not hazardous to themselves or their property. Because the senses are important to the exercise of power, Anaconda hoped to maintain control over the situation.[73]

Anaconda's experts started by thoroughly investigating air concussions, the energy from a blast that did not go into shattering rock in the pit or making ground vibrations. Glass was most susceptible. Windows could break if the federal safe limit of 140 decibels for airwaves was exceeded. (By 1980, the Bureau of Mines had lowered the official limit to 133 decibels.)[74] Blasts at the Berkeley Pit often were as low as 90 decibels and rarely exceeded 105. Occasional mishaps did occur. In one case, two kitchen windows shattered in one man's house, which "frightened his Mother who screamed when the windows broke."[75] If Anaconda received many damage complaints at once, if the decibel measurement was high, or if the damage looked typical of concussion—with windows blowing out due to sudden pressure change—the company would settle.[76] Concussion claims never involved major structural harm. Anaconda's engineers did not see readings above a rather high 173 decibels, which was the point at which they believed that anything beyond glass damage would occur.[77]

Air concussions had their greatest impacts on people's sensory perceptions. Lewis Oriard wrote, "Residents of the Butte area can be expected to respond more adversely to those blasts which generate the stronger air waves." According to him, "People living beyond the range of detection of ground motion will often testify to 'violent shaking' because of some effect produced by an air wave. This could be the sound of a loud blast, or the sound of rattling of a loose door or window, the 'thud' produced by an air wave striking the building walls, or merely the pressure-rarefaction variations against the person's ears."[78]

Explanations by engineers that the affected houses were too far from the blast to be damaged did not convince many regular citizens, who frequently confused vibrations from air concussions with ground tremors, even though the two effects were often separated by ten or more seconds.[79] When McQueen's Holy Savior School registered a complaint, Anaconda lawyer J. J. Picarelli tried to explain that "the occurrence of noise and window rattling from the blasting is unavoidable, but the school should, otherwise, remain unaffected by the present mining operations."[80]

Decreasing the noise coming out of the blast thus became engineers' prime objective. Even in their original 1960 tests, Anaconda's Mining Engineering Department "found that the noise level is directly related to the vibration complaints." Dave Piper reported, "The average layman associated damage with the volume of sound. The louder the noise, the more damage he believes his property has sustained. It is not true; there is no direct relationship between the volume of sound and the amount of damage."[81] Consultant Lewis Oriard confirmed Piper's research, but annoyed laypeople refused to believe this "truth."[82] To better understand how noise worked, Anaconda's engineers began to collect articles from other industries that worked with explosives and from the national press.[83]

Many people at Anaconda were soon focusing in particular on what Lewis Oriard called the "human response problem." As Oriard put it in 1970, "People are extremely sensitive to any type of sound or vibration. This sensitivity has a complex relationship to physiological, psychological and socio-economic factors."[84] Dave Piper had long touted the idea that people were simply too sensitive. On Anaconda's early blasting tests in 1960, he wrote, "Although the human body is quite susceptible to the effects of vibrations, most people cannot judge these effects reliably without accurate instruments. Tests have demonstrated that some people can detect vibrations that are no more than 1/100 the strength of those that are capable of causing damage to buildings." Oriard concurred with Piper, claiming that the ratio of perception to damage could sometimes fall down to 1/1,000.[85]

Oriard, Anaconda's engineers, and other people in the mining industry attempted to measure not just the sound and ground waves but also human perceptions of them. Oriard developed figures based on his research into human sensitivity in the late 1960s and early 1970s. In a presentation to the Association of Engineering Geologists, he suggested that at 0.02 inches per second, "biased observers" will possibly complain; at 0.2 inches per second, they already believe blasting to be "severe." "Impartial observers" notice blasting at 0.06 inches per second, find it disturbing at 0.4 inches per second, and severe at 1.2 inches per second, a full inch per second after the biased observers were panicking. In fact, all of the numbers for impartial observers were far higher than the numbers for "biased observers," but also still below the then-extant standard of 2 inches/second for actual damage.[86] Later research at the U.S. Bureau of Mines agreed that even 0.5 to 0.75 inches per second would feel "less than acceptable" to most residents.[87]

As Oriard's report shows, attitudes toward open-pit mining were increasingly seen as a pertinent factor. Those viewing the blasting as necessary for Butte's economic vitality wanted it reasonably controlled but not stopped; a 1980 study by the Bureau of Mines found that people accepted loud noises if the economy was closely tied to the local mines, as had been the case in Butte.[88] Those who instead connected blasting with "danger, cost, intrusion or pollution," as Oriard put it, were much more likely to oppose it.[89] Such had become the case for many near the Berkeley Pit.

Factors beyond peak particle velocity and attitude also affected "the threshold of perceptibility" Oriard was attempting to measure. The duration of blasts as well as differences of physiology among people had an impact.[90] Anaconda therefore faced a difficult perception problem that could not simply be solved through technological measurement.

Residents struggled to describe their experiences with blasts to company personnel. Recognizing that they would need proof of open-pit infractions, some kept lists of blasts that they had felt over months or even years. One man gave a list to the engineers of the "bad shakes" versus the "real bad shakes." Residents like Mrs. Whalen of Butte's East Side commonly spoke of the "earthquake type" of blast.[91]

Oriard noted that residents had another blind spot: a "lack of knowledge of the normal static (non-vibratory) physical forces which are involved in their daily lives and act on the buildings in the community." They tended to blame mining for property deterioration that occurred simply with the passage of time.[92] The logical assumption that a householder would have a "reasonably accurate idea of the number of cracks, their location and size, that exist in his home" was unfounded, he explained. Old paint covering cracks in plaster and grass growing in concrete splits revealed that damage discovered in the wake of a blast often was evidence of older problems. In contrast to a single, immediate loud blast, such changes happened too slowly for a person to sense. In his view, most residents were not trying to trick the Anaconda Company; they just had faulty sensory capabilities.[93]

Under the guidance of Oriard, Anaconda started to deal aggressively with what they called the human response problem. The company learned that supposed improvements in their blasting technique, introduced for safety, unfortunately aggravated the noise issue. In order to reduce the danger of premature explosions set off by stray currents from lightning, radio transmitters, drills, and shovels, engineers in the late 1950s switched from silent, electric blast initiation devices to nonelectric primacord, which produced a sharp report. Primacord was similar to a safety fuse, but instead of containing a core made of gunpowder, it contained powerful, high-velocity explosives that detonated quickly. In 1960 Piper reported, "The present animosity to the eastward expansion of the Berkeley Pit by the residents of East Butte and McQueen may in part be influenced by the excessive noise created by primacord blasting."[94]

In 1960 Anaconda experimented with low energy detonating cord (LEDC). Produced by DuPont, LEDC contained much less explosive per foot than normal cord and proved to reduce "excessive noise and concussion" from blasts. In Anaconda's tests, LEDC shots produced few complaints, even though they were triggered on the edge of the Berkeley Pit, only three hundred feet away from occupied dwellings. Engineers decided to try the LEDC system "in selected areas of the pit where noise may be objectionable," but they still used basic primacord for other blasts.[95] Oriard also suggested covering the cord with material, which could reduce some of the noise.[96]

LEDC could not solve complaints about multiple blasts at once. Mrs. Roland Navarro, a neighbor to the Alice Pit, noted, "We have noticed short, fast blasts more—the ones that go 'boom, boom, boom!'"[97] Seriatim explosions meant less vibration, which is why Anaconda had introduced interval delay, but these explosions protracted the blasting time. Subsequently, each blast became more noticeable to people in the community. Throughout the 1960s and 1970s, the company investigated different delay systems, hoping that each would fix the noise problem. In the 1970s the company was back to using an improved kind of electric initiation, but by the late 1970s it was testing, and then using, a new nonelectric (NONEL) delay system. NONEL kept the improvements in damage creation that primacord had brought, but it also produced a soft-enough noise that seriatim explosions caused less concern. Engineers' final tests suggested that this system had successfully "reduced surface noise from blasting" and so they continued using it until the end of the Berkeley Pit.[98]

Refinements in the technique known as stemming became another of Anaconda's noise abatement efforts. Stemming is inert material, often made of previously drilled rock, used to plug the surface end of the drill hole in which the charge sits. Plugging the drill hole directs more of the energy into breaking the rock and lessens the blowout that makes loud noise. When he observed that blasting noisily ejected much of the rocky stemming known as drill cuttings, Oriard instituted trials of substitute materials.[99] He also suggested drilling holes deeper so that they could be stemmed more thoroughly.[100]

Extra care during each push back of the pit's boundaries became another, less successful, part of the company's technical solution. When crews blasted on the edges of the Berkeley Pit, thus closer to homes, they lessened the amount of explosives per hole. Changing the amount of explosives per hole, however, proved to have little effect on the number of complaints Anaconda received.[101]

By the late 1960s Oriard was suggesting a much more successful tactic: less-frequent and better-planned blasts. Oriard wanted the

DETONATION IN THE BERKELEY PIT, C. 1972. Photo courtesy of Butte-Silver Bow Public Archives, Anne Dobney Collection, PH227.

company to detonate a small number of bigger blasts rather than a big number of smaller blasts, thus creating "fewer events to be perceived by the public."[102] Following his advice, the company reduced the daily frequency of blasting. Likewise, to eliminate surprise, the company accepted Oriard's suggestion that it establish a "fairly regular routine of blasting" in lieu of its previous practice: intermittent detonations throughout the day.[103] As Oriard wrote in 1968, "People tend to become accustomed to routine procedures, and will pay less attention to something expected and familiar.... People are more sensitive to sounds when they are at rest and when the surroundings are quiet. For that reason, the night and early morning hours are the times when sounds seem to be the most disturbing." Anaconda therefore restricted its big blasts to the noon hour and four in the afternoon, when people were busy at work and not resting at home.[104]

Dave Piper and Lewis Oriard also investigated the effects of wind and of temperature inversions, factors not part of the underground blasting environment. They learned that weather conditions did not produce additional energy from the blast but often altered the pattern of energy dispersion. Tests showed that, if "wind conditions were right," sounds of blasts increased by 10 to 15 decibels thousands of feet downwind, when compared with the levels in no-wind or crosswind situations. To compensate, Piper suggested that crews blast when the wind direction would cut off the sound from populated areas rather than send it toward them.[105]

Engineers also watched for temperature inversions, during which cool air does not rise, but becomes trapped by overlying warm air. Under such abnormal conditions, air blasts do not refract from the surface of the earth but bend toward it, with the result that the wave ricochets, landing back on the ground, sometimes in places far from its origin in the mine.[106] When the company received complaints of air blast damage, including broken windows, it would often settle if temperature inversions had been a factor.[107]

Time of day affected the likelihood of inversions. Both the ground and the air above it heat in the morning and cool in the evening, but at different rates. Hence, inversions are more likely to happen during the mornings and evenings.[108] Avoiding inversions therefore became another reason to schedule most blasts during the afternoon. Anaconda also became more wary of blasts during the winter, when temperature inversions were more common.[109]

Assumptions by residents that damage from blasting accumulated over time was a final obstacle for Anaconda to overcome. Even homeowners who did not initially complain about the blasts gradually became upset about the "eventual effects" that blasts could have on their property.[110] Spots near the Berkeley Pit, like Harrison Elementary School, were believed by the public to be gradually "shaken apart by shock waves from blasting at the Berkeley Pit."[111] Investigations between 1968 and 1971 reassured the company that blasting did not have strong cumulative effects. Dave Piper claimed that most damage instead resulted from an "accumulation of a poor repair program," not blasting.[112]

After its engineers reached agreement that the cumulative effects of blasts did not cause damage to buildings, Anaconda developed a program to educate homeowners. Oriard initially wanted to conduct preblast surveys of homes that might be affected. If people could just see what their home looked like before blasting, then they could finally understand how minimal pit mining's effects were. To make this kind of investigation easier, Oriard suggested that Anaconda designate test houses. He wrote to Nelson Jones: "When you acquire ownership of some of the buildings around the rim of the pit, look for an opportunity to preserve some of these for long-range observation of vibration effects." The company could then create its own data on the issue, showing people that most property defects were simply caused by age and other nonmining forces.[113] The use of test houses and model towns was common for blasting research, including those for atomic weapons, so it is unsurprising Anaconda used them in its investigations too.[114]

Anaconda was soon using test houses. In the early 1970s, before company operations began at the Continental East Pit, adjacent to the Berkeley Pit, Nelson Jones and Lewis Oriard inspected an occupied residence in the Columbia Gardens area, taking multiple photos. During each of Oriard's quarterly visits to Butte, he returned to the house, which sat 1,600 feet from the open pit. Taking additional photos and filing multiple reports, Oriard verified no change in the structure's condition. Oriard repeated this initiative when Anaconda started a new expansion near the Hillcrest neighborhood later in the decade. He believed that obtaining a "long-term history" for a house could finally prove his point to residents; their senses were deceiving them.[115] It is hard to tell how useful these changes proved. Even as the Berkeley Pit encroached on increasing numbers of homes, the number of settled complaints maintained a low rate through the 1970s, but that result was due to a related Anaconda program: the buffer zone.

## THE BUFFER ZONE

Educating the public about mining's sensory experience was helpful, but Anaconda took an even bigger step to improve both its liability and its public relations issues. Proximity to mining operations encouraged people to move—an act that was often helpful to Anaconda—but the company decided that a large buffer zone between industrial work and home life was a more effective approach to the community. They knew that urban planners looked at industrial zoning in a similar way, seeking to make people's lives free of pollution through a spatial arrangement. Putting industrial zones away from residential areas would protect residents from industrial hazards, save the company from financial claims, and convince Butte to be more accepting of open-pit mining.

Anaconda had used this kind of a pollution solution, which environmental historian Franz-Josef Brüggemeir has called the regionalization of pollution, numerous times in the past.[116] Like other mining companies in the American West, it often simply bought the property on which it was causing damage. In the 1940s Anaconda bought a lot of the property in central Butte most affected by subsidence and obtained easements and liability releases on much of the rest. The same held for the area along Silver Bow Creek, where Anaconda had long dumped mine waste. Earlier in the century, to placate farmers in the Deer Lodge Valley affected by arsenic-laden smoke, the company similarly bought property in the path of smelter emissions. In 1927 Anaconda even negotiated a swap with the National Forest Service of 22,000 acres of land it already owned for forest that it was polluting. As historian Fred Quivik puts it, the Anaconda Company had gradually "come to own the right to pollute nearly all the property across which it spread its waste products."[117]

In the 1960s and 1970s, Anaconda likewise bought Butte property on which it was alleged to be causing damage. In one case, the company entered "an agreement to buy the premises rather than have the Stapletons pursue their allegations that the damage to the west part of 410–416 East Park Street was caused by continuous noon time blasting of The Anaconda Company in its Berkeley Pit operations." This agreement even included a requirement that the owners file a

$100,000 bodily injury policy with Anaconda.[118] Engineers frequently recommended that the company purchase properties where inhabitants were likely to continually feel, hear, or smell the effects of mining. Calling for the purchase of four homes on Anaconda Road, which sat just to the northwest of the Berkeley Pit, Chief Engineer Dave Piper wrote that one elderly resident "had become quite nervous and frightened" by pit blasting, which was happening only three hundred feet from his house. Piper noted, "The four properties in this area are in such location that they will be subjected to further vibration, and, although not damaging, this will likely cause considerable complaint."[119]

A buffer zone had been absent during Anaconda's pit work in the 1950s. People in Walkerville and Meaderville did not receive such a benefit, and hence mining's effects and related complaints grew, but the company began to fashion a zone between the Berkeley Pit and residences at the end of the decade. By the early 1960s, years before expansion reached Butte's East Side neighborhood, Anaconda had already begun buying property there. Little by little, the Land Department purchased even more property during the rest of the 1960s and 1970s. The company never performed much mining work in the area once taken up by East Side houses; the emptied area instead functioned simply as a buffer.[120]

Open-pit mining had created a new set of hazards for community members and the company. The pit spread risks widely, causing complaints across Butte; but their wide dispersal, coupled with the company's legal, technological, and procedural changes, meant that it also became increasingly difficult to prove actual damage from mining. The biggest of those changes, the buffer zone, suggests that Anaconda's talented lawyers and engineers had managed what they called the human response problem with efficacy. As figures cited earlier show, Anaconda settled around 45 percent of the more than 1,500 underground subsidence complaints it received, but settled only 7.5 percent of nearly five hundred blasting complaints.

Yet this apparent success covers up the real costs of open-pit mining to the company. Such costs do not show up in engineers' files, but instead are part of the Land Department, whose records are mostly absent from the preserved record. The land records that do exist, though, suggest that many resources, including significant amounts of money, became devoted to the buffer zone project, which necessarily grew as the Berkeley Pit itself grew. Because Butte residents were often fairly savvy negotiators, relocating people took tact and money. The process of encouraging Butte residents to become more accepting of open-pit mining was therefore expensive.

Figures also cannot account for the worry felt by residents who disliked the open pit's invasiveness, but who believed, because of the dispersed nature of open-pit hazards, that they had no way to make a viable claim for damages. Many residents hence began to relocate themselves away from the pit, causing a cascade of additional problems for the city of Butte. Life at the edge of an expanding industrial zone was a risky proposition for residents, the company, and, eventually, the city. Relocation soon seemed to be the most viable way to answer what historian Jessica Van Horssen has called an open-pit mine's "competing demands for life and work space."[121]

## Notes

1. Complaint numbers are based on the company's complaint files in the twenty-three Boxes of the Dave Piper Records, MC388, BSBPA.

2. For example, see J. F. Harvey to John A. Frankovich, "Initial Inspection of Property Owned by John Brozovich," July 10, 1964, Fldr. "Watson St.," Box 5B, MC388, BSBPA.

3. Parr, *Sensing Changes*. On noise as a way to sense environmental change, see Peter A. Coates, "The Strange Stillness of the Past: Toward an Environmental History of Sound and Noise," *Environmental History* 10, no. 4 (Oct. 2005): 636–65; Raymond Wesley Smilor, "Toward an Environmental Perspective: The Anti-Noise Campaign, 1893–1932," in *Pollution and Reform in American Cities, 1870–1930*, ed. Martin V. Melosi (Austin: University of Texas Press, 1980), 135–51; Raymond Wesley Smilor, "Personal Boundaries in the Urban Environment: The Legal Attack on Noise: 1865–1930," *Environmental Review* 3 (1979): 24–36.

4. Linda Soneryd, "Hearing as a Way of Dwelling: The Active Sense-making of Environmental Risk and Nuisance," *Environment and Planning D: Society and Space* 22 (2004): 737–53; Michelle Murphy, *Sick Building Syndrome and the Problem of Uncertainty: Environmental Politics, Technoscience, and Women Workers* (Durham: Duke University Press, 2006); Joy Parr, "Local Water Diversely Known: Walkerton, Ontario 2000 and After," *Environment and Planning D: Society and Space* 22 (2004): 347–48; Paul Slovic, "Perception of Risk," *Science, New Series* 236, no. 4799 (Apr. 17, 1987): 280–85.

5. Ulrich Beck, *Risk Society: Towards a New Modernity*, trans. Mark Ritter (Newbury Park, CA: Sage Publications, 1992), 10–90; Mads P. Sorensen and Allan

Christiansen, *Ulrich Beck: an Introduction to the Theory of Second Modernity and the Risk Society* (New York: Routledge, 2013), 8–25.

6. Trux Fisher in Fisher and Corcoran interview; "World's Most Modern Concentrator," 1965; "World's Most Modern Concentrator," 1967; "Welcome to the Weed Concentrator," file 29906.01A, Fldr. "Butte, 1961–1985, Mine Operations," AGDC, AHC.

7. Fisher and Corcoran interview; "World's Most Modern Concentrator," 1965.

8. Letters in Fldr. "Caprara, Frank," Box "Cabinet 1, Drawer 3 (Box 1)," MC388, BSBPA: B. T. McDonald to M. K. Hannifan, "Attached Memo," Apr. 28, 1970, Bernard P. Harrington to Frank Laird, Apr. 21, 1970, and Joseph J. Picarelli to Merle Emmert, "Blasting Complaint," Feb. 21, 1973; John W. Warren to John A. Frankovich, Sept. 23, 1968.

9. For examples, see John A. Frankovich to R. D. Piper, Aug. 17, 1965; Nelson A. Jones to John A. Frankovich, Aug. 24, 1965, in Fldr. "Watson St.," Box 5B, MC388, BSBPA; and M. W. Emmert to J. J. Picarelli, Sept. 21, 1972, in Fldr. "Roy, Ray," Box 4C, MC388, BSBPA.

10. A. Guidoni to Clyde Weed, Dec. 30, 1964, in Fldr. "Spruce Street," Box 5B, MC388, BSBPA; Nelson A. Jones to J. J. Picarelli, "Re: Complaint from Arthur Martinetto," Dec. 20, 1972, Fldr. "Spruce Street," Box 5B, MC388, BSBPA; "McQueen Clean Up," *Mont. Std.*, Feb. 10, 1973.

11. R. D. Piper to M. K. Hannifan, June 4, 1965, Fldr. "Willow St.," Box 5B, MC388, BSBPA; Trux Fisher in Fisher and Corcoran interview; Quivik, "Smoke and Tailings," 497–505.

12. Dead zones in reference to mining comes from LeCain, *Mass Destruction.*

13. N. A. Jones to Merle Emmert, "Test House for Southeast Berkeley Pit Operations," Oct. 25, 1979, Fldr. "Test House SE Cont. Pit," Box 2A, MC388, BSBPA.

14. First quote is M. W. Emmert to J. J. Picarelli, "Damage to Siding of House at 2303 Yew St.," Fldr. "Jaksha, John W.," Box "Cabinet 1, Drawer 2 (Box 1)." Second is John A. Frankovich to R. D. Piper, re: 1259 E. Mercury St., July 22 1964, Fldr. "Mercury St.," Box 3C, MC388, BSBPA. Other examples include Nelson A. Jones to J. J. Picarelli, "Blasting Complaint at 26 North Drive, Columbia Gardens," Jan. 29, 1974, Fldr. "Caress, Mrs. H. C.," Box "Cabinet 1, Drawer 3 (Box 1)," MC388, BSBPA; Mar. 8, 1960 handwritten note, Fldr. "Indiana," Box 3B, MC388, BSBPA; J. J. Picarelli to N. A. Jones, Sept. 19, 1974, Fldr. "Aberdeen/Agate," Box 3A, MC388, BSBPA.

15. For examples of this complaint type, see R. D. Piper memorandum, Sept. 22, 1970, Fldr. "Notes, Maps, Clippings," Box 1; R. D. Piper to P. J. Berget, Apr. 12, 1956, Fldr. "Galena," Box 3B; handwritten note about Mrs. Mansanti, Fldr. "Daly Street," Box 3B; all in MC388, BSBPA.

16. Joseph J. Picarelli to Merle Emmert, "Blasting Complaints," Mar. 28, 1973, Fldr. "Rooney, Mrs. Delores," Box 8, MC388, BSBPA.

17. Sig. J. Dahlquist to Largey P. MacDonald, Dec. 31, 1969, Fldr. "Finlen Hotel," Box "Cabinet 1, Drawer 3 (Box 2)," MC388, BSBPA.

18. Cecilia S. Hakala and William H. Hakala to the Anaconda Company, May 5, 1972, Fldr. "2nd Street," Box 5B, MC388, BSBPA.

19. In Fldr. "Ledling, Mrs. J," Box 4C, MC388, BSBPA: John F. Harvey, Asst. Chief Mining Engineer, to John A. Frankovich, Legal, Dec. 17, 1963; Mrs. John Ledling to

E. D. Renouard, Apr. 14, 1965; Mrs. John Ledling to Anaconda Company Land and Tax Department, Sept. 19, 1972.

20. Leeper, "Changing Urban Landscape," 86–87.

21. Nicholls interview; Holter interview, by Hayward; Ralph interview; Wayrynen interview; Kearney, *Butte Voices*, 211; McCormick, "East Park Survivors"; McCormick, "Old Drug Store"; McCormick, "Finn Stories."

22. Van Horssen, *A Town Called Asbestos*, 127.

23. N. A. Jones to J. J. Picarelli, "Blasting Complaints from East Butte Area," Feb. 12, 1973, Fldr. "Plum Street," Box 3C, MC388, BSBPA.

24. On blockbusting, see Dmitri Mehlhorn, "A Requiem for Blockbusting: Law, Economics, and Race-Based Real Estate Speculation," *Fordham Law Review* 67 (Dec. 1998): 1145–61; Beryl Satter, *Family Properties: How the Struggle Over Race and Real Estate Transformed Chicago and Urban America* (New York: Henry Holt and Co., 2009); Amanda I. Seligman, *Block by Block: Neighborhoods and Public Policy on Chicago's West Side* (Chicago: University of Chicago Press, 2005); Gerald Gamm, *Urban Exodus: Why the Jews Left Boston and the Catholics Stayed* (Cambridge, MA: Harvard University Press, 1999).

25. Cecilia S. Hakala and William H. Hakala to the Anaconda Company, May 5, 1972, Fldr. "2nd Street," Box 5B, MC388, BSBPA. On the Federal Housing Authority, see example in John F. Harvey to P. J. Berget, "Initial Inspection of a New Brick and Wood Home," Nov. 6, 1963, Fldr. "Fagan, W.," Box "Cabinet 1, Drawer 3 (Box 2)," MC388, BSBPA.

26. Leeper, "Changing Urban Landscape," 87.

27. "Gardens Area Residents Meet," *Mont. Std.*, Oct. 9, 1973. See also Vidrich interview; Petroni and Petroni interview.

28. R. D. Piper, "Report on Vibrations Caused by Blasting in the Vicinity of the Berkeley Pit," Anaconda Company, June 3, 1960, unmarked Fldr., Box 3B, MC388, BSBPA, 1–2.

29. C. J. Hansen, "Potential Liability of Syndicate Pit Operations," Dec. 18, 1963, Fldr. "Syndicate Pit Problems," Box 2C, MC388, BSBPA.

30. Ibid.

31. R. D. Piper, "Report on Vibrations," 5. On flyrock problems, see examples in R. D. Piper, "Damage to Property by Blasting," May 29, 1968, Fldr. "Seismograph Correspondence General," Box 3A, MC388, BSBPA; Jones to Picarelli, "Blasting Complaints from East Butte Area"; "Berkeley Blast Throws Rocks," *Mont. Std.*, Aug. 18, 1970; "Flying Rock Jolts Area Close to Pit," *Mont. Std.*, Feb. 3, 1973.

32. Urban L. Roth to Dave Hardy, Jan. 6, 1972, in Fldr. "Seismograph General Correspondence," Box 3A, MC388, BSBPA.

33. James A. Robischon to M. K. Hannifan, "Blasting Activity—Berkeley Pit," Jan. 13, 1972, in Fldr. "Seismograph General Correspondence," Box 3A, MC388, BSBPA.

34. Piper, "Damage to Property by Blasting."

35. R. D. Piper, "Report on Vibrations," 1–2.

36. Lewis L. Oriard, "Drilling, Blasting, and Dredging Techniques in the Panama Canal," *World Dredging and Marine Construction* (June 1980): 24–25; Lewis L. Oriard

resume, Fldr. "Seismograph Correspondence-General," Box 3A, Fldr. "Blast Patterns—Oriard," Box 2A, MC388, BSBPA; Lewis L. Oriard to Dave Walker, Dec. 26, 1977, Fldr. "Oriard Reports, Statements," Box 2A, MC388, BSBPA; Gardiner interview.

37. W. R. Gibson to K. F. Beyer, Nov. 7, 1968, Fldr. "Seismograph General Correspondence," Box 3A, MC388, BSBPA; M. K. Hannifan to Woodward-Clyde & Associates, Attn: Lewis L. Oriard, July 31, 1968, Fldr. "Seismograph Correspondence—Oriard," Box 3A, MC388, BSBPA; R. D. Piper to M. K. Hannifan and D. Dasinger, "Inspection of Blasting Complaint at 310 E. Woolman," Jan. 19, 1971, Fldr. "McGonigle, Daniel C.," Box 2C, MC388, BSBPA.

38. John G. Hall to M. K. Hannifan, "Blasting Complaints—Berkeley Pit," June 26, 1968, Fldr. "Seismograph General Correspondence," Box 3A, MC388, BSBPA.

39. Lewis L. Oriard to J. A. Robischon, R. D. Piper, and P. J. G. duToit, Sept. 9, 1971, Fldr. "Seismograph Correspondence," Box 3A, MC388, BSBPA. See also Lewis L. Oriard to Nelson Jones, June 18, 1973, Fldr. "Seismograph Correspondence—Oriard," Box 3A, MC388, BSBPA.

40. Piper, "Damage to Property by Blasting."

41. David E. Siskind, Mark S. Stagg, John W. Kopp, and Charles H. Dowding, "Structure Response and Damage Produced by Ground Vibration from Surface Mine Blasting" (Bureau of Mines, Report of Investigations 8057, 1980, 1–59); Lewis L. Oriard, "Report of Investigations: Seismic Effects of Blasting Operations Berkeley Pit, Butte, Montana," Prepared for the Anaconda Company, Sept. 13, 1968, Fldr. "Seismic Study: Berkeley Pit," 1, Box 2C, MC388, BSBPA; Lewis L. Oriard, "Blasting Operations in the Urban Environment," Presentation at the Association of Engineering Geologists, 1970 Annual Meeting, Washington, DC, Oct. 20–24, 1970, 4, unmarked Fldr., Box 3B, MC388, BSBPA; U.S. Department of the Interior, Bureau of Mines to Nelson A. Jones, Anaconda Company, 1980, Fldr. "Test House SE Cont. Pit," Box 2A, MC388, BSBPA.

42. N. A. Jones to Stephen Foster, "Alleged Blast Damage, 449 E. Park," May 2, 1979, Fldr. "Park St.," Box 5B, MC388, BSBPA; Oriard, "Report of Investigations," 6.

43. Piper, "Damage to Property by Blasting"; Lewis L. Oriard to Nelson Jones, Sept. 5, 1972, Fldr. "Granite St.," Box 3B, MC388, BSBPA; R. D. Piper to P. J. Berget, "Inspection of Property at 510 E. Park," Mar. 17, 1965, Fldr. "Park Street," Box 5B, MC388, BSBPA.

44. Piper, "Damage to Property by Blasting."

45. S. W. Hurlbut, Anaconda Research, to Frank S. Wesson, U.S. Air Force Claims, Jan. 27, 1966, in unlabeled Fldr., Box 3B, MC388, BSBPA.

46. On Butte's blasting procedures, see Gardiner interview; Shepich, "Health and Safety Inspection," 2; McWilliams, "Mining Methods and Costs," 24–26.

47. H. J. Poel, DuPont, "New Developments in Initiation of Blasts in Mining," Oct. 13, 1960, unlabeled Fldr., Box 3B, MC388, BSBPA, 2–3; Lewis L. Oriard to R. D. Piper, Feb. 18, 1971, Fldr. "Seismograph General Correspondence," Box 3A, MC388, BSBPA; Lewis L. Oriard to Nelson Jones, Dec. 8, 1972, Fldr. "Seismograph Correspondence—Oriard," Box 3A, MC388, BSBPA; Lewis L. Oriard to Nelson Jones and Dave Walker, Aug. 30, 1978, Fldr. "Oriard Reports, Statements," Box 2A, MC388, BSBPA.

48. N. A. Jones to R. D. Piper, "Berkeley Pit Blasting," Dec. 15, 1971, Fldr. "Seismograph General Correspondence," Box 3A, MC388, BSBPA.

49. "Pit Blasting Safe," *Mont. Std.*, Jan. 23, 1972.

50. Example in F. Lee Williams to Roy Ray, Apr. 18, 1974, Fldr. "Walnut Street," Box 5B, MC388, BSBPA.

51. M. K. Hannifan to A. J. McDonell, "Berkeley Pit Blasting Nuisance Suit," June 25, 1970, Fldr. "Seismograph Correspondence General," Box 3A, MC388, BSBPA. See also Gardiner interview.

52. Hannifan to McDonell, "Berkeley Pit Blasting Nuisance Suit."

53. "In Anticipation of Damage Claims Associated with More Moves Away from Butte Hill," n.d., Fldr. "Notes, Maps, Clippings," Box 1, MC388, BSBPA. See also Dasinger to Cyr et al., "Subsidence Claims and Procedures"; James A. Robischon to W. R. Gibson, "Subsidence-Blasting-Property Damage-Claims," Jan. 26, 1972, Fldr. "Seismograph Correspondence General," Box 3A, MC388, BSBPA; William P. Mufich to P. W. Murray, "Subsidence, Blasting, Property Damage Claims," Feb. 4, 1972, Fldr. "Notes, Maps, Clippings," Box 1, MC388, BSBPA.

54. W. J. Fogarty to P. J. Berget, Feb. 15, 1965, Fldr. "Fir Street," Box 3A, MC388, BSBPA.

55. Note re: Lydia Curran, May 2, 1956, Fldr. "Property Damage: Miscellaneous Claims," Box "Cabinet 1, Drawer 4, (Box 1)," MC388, BSBPA.

56. Nelson A. Jones to Doug Dasinger, "Blasting Complaint at 54 Anaconda Road," July 7, 1969, Fldr. "Anaconda Road," Box 3A, MC388, BSBPA; Con J. Kelly and P. J. Berget, "Surface Blasting," June 10, 1952, Fldr. "Casagrande, Mrs. Evelyn," Box "Cabinet 1, Drawer 3 (Box 1)," MC388, BSBPA.

57. See examples in R. D. Piper to P. J. Berget, "Examination of Alleged Damage by Blasting to the Dwelling at 835 East Park Street," Dec. 26, 1962, Fldr. "Park Street," Box 5B, MC388, BSBPA; John F. Harvey to P. J. Berget, "Initial Inspection of Luther Apostilic Church," June 15, 1964, Fldr. "Park Street," Box 5B, MC388, BSBPA; Jones to Picarelli, "Blasting Complaint at 26 North Drive, Columbia Gardens."

58. Oriard to Jones, Sept. 5, 1972.

59. Frank H. Fogliani to Carla M. Gray, May 4, 1981, Fldr. "Elm Street," Box 3B, MC388, BSBPA.

60. Hall to Hannifan, "Blasting Complaints—Berkeley Pit," June 26, 1968.

61. Lewis L. Oriard to James Robichon, June 19, 1970, Fldr. "Seismograph Correspondence General," Box 3A, MC388, BSBPA.

62. Examples in Anna E. Jursnich to P. W. Murray, Feb. 14, 1972, Fldr. "Jursnich, Anna E.," Box "Cabinet 1, Drawer 2 (Box 2)," MC388, BSBPA; Joe McCarthy to Ed Bartlett, "Building Inspection—445 E. Park Street," Oct. 9, 1981, Fldr. "Park Street," Box 5B, MC388, BSBPA.

63. Examples in R. D. Piper to P. J. Berget, "Inspection of a Brick Veneer 8-Room Apartment Building at 24 North Atlantic," Nov. 29, 1962, Fldr. "Atlantic Street," Box 3A, MC388, BSBPA; Bernard P. Harrington to W. R. Gibson, Oct. 28, 1969, Fldr. "Diversified Realty, Inc.," Box 4B, MC388, BSBPA.

64. For examples, see Fldr. "Caprara, Frank," Box "Cabinet 1, Drawer 3 (Box 1)," MC388, BSBPA: Bernard P. Harrington to Frank Laird, Apr. 21, 1970; Joseph J. Picarelli to Merle Emmert, "Blasting Complaint," Feb. 21, 1973; John W. Warren to John A. Frankovich, Sept. 23, 1968. Among others, see John F. Harvey to John A. Frankovich, "Initial Inspection of Property Owned by Mrs. Piskolich," July 8, 1964, Fldr. "Park St.," Box 5B,

MC388, BSBPA; A. C. Ducich to Bernard P. Harrington, "Property at 101–5 East Broadway," Apr. 17, 1974, in Fldr. "East Broadway St.," Box 3A, MC388, BSBPA; Chuck Krause, memo on "Mrs. Zinnie," Sept. 7, 1977, Fldr. "Wyoming Street," Box 3C, MC388, BSBPA.

65. Mrs. Costello in J. F. Harvey to John A. Frankovich, "Initial Inspection of Property Located at 418 East Broadway," Feb. 8, 1965, Fldr. "East Broadway St.," Box 3A, MC388, BSBPA; Nelson A. Jones to John A. Frankovich, "Second Inspection of Property Located at 418 East Broadway," Aug. 18, 1965, Fldr. "East Broadway St.," MC388, BSBPA.

66. Rush Muse to E. O. Bonner, May 28, 1968, Fldr. "Seismograph General Correspondence," Box 3A, MC388, BSBPA. On the New York Trap Rock Company's blasting, see J. R. Kringel, "Control of Air Blast Effect Resulting from Blasting Operations," *Mining Congress Journal* 46, no. 4 (Apr. 1960): 51–56. Even explosives companies saw the need to engage in public relations campaigns. See DuPont Company, "Let's Reduce the Noise from Blasting," n.d., in blank Fldr., Box 3B, MC388, BSBPA.

67. First quote is in Rush Muse to E. O. Bonner, May 28, 1968, Fldr. "Seismograph General Correspondence," Box 3A, MC388, BSBPA. Second is Hall to Hannifan, "Blasting Complaints—Berkeley Pit," June 26, 1968.

68. Joseph J. Picarelli to Merle Emmert, "Blasting Complaints," Mar. 28, 1973, Fldr. "Rooney, Mrs. Delores," Box 8, MC388, BSBPA.

69. J. J. Picarelli to Nelson Jones, "Blasting Complaint," July 25, 1972, Fldr. "Lee, Bob," Box 4C, MC388, BSBPA.

70. Joseph J. Picarelli to William P. Mufich, "Complaints from Columbia Gardens Residents—Continental-East Operations," Aug. 23, 1973, Fldr. "Jenkins, Kathy—Columbia Gardens," Box "Cabinet 1, Drawer 2 (Box 1)," MC388, BSBPA.

71. Oriard, "Blasting Operations in the Urban Environment," 24.

72. Robert M. Cox, "Rock Blasting in the Urban Environment—Legal and Technical Problems," *Proceedings of the 18th U.S. Symposium on Rock Mechanics*, June 22–24, 1977, Colorado School of Mines, Golden, Colorado.

73. On the senses and power, see Sarah Keyes, "'Like a Roaring Lion': The Overland Trail as a Sonic Conquest," *Journal of American History* 96, no. 1 (June 2009): 19–43; Andrew J. Rotter, "Empire of the Senses: How Seeing, Hearing, Smelling, Tasting, and Touching Shaped Imperial Encounters," *Diplomatic History* 35, no. 1 (Jan. 2011): 3–19.

74. David E. Siskind, Virgil J. Stachura, Mark S. Stagg, and John W. Kopp, "Surface Response and Damage Produced by Airblast from Surface Mining" (Bureau of Mines, Report of Investigations 8485, 1980), 1–9, 66–67; Piper, "Damage to Property by Blasting"; N. A. Jones to John A. Frankovich, "Blasting Complaint at 449 East Park," Sept. 13, 1968, Fldr. "Park St.," Box 5B, MC388, BSBPA.

75. M. W. Emmert to J. J. Picarelli, "Alleged Blast Damage, 'Club 156,'" Mar. 1, 1973, Fldr. "Gergurich, Mrs. Mary," Box "Cabinet 1, Drawer 3 (Box 2)," MC388, BSBPA.

76. Merle W. Emmert to J. J. Picarelli, "Alleged Blast Damage, Uptown Butte and E. Park Street," Aug. 23, 1972, Box 3B, Fldr. "Granite St.," MC388, BSBPA.

77. David D. Walker to P. J. Berget, "Examination of Property at 510 E. Park," Fldr. "Park Street," Box 5B, MC388, BSBPA.

78. First quote is Oriard to Piper, Feb. 18, 1971, Fldr. "Seismograph General Correspondence." Second is Oriard, "Report of Investigations," 14.

79. See descriptions of this phenomenon in Oriard, "Report of Investigations," 10; Siskind et al., "Structure Response," 1.

80. Joseph J. Picarelli to Jean Finley, Feb. 28, 1973, Fldr. "Holy Savior School," Box "Cabinet 1, Drawer 2 (Box 1)," MC388, BSBPA.

81. R. D. Piper, "Report on Vibrations," 2–3.

82. Oriard, "Blasting Operations in the Urban Environment," 15, 20.

83. Examples include "Muffling the Clamor of Urban Construction," *Businessweek*, Dec. 14, 1968; James Stewart-Gordon, "We're Poisoning Ourselves with Noise," *Reader's Digest*, Feb. 1970, Box 3B, MC388, BSBPA.

84. First quote is Oriard, "Blasting Operations in the Urban Environment." Second is Lewis L. Oriard to publicity chairman, Northwest Mining Association, Oct. 29, 1970, Fldr. "Seismograph Correspondence—Oriard," Box 3A, MC388, BSBPA.

85. Piper, "Report on Vibrations," 4; Oriard, "Report of Investigations," 11.

86. Oriard, "Blasting Operations in the Urban Environment," 18.

87. Siskind et al., "Structure Response," 1, 67.

88. Siskind et al., "Surface Response," 64.

89. Oriard to publicity chairman, Oct. 29, 1970. Later research concurred. See Siskind et al., "Surface Response," 64.

90. Oriard, "Blasting Operations in the Urban Environment," 12–14.

91. First quote is untitled handwritten list, n.d., Fldr. "Copper Street," Box 3A, MC388, BSBPA. Second is in J. F. Harvey to P. J. Berget, "Inspection of 439, 441 East Broadway," May 16, 1961, Fldr. "East Broadway," MC388, BSBPA. See also Jones to Frankovich, "Second Inspection of Property Located at 418 East Broadway," Fldr. "600 W. Gold St.," Box 3A, MC388, BSBPA; Robert B. Kroeze to Pat Monninger, June 8, 1972, Fldr. "600 W. Gold St.," Box 3A, MC388, BSBPA.

92. Oriard, "Report of Investigations," 10.

93. Oriard, "Blasting Operations in the Urban Environment," 17–23.

94. R. D. Piper to Martin K. Hannifan, "Berkeley Pit Blasting," Mar. 30, 1960, Fldr. "Seismograph Correspondence General," Box 3A, MC388, BSBPA. On primacord, see Twitty, *Blown to Bits in the Mine*, 157.

95. W. C. McLaughlin, "Test of Low Noise Level Trunk Line Delay Blasting System—DuPont," Anaconda Company Mining Research Department, May 31, 1960, unlabeled Fldr., Box 3B, MC388, BSBPA.

96. Oriard, "Blasting Operations in the Urban Environment," 20, 23.

97. Mrs. Roland Navarro quoted in note to Dave Piper, Feb. 24, 1959, Fldr. "North Walkerville," Box 5B, MC388, BSBPA.

98. Lewis L. Oriard to The Anaconda Company, Feb. 23, 1981, Fldr. "Oriard Reports, Statements," Box 2A, MC388, BSBPA. On testing systems, see N. A. Jones to P. Balentine, "Pit Blasting," Oct. 28, 1969, Fldr. "Seismograph General Correspondence," Box 3A, MC388, BSBPA; Oriard to Walker, Dec. 26, 1977.

99. On stemming, see Twitty, *Blown to Bits in the Mine*, 158. On Butte's improvements in stemming, see Oriard to Jones, Sept. 5, 1972.

100. Lewis L. Oriard to Dave Walker and Nelson Jones, June 3, 1978, Fldr. "Oriard Reports, Statements," Box 2A, MC388, BSBPA.

101. R. D. Piper to F. J. Balentine, Emmet Murphy, Frank Gardiner, and Gordon Lyda,

"Memorandum: Berkeley Pit Blasting," Fldr. "Seismograph Correspondence," Box 3A, MC388, BSBPA; Rush Muse to E. O. Bonner, May 28, 1968, Fldr. "Seismograph General Correspondence," Box 3A, MC388, BSBPA; Lewis L. Oriard to R. D. Piper, Nov. 27, 1971, Fldr. "Seismograph Correspondence—Oriard," Box 3A, MC388, BSBPA; Oriard to Jones, Sept. 5, 1972.

102. Lewis L. Oriard to The Anaconda Company, July 29, 1980, Fldr. "Oriard Reports, Statements," Box 2A, MC388, BSBPA. See also Oriard, "Blasting Operations in the Urban Environment"; Oriard, "Report of Investigations," 7.

103. S. W. Hurlbut, Anaconda Research, to Frank S. Wesson, USAF Claims, Jan. 27, 1966, unlabeled Fldr., Box 3B, MC388, BSBPA.

104. Oriard, "Report of Investigations," 7, 13.

105. Piper to Hannifan and Dasinger, "Inspection of Blasting Complaint at 310 E. Woolman," Jan. 19, 1971. See also R. D. Piper to J. A. Rogischon, "Berkeley Pit Blasting," July 17, 1970, Fldr. "Seismograph—General Correspondence," Box 3A, MC388, BSBPA; Oriard, "Blasting Operations in the Urban Environment," 20, 23.

106. David E. Siskind, Steven V. Crum, and Matthew N. Pils, "Blast Vibrations and Other Potential Causes of Damage in Homes Near a Large Surface Coal Mine in Indiana" (Bureau of Mines, Report of Investigations 9455, 1993), 8.

107. For examples of inversions causing damage, see Oriard to Jones, Sept. 5, 1972; Emmert to Picarelli, "Alleged Blast Damage, 'Club 156.'"

108. Siskind, Crum, and Pils, "Blast Vibrations," 5.

109. Piper, "Report on Vibrations," 2; Oriard, "Report of Investigations," 14; Oriard to Anaconda Company, Feb. 23, 1981.

110. Kroeze to Monninger, June 8, 1972.

111. Rick Foote, "Harrison School Is Falling," *Mont. Std.*, Dec. 10, 1971.

112. R. D. Piper to J. A. Robischon, "Blasting Vibrations—Harrison School," Jan. 3, 1972, Fldr. "Harrison School-Fir Street," Box "Cabinet 1, Drawer 2, (Box 1)," MC388, BSBPA.

113. Oriard to Jones, Sept. 5, 1972. See also Oriard, "Blasting Operations in the Urban Environment," 19.

114. On nuclear blasting tests and model houses, see *Cue for Survival: Operation Cue, A.E.C. Nevada Test Site, May 3, 1955; a Report* (Washington : U.S. Government Printing Office, 1956).

115. Oriard to Anaconda Company, Feb. 23, 1981. On Anaconda's test houses, see N. A. Jones to Merle Emmert, "Test House for Southeast Berkeley Pit Operations," Oct. 25, 1979, Fldr. "Test House SE Cont. Pit," Box 2A, MC388, BSBPA; G. W. Thompson to Mr. and Mrs. Steve Matule, Jan. 8, 1980, Fldr. "Test House SE Cont. Pit," MC388, BSBPA; Nelson A. Jones to B. J. Holdsworth, "Visit to 'Test Houses,' Reimbursement," June 3, 1981, Fldr. "Test House SE Cont. Pit," MC388, BSBPA; Gardiner interview.

116. Franz-Josef Brüggemeir, "The Ruhr Basin 1850–1980: A Case Study of Large-Scale Environmental Pollution," in *Silent Countdown: Essays in European Environmental History*, ed. Christian Pfister and Peter Brimblecombe (Berlin: Springer-Verlag, 1990), 210–27. On spatial solutions to industrial noise pollution, see Karin Bijsterveld, *Mechanical Sound: Technology, Culture, and Public Problems of Noise in the Twentieth Century* (Boston: MIT, 2008), 53–90.

117. Quivik, "Smoke and Tailings," 11; LeCain, *Mass Destruction*, 64–107. See a similar approach in Huggard and Humble, *Santa Rita Del Cobre*, 69.

118. Bernard P. Harrington to William P. Mufich, "Old C.O.D. Laundry Building," Nov. 26, 1974, Fldr. "Stapleton, N. H.," Box 1B, MC388, BSBPA.

119. Dave Piper to P. J. Berget, "Inspection of Ryan Property," May 26, 1964, Fldr. "Anaconda Road," Box 3A, MC388, BSBPA.

120. See home-purchase files for the East Side in Boxes 1A and 1B of MC388, including Bernard P. Harrington to J. P. Hagerty, July 19, 1968, Fldr. "Alto, Sandra, East Side," Box 1A, MC388, BSBPA; Dennis M. Shea to W. R. Gibson, Apr. 17, 1968, Fldr. "Egan, Maurice & Catherine (Dave Holland)," MC388, BSBPA.

121. Van Horssen, *A Town Called Asbestos*, 124.

## CHAPTER 6

# ACQUISITION

BUTTE RESIDENTS REFLECTING on the expanding Berkeley Pit agreed on one topic: relocation hurts. Some, like Vincent Ciabattari, believed that protest against relocation was a necessary tactic, but that people, sadly, had abandoned it. In a 1974 interview, he related that his once close-knit group of neighbors had deserted him among the waste dumps. As he put it, "People said they'd fight, but then they jumped."[1] Others, like long-haul truck driver Bill Long, found the relocation process depressing but necessary. Just a few years after Ciabattari's interview, Long pulled up next to the site where the Anaconda Company dump man directed him to release a load of waste from the Berkeley Pit. As he looked in his rearview mirror, he saw rocks tumbling out of his truck and down onto Holy Savior School, a grade school for the Catholic residents of two Butte neighborhoods: Meaderville and McQueen. The residents were now gone—the company had relocated them. The parish buildings were all that remained. Covering up the still-standing school, he later wrote, "left an indelible mark in my memory. I felt like a murderer. Progress does have its price."[2]

In exploring the process of neighborhood relocation, this chapter will attempt to explain why people like Ciabattari and Long found pit expansion demoralizing but also increasingly inevitable. The decision to shift Butte's mining to open pit was driven by the need for efficiency, with increased safety a pleasant benefit; the pit's continued development initiated a process only partly foreseen by early planners, however. The relocation of homes, businesses, and, eventually, whole neighborhoods, soon became a necessity for the Anaconda Company and an accepted program by locals. Given residents' early protests against open-pit mining, it seems surprising at first that a large-scale relocation program could become normalized.

Bill Long's above comment about the price of progress suggests an initial answer to this conundrum. Access to good jobs for Butte's residents increasingly overcame locals' earlier resistance to open-pit mining. Worried about Butte's smaller role in the company's copper empire, Anaconda's workers as well as other residents came to believe that open-pit mining's productivity represented necessary technological progress. Journalists commented on residents' love-hate relationship with Anaconda; as much as they resented being pushed around by it, "the people of Butte feel they owe the company everything," as one paper put it.[3] Anaconda had created Butte and it still employed almost one-third of the city's workforce. Its workers had much pride in the industry—they had produced metals that had been significant in two world wars and in the Cold War.[4] As horrible as Bill Long felt about relocation, he still believed that pit expansion equaled a healthy economy.[5] One reporter described the situation rather poetically: "The Pit's neighbors generally appear resigned to live with it. The Pit contains Butte's life blood—copper. The Pit is life in Butte."[6] Another journalist was more blunt: "Rebellion has been dissipated by the economics of a one-industry town."[7]

Employment concerns and pride in the industry might be the most obvious answers, but they are not the only ones. In the postwar era, Anaconda and other mining companies across the American West were working to rid themselves of an image that they were colonial overlords. Anaconda's crude but effective methods had once made Butte the quintessential company town.[8] By the 1970s, however, the days when western companies like Anaconda used gunplay, union infiltration, politician buying, and press ownership to get its way were gone.[9]

Anaconda's approach to exerting its corporate influence was in flux between the 1950s and 1970s. The company initially turned back to an older tactic: welfare capitalism. After World War II it built more housing, a workers' club, a civic auditorium, and a hospital.[10] Yet the company also began to divest itself of weapons that had made itself such a political force. Formerly the owner of all but one of Montana's major dailies, Anaconda sold its papers to the Lee Corporation.[11] Even

Anaconda's harshest critic, historian K. Ross Toole, recognized that the company lost its unfettered power during this period.[12] Having disavowed its most blunt tactics, Anaconda then pulled back from overt welfare capitalism, withdrawing support from the hospital it built. The company instead tried what had emerged as a more modern approach to community relations. It hoped to be "more sensitive to people," in the words of one executive.[13]

To win popular support, Anaconda attempted to carefully and gradually break the link between home space and workspace. Its personnel sought to make pit expansion smooth and admissible by employing a sophisticated property acquisition system and using a modern, gentle form of public relations management. Anaconda's approach meant acquiring new legal powers, paying decently for property, and gently leaning on holdouts who refused to move. Some things the company did not control, like neighborhood deterioration, residents' social divisions, and even zoomorphism, in which people began to explain the mine as if it were an organism, also contributed to Butte residents' acceptance of open-pit mining, even if those same factors could make displaced people depressed and sometimes resentful.

## Open-Pit Planning

During pit expansion, Anaconda typically kept quiet about exactly why it wanted to buy in specific areas and when it would expand to them. The approach provided the company with benefits, including the fact that residents could not easily join together to drive up purchasing prices, but it was also likely a mistake because it generated community suspicion. Records from the company's Geological Department suggest that Anaconda was less devious than many assumed. The company wanted to keep its property acquisition costs down, so pit planners did pay attention to where people lived, sometimes cutting back on the amount of neighborhood property that they purchased, but planners also did not target specific neighborhoods because they lacked the power or status to fight back; geological data for the ore body and economical engineering decisions were the true shapers of expansion, even if surface features, including residential areas and company facilities, often factored into the perimeter's amoebic shape.

In the early days of the Berkeley Pit, geology drove development. Placement of the Berkeley-Rarus ore body determined why Meaderville property was the first threatened. In the area originally planned for the Berkeley Pit, geologists expected to find 150 million tons of ore at 0.75 to 0.90 percent copper.[14] Stripping 250 feet of cover rock that lay over the extensive low-grade copper took several years. Further geological exploration led engineers to direct development toward the Meaderville neighborhood. The perimeter expanded gradually to the northeast, up to the Butte–Helena highway and then against Meaderville by the late 1950s.[15]

Despite the priority of geological considerations, other factors gained increasing importance. Even though many residents claimed that they were too often an afterthought, discussions between the Geological Department and Mining Engineering Department usually involved surface or physical features during planning decisions.[16] As Anaconda executive Robert Corbett later explained, the company's biggest problem was that "the community of Butte is perched right on top—not really on the ore body but close enough."[17] Both existing company facilities and residential areas sat on every side of the new pit. During the Berkeley Pit's early years, Anaconda worried about expanding to the northeast because the Leonard and East Colusa shafts there were vital to the existing underground ventilation system. Private homes and businesses also hampered pit expansion to the east until the company could relocate Meaderville's inhabitants. To the northwest, the existence of additional underground mines and a dramatic rise in elevation meant the Berkeley Pit could only gradually expand only in that direction. Expanding southward would also require a lot of effort and money because the Great Northern Railroad and the Northern Pacific Railroad tracks were there. To the west sat the residential districts of Butte's East Side, particularly Finn Town. This area had long been heavily populated, because it adjoined the CBD. For the company, expansion in that direction would therefore be very expensive.[18] Planners thus quickly recognized that surface features had the ability to interrupt pit expansion in virtually every direction.

To solve this problem, pit planners crafted a new process. By the 1960s the considerations of economic geology continued to drive company plans, but now they were typically in concert with the rising surface features issue. Because engineers checked ore grades and volume daily, much more became known about the shape of the economically viable ore body as mining development progressed.[19] Geologists continually checked these data and revised projections. Anaconda's vice president and chief geologist, Vincent ("Vin") D. Perry, oversaw early pit expansion from his office in New York. His trusted lieutenants, Edward Shea, the chief geologist in Butte, and fellow geologist Murl Gidel, provided geological, day-to-day guidance, while Chester Steele, the vice president in charge of western operations, gave the project its overall direction. To provide Berkeley Pit expansion with further geological input, Perry asked geologist Wallace O'Brien to explore possibilities for low-grade ore that were not initially part of company plans. Under a long-standing Geological Department procedure, the Ton-for-Ton Program made it "mandatory that each ton of ore mined was replaced with a ton of new ore reserve."[20]

O'Brien submitted recommendations in 1958 to drive the Berkeley Pit much deeper. O'Brien thought Anaconda was missing out on enormous quantities of ore far below the current pit.[21] Unfortunately for Anaconda, the process of deepening an open pit required workers to broaden it out as well. Haul trucks could access deeper ore only by spiraling downward in a circular manner. During each pit limit pushback, the benches on which the trucks drove up and down had to be removed from the top down, then replaced with new benches farther out from the pit's center.[22] To go downward therefore always meant to expand outward as well.

In examining the outer perimeter, O'Brien knew that ore continued eastward, which was the current direction of the pit's axis, but he also saw lots of ore left in higher-grade areas to the west. That area had housed a number of old underground mines, whose ore remnants whetted executives' appetites, because they would greatly improve the project's ore-to-waste ratio. Hence, O'Brien's plans also suggested moving the pit's boundaries westward.[23]

By 1961 O'Brien had formulated five different plans to expand the larger Berkeley enterprise, each of which took into account the shape of the ore body, the ore-to-waste ratio, and property-acquisition costs. Plan 2 gained some original attention, but it had

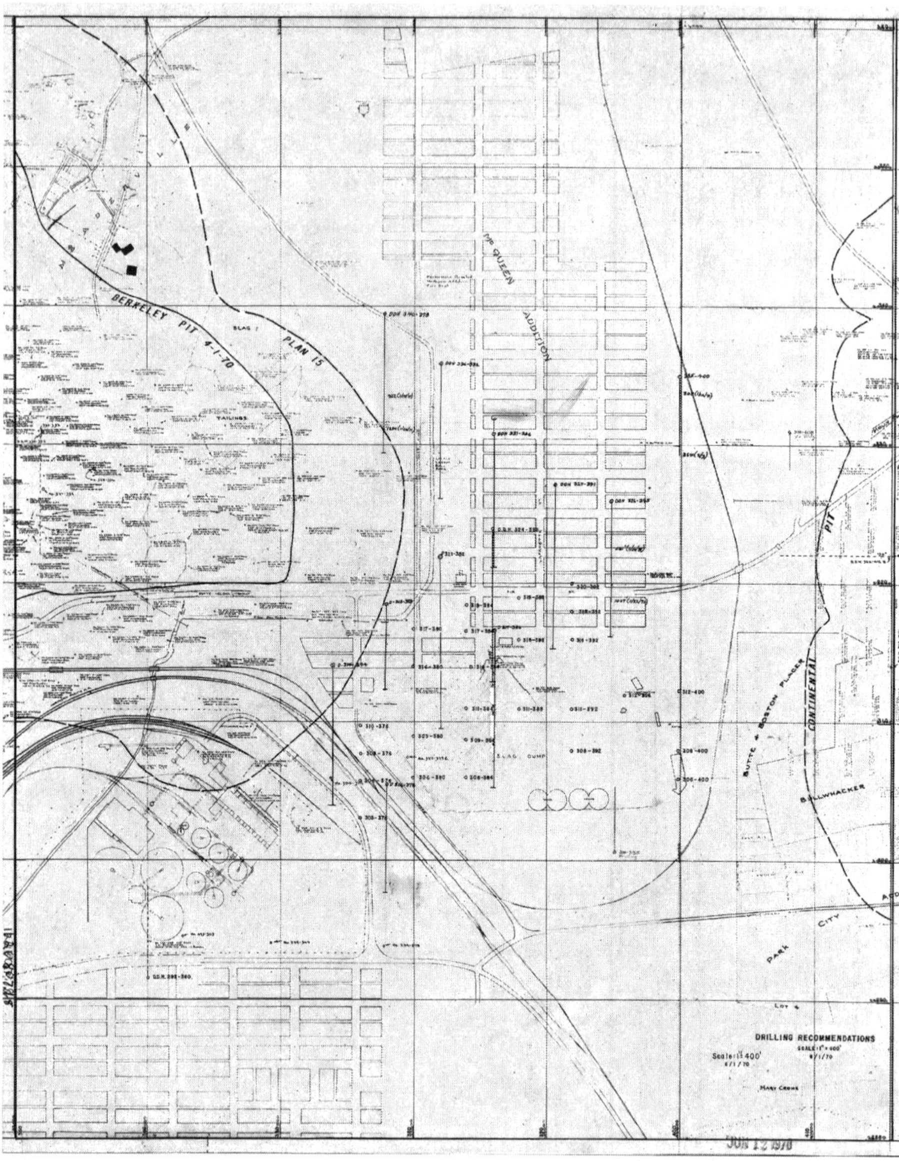


ANACONDA OPEN-PIT PLANNING MAP, 1970. This Anaconda Company map was made to suggest recommendations for holes to be drilled during geological exploration. It also shows mine expansion plans as of 1970. The Berkeley Pit, on the left of the map, was planned to expand farther to the east. This plan is labeled "Plan 15." Outlines for a new Continental Pit—a pit that would later border the Columbia Gardens—are on the right of the map. Photo courtesy of American Heritage Center, University of Wyoming, Laramie, Wyoming, Anaconda Geological Documents Collection, File 120807.18.

an unacceptable ore-to-waste ratio and it extended well into the East Side neighborhood, which would mean a significant financial outlay for Anaconda. Plan 5 interfered less with the residential area to the west, although it still would necessitate a fair amount of property purchase. Officials also liked Plan 1: this plan intruded less than others into the neighborhood west of the pit, and instead moved the pit's boundaries farther to the east, encompassing the rest of Meaderville and touching up against McQueen. Plan 1 meant property acquisition costs, but there were fewer large purchases than going far to the west. In addition, the projected ore-to-waste ratio looked quite positive.[24] By 1962 Plan 1, with greater extensions to the east, and Plan 5, with some extensions to the west, combined to become Plan 15.[25]

Following Plan 15 during the 1960s and early 1970s meant that the Berkeley Pit expanded primarily to the east. It enlarged northeast, past a soon-obliterated Meaderville and toward the neighborhood of McQueen, as well as southeast, toward the neighborhood of East Butte. Anaconda therefore began to relocate residents from both areas as quickly as possible. This eastward development, which involved financial costs to the company and social costs to residents, was based primarily on ore calculations.[26]

One reason the perimeter never significantly expanded to the south had little to do with geology. In the early 1960s officials sited the newly built, costly, Weed Concentrator on the southern edge of the pit, much to the consternation of some Anaconda geologists. Vin Perry later bemoaned the fact that some good ore lay under the concentrator, never to be had, simply because, in his view, executives had listened too much to engineers and not enough to geologists.[27]

Therefore, both geological reasons and reasons related to already-extensive surface features meant that the McQueen/East Butte area, sitting east of the pit's center, joined the East Side, which sat west of the pit's center, as Anaconda's two property acquisition targets during the late 1960s and the 1970s. Acquiring property south of the pit was out due to a lack of viable ore and the existence of a pricey concentrator facility, while the Land Department also did not need to acquire property north of the pit's center, because expansion in that direction ran into ground the company already owned.[28]

Land needs beyond the Berkeley Pit itself also drove property acquisition. Even though many people today erroneously assume that Butte's vanished neighborhoods disappeared straight into the Pit, most of the land taken by Anaconda during expansion did not actually become the Berkeley Pit. Unlike in Meaderville's case, the borders of the Berkeley Pit never reached the McQueen/East Butte area, nor did they touch much of the East Side. At the same time, though, the company still claimed large amounts of land in both of these areas.

The reasons for these purchases were simple. As historian Timothy LeCain has made clear, an open-pit system generates an enormous hunger for land.[29] Some land becomes subtracted, creating what geographer Richard Francaviglia describes as the "deep, scooped out depression" of an open-pit mine. Francaviglia notes that wherever subtraction exists, however, so does addition.[30] Not too much of what came out of the ground contained ore. Where to put everything else thus became an important concern.

Anaconda's operation therefore required a lot of additional space. To separate ore from waste, Anaconda needed the nearby concentrator and a smelter that was farther away. An open-pit copper operation also requires a tailings pond into which slimy wastes, now ridden of their metals, are emptied from the concentrator in order to evaporate. Maps show that Butte's massive Yankee Doodle Tailings pond, which still sits to the north of the pit, actually takes up more space than the Berkeley Pit mine itself. Besides sites for all of the processing work, Anaconda needed locations for mined material that did not carry much ore and therefore was not worth putting through the concentrator. Butte's open-pit operation thus demanded a place to deposit waste rock that had no obvious value, and a large area for leach dumps, where Anaconda could pump acidic water through slightly more valuable materials than waste rock. The water precipitated out in metallic form, providing a small amount of metal. Company engineers worked to keep all these areas as close as possible to one another to minimize fuel costs as trucks transported rocks from site to site.[31]

The entire area around the mine therefore transformed into an industrial factory. As the pit grew, additional space was added to this industrial system. Space that Anaconda acquired in Meaderville became

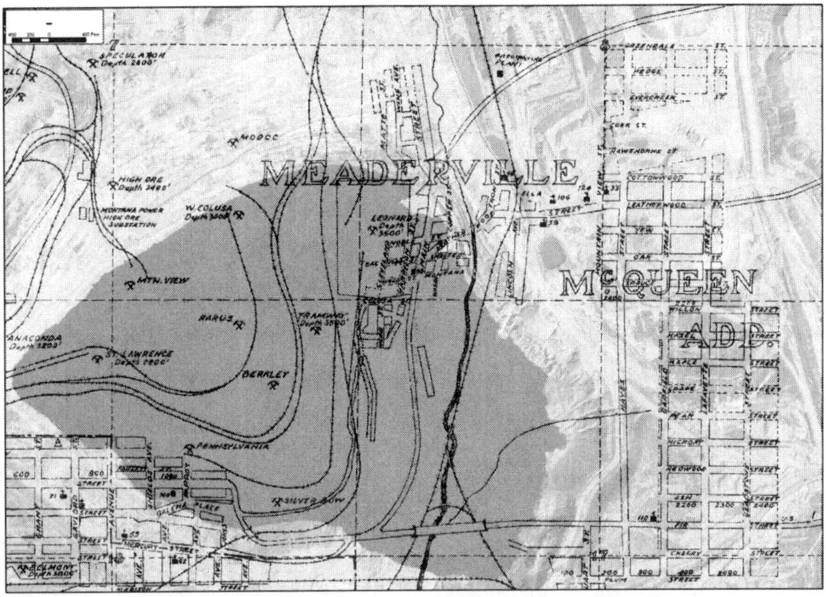


Berkeley Pit Map Overlay, 1951 and 2010. This map is a close-up of a 1951 Butte street map overlaying today's Butte landscape, with the Berkeley Pit prominently featured. New map made by the Butte–Silver Bow GIS Department for the author, 2010. The 1951 map courtesy of Butte–Silver Bow Public Archives, "Detailed Map of Butte: Heart of Montana's Magicland," Butte Chamber of Commerce, 1951, Butte Street Maps, Vertical File 1361.

part of the mine, while additional space in Meaderville, McQueen, and East Butte became necessary for dumps, as Anaconda sought the cheapest and most efficient disposal of waste. The rest of McQueen, East Butte, the East Side, and what was left of Dublin Gulch, on the other hand, became buffer zones between industry and neighborhood life.[32] By the late 1960s, Meaderville had completely vanished. By the late 1970s and early 1980s, both McQueen and East Butte had joined it. Meanwhile, by the early 1970s many people had abandoned the far reaches of the East Side.

Because most of this land acquisition clearly followed geological data for the ore body and the logic of an open-pit operation, the Berkeley Pit's story does not neatly fit into traditional tales of environmental justice. The environmental justice community has shown that marginalized communities became the sites for a disproportionately large share of the world's environmental hazards, but, as Arn Keeling and

John Sandlos point out, "Mines are sited where viable ore deposits are found, not necessarily where pliant or disenfranchised communities are located."[33] Certainly, it was often the poorest families in the Butte area who suffered from open-pit expansion, but the reasons for their distress had less to do with direct targeting of toxic wastes and more to do with the history of larger structural imbalances, as environmental justice advocates are now recognizing. Ethnic ties, cheap housing, and proximity to workplaces had determined where working-class families lived. Yet when much of this underground mining area became an aboveground mining area, living above rich veins of ore became an increasing liability.

## Bargaining for Property

The Berkeley Pit thus required increasing amounts of property. Residents of Meaderville, where the Berkeley Pit began, watched with trepidation. The owner of a Meaderville gas station quickly understood that "as the hole got deeper, they needed more width and breadth for the side, they needed more room on the top that did not have structures on it."[34] Residents who were farther away, however, did not foresee that the pit would eventually come to border the East Side, a mile away. As Art Kesikitalo remembered, "We never thought about it. I guess hardly anybody knew the pit was going to be as big as it is."[35]

When Anaconda's land agents called, residents suddenly found themselves compromised by decisions their parents and grandparents had made decades earlier. In those times, building homes close to the underground mines where they worked made sense, especially for poor families that could not afford daily transportation. The cheap cost of housing next to running mines and smoke-spewing smelters also played a role in moving decisions, as did living in the right parish for many Catholic migrants.

Descendants of the later-arriving southern and eastern Europeans, who settled in unincorporated parts of the county like Meaderville, McQueen, and East Butte, were especially disadvantaged. A simple comparison of Walkerville, which staged a relatively successful protest, with the neighborhoods that later faced pit expansion, proves instructive. Walkerville residents were descendants of early Irish and

Cornish arrivals, and their long-standing community had become an incorporated city with various legal protections.[36] The Walkerville City Council was able, for example, to add to the bargaining power of its citizens by turning debates over expansion into discussions about public roads.

Compared with residents of Walkerville, those in McQueen, East Butte, and Meaderville had less leverage. Many did not own the land on which their houses stood and instead rented it from the company, often for a nominal fee of $1 per month. When mining operations began to encroach on Meaderville, Anaconda needed to acquire only the house, without spending any money on land.[37]

Anaconda had three options when it acquired a home. The first was to trade the home for a house already owned by its housing affiliate, Kenwood Realty, in another part of town. The second was to buy the home and demolish it, while the resident used the settlement money either to build a new home or to purchase an existing one. These two options were the most common choices early in the relocation process, especially in Meaderville. The final option meant the company bought the home and then sold it back to the owner for $1. That choice saw more frequent use in McQueen and East Butte, where residents typically owned larger, newer homes that they often sought to keep and then move to a new location.[38]

Neither owning land nor valuable houses, Meaderville's residents often had no choice except to trade for a home owned by the company. Many were dissatisfied with the outcome and expressed more resentment than people from neighborhoods that later faced expansion. Residents of the East Side, McQueen, and East Butte often at least owned the surface, so they received a better appraisal from real estate agents and hence a better final deal—usually between $14,000 and $20,000. McQueen resident Tom Holter said that, in his area, "You made out well. They didn't cheat you out of anything. You got a nice home." In Meaderville, however, his grandfather and uncle suffered from the land ownership issue. They received very little for their store, which was built on Anaconda ground.[39]

Years later, Butte-area residents still disagree over how well Anaconda handled prices during the relocation process. Bob Kovacich, for example, claims that the "company was fair about relocation, more than fair." His mother received $22,000 for a home that she bought for $500 and he received $50,000 for his father's home, which had been built almost entirely out of stolen company lumber.[40] Some people thus believed the company provided "kind of a tight deal. They would swap homes. You would get practically a new home."[41] Trading a house based on a log cabin for a dwelling with modern conveniences did tempt a number of people. Those leaving McQueen, for example, mostly gained "lovely homes, just lovely homes," in the words of one former resident. Some of McQueen's former residents complained that the taxes were higher on the Flats, which sat below McQueen, but such complaints were few.[42]

A geographer's examination of Berkeley Pit expansion during the 1960s and early 1970s confirms that most residents received relatively new, nice housing from Anaconda. The average assessed valuation of a home in Meaderville was $4,000 in 1955, but the valuation of the homes that people received in trades during the 1960s was two to three times as high. Some Butte residents even claimed that the removal of Meaderville's homes improved civic pride in Butte due to the neighborhood's purportedly ramshackle appearance.[43] The general manager for Anaconda's Montana operations agreed the company was performing a community service by demolishing many of the substandard homes that it had purchased. "If the city had a demolition program equal to ours we wouldn't have the eyesores we have in the community," he explained.[44] Proponents for large cities' urban renewal projects often propagated similar opinions about otherwise vibrant areas targeted for destruction.[45]

Despite such evidence, relocation sometimes involved lengthy price disputes. Leaders of the McQueen Athletic Club met with Anaconda as often as twenty-three times to resolve a dispute over both the price Anaconda offered and the construction company it recommended before the group finally moved in 1978.[46] Although Anaconda offered residents market prices for their homes, many residents insisted that only replacement cost would let them find suitable substitutes, because mining operations had already lessened their property values.[47]

The land ownership issue made dissatisfaction especially high in Meaderville. Angelo Petroni, the community's former fire chief, remembered that a number of widowed miner's wives received only $1,500 for their homes, which simply was not enough money to live on when "they didn't have nothing."[48] Learning from such complaints, Anaconda's Land Department became much better about the prices it offered in neighborhoods that later faced expansion.[49]

Residents ultimately were not in a strong bargaining position. Whether or not they owned the land on which their homes rested, they almost never owned rights to the minerals located below the surface. In a situation known now as a split estate, even if a resident owned the top few feet of ground, the mining company held the mineral rights farther below. The Homestead Acts had encouraged the splitting of property rights and mineral rights, with mineral rights considered dominant over surface rights. This split situation remains common in the region. Having spent past decades buying up what other mining companies in Butte once owned, Anaconda therefore had access rights to minerals throughout the Butte area.[50]

### The Power of Condemnation

State legislation reinforced Anaconda's position. In January 1961 Butte's sole state senator, Frank Reardon, joined with state Senator John McKeon of the city of Anaconda and senators from nearby counties to introduce a bill prepared by the Anaconda Company's general counsel. Senate Bill 112 extended the right of eminent domain to the mining industry so that it could condemn private property for open-pit purposes.[51] Eminent domain allows the state to seize a property for public use without the owner's consent, provided due monetary compensation is given. Courts in the East narrowly defined public use, but legal scholars like Alexandra Klass have shown that states in the interior West routinely used this power on behalf of private entities that would develop natural resources. Constitutions and statutes in Arizona, Colorado, Idaho, Nevada, North Dakota, Oklahoma, South Dakota, Utah, and Wyoming all granted eminent domain authority to private businesses engaged in oil, gas, mining, and other extractive work.[52]

In Montana the development of mining resources by private companies had long been seen as fostering the common good. The legislature fashioned similar laws for underground mining in the 1870s, and the Montana Supreme Court recognized the principle in 1910. By midcentury, companies already had extensive rights to deposit mine debris or tailings. The 1961 bill enhanced company rights, granting full condemnation powers to any company that needed land "for the mining and extracting of ores, metals, or minerals when the same are owned by the plaintiff but located beneath or upon the surface property where the title to said surface vests in others."[53]

With operations at Walkerville's controversial Alice Pit suspended by 1961, Meaderville, across the road from the Berkeley Pit, seemed to be most threatened by the new bill. Protests from Walkerville, however, proved just as loud as those from Meaderville. Walkerville's mayor, Jimmy Shea, immediately issued a newspaper advertisement that the bill meant nothing but "special privilege" for the mining industry.[54] Fearing that the Alice Pit could be reactivated, one hundred people packed into Walkerville's tiny firehouse to argue against the bill. State Senator Reardon defended the bill in front of a hostile crowd. Using a familiar employment argument, he insisted that "mining operations must expand both east and west of the Berkeley pit and also in other areas in and adjacent to Butte if jobs are to be provided.... My sole purpose in introducing this measure...is to keep the Berkeley pit in operation." Without this special power, he implied, the company might have to shut down its open pits, leaving Butte's economy in the lurch. Reardon attempted to show a brighter side to this bill, stating that eminent domain did at least ensure that property owners would get paid market value, giving each person "a stronger bargaining position."[55]

Using precedent to reassure those at the meeting, Reardon recalled Anaconda's work to compensate residents for subsidence that had damaged homes in central Butte near the Emma Mine. He claimed, "I have heard no complaints by anyone that they were not satisfied with the transactions completed in that area." Reardon was correct that the company had paid owners a decent price, although he overplayed recipients' satisfaction. Reardon remained confident that Anaconda would similarly provide decent payment to homeowners who faced open-pit expansion.[56]

Mayor Shea offered a scathing rebuttal to Reardon's remarks. Discussing the travails Walkerville's residents had experienced with the Alice Pit, Shea declared that this bill seemed like nothing more than an attempt to rob people of their rights. He fired back at Reardon that he was too far removed from the threatened people in Walkerville, Meaderville, and McQueen. Meanwhile, members of the audience yelled out the name of other potentially affected areas, including East Butte and the East Side.[57]

A similar meeting took place in Meaderville's volunteer fire station. Representatives from many areas threatened by the Berkeley Pit, including McQueen, East Butte, and the East Side, were present. Fireman Angelo Petroni declared that residents were upset about the bill and about the low prices Meaderville residents were receiving for their homes. Community leaders immediately had encountered problems. Petroni explained, "We invited several of the senators and that to come to Meaderville to see the situation and we would kind of escort them around. But, hell, they, uh, they had the Highway Patrol—the company had the Highway Patrol meet them at the top of the hill before they got to Meaderville. And then they took them onto their version of everything."[58]

Residents of Meaderville, Walkerville, and other threatened areas sent a delegation, including Mayor Shea, to Helena, the state capital. The labor-oriented and Helena-based *People's Voice* supported their side, calling the bill a "dangerous precedent" that endangered individual property rights. The paper suggested that the bill "opens the way wide for not a single Montanan to be secure in his home or business."[59] The *Helena Independent Record* offered a more balanced view. It first listed claims by the bill's opponents, including the following: A mining company was still a private enterprise, even if it provided employment to the area. Only the government and public utilities had the privilege of eminent domain, as they could claim to be acquiring property in the public interest. Unlike a public utility, Anaconda could simply fire workers without reproach, even if the bill passed in its favor. Just as importantly, the "average person could not afford the expense of litigation" if he or she needed to challenge a condemnation order.[60]

The paper then put forward arguments by the bill's proponents: New methods of open-pit mining necessitated this bill. Without the legislation, "the Anaconda Company is in a strait-jacket as far as expansion is concerned," because residents could hold the company hostage over property prices if they wanted to do so. In sum, "If Butte is to continue to exist, mining must continue and without SB112 this will be impossible."[61]

Despite residents' efforts, the bill easily passed. The state senate approved it unanimously.[62] The house provided only twelve nays to seventy-nine ayes. Butte's representatives were evenly divided. Torn between assisting the industry that sustained Butte and supporting threatened working-class residents, four of Silver Bow County's members of the house voted for eminent domain, while the other four voted against it.[63]

Later in the session, legislators passed a bill that was of more obvious benefit to working-class residents near mining operations. Even Walkerville's mayor, Jimmy Shea, supported it. Senate Bill 181 directed mining companies using eminent domain to take surface property for open-pit work to protect the public in the immediate area from industrial hazards like subsidence and blast damage. To offer this protection, the mining company would also have to "purchase all residential properties within three hundred feet of the surface tract condemned."[64] Although the bill did look out for residents' interests, it also helped the company by supporting Anaconda's efforts to separate expanding industrial operations from residential areas.

After gaining them in 1961, Anaconda exercised its condemnation rights only three times.[65] The simple existence of eminent domain power made its use unnecessary. As one journalist wrote, "It works as a very effective lever against any potential holdout." The threat was the most important aspect. Actual legal challenges generated negative publicity, but fear and uncertainty were both quiet and effective.[66]

## The Relocation Routine

Despite rebuke by the legislature, a number of residents continued to fight. In Meaderville, the first neighborhood to face Berkeley Pit expansion, residents formed a committee to meet with the Anaconda

Company. Seeking fair treatment for all, a number of people even tried to convince Anaconda to relocate most of the neighborhood's population as a group. Despite frequent meetings with Anaconda in its uptown Butte headquarters, the committee was not able to get the company to budge from its intention to deal individually with each landowner.[67]

The intentions of this company policy were primarily to keep costs down, but dealing with residents individually also divided residents from each other, making relocation both easier to do for the company and harder to get over for residents. When Anaconda first purchased property in Walkerville and Meaderville, it tried to buy large sections of property at once. That approach created the sense of an immediate threat in the entire neighborhood and convinced residents to fight back as a group. In later acquisitions, the company resorted to spot buying. When encountering a stubborn group of landowners, the land department instead purchased houses one by one. The policy came from the top. Company vice president R. S. Newlin, for example, instructed Butte management in 1961, "As a final admonition: you should follow a program of picking up property all the time, little by little as it might become available under favorable terms, in the areas which may come within the expanded pit limits."[68] Under these terms, all 206 homes in Meaderville were removed or demolished in the early 1960s.[69]

The company continued a policy of individual negotiations throughout the relocation process. In 1976 Anaconda was expanding its operations into and beyond the community of McQueen. While conducting an environmental assessment of the area, the Montana Department of State Lands received multiple letters from residents who wanted the government to convince the company to acquire all of the property in the area at once so that they could know what to expect. Acting commissioner Leo Berry Jr. replied, "The acquisition of property is a private matter between the Company and the landowner." The fact that residents appealed to the state for help suggests that they believed the individual purchases were not in their best interest.[70]

Anaconda developed a standard routine for acquiring land. Personnel from the Engineering Department and the Geological Department would ask the Land Department to look into a piece of property. The Land Department would contact the Abstract Department, whose few employees would scour the county courthouse for records about ownership, liens, taxes, mining claims, and more. Tracking down owners could involve scouring old city directories and even death records. The abandonment of many mining claims from Butte's early boom complicated the employees' searches. With information received from Abstracts, the Land Department would ask the Butte Board of Realtors to appraise a property, for which it would then make an offer. Anticipating later expansion of mining operations, the company often bought properties that it did not immediately need. In a few cases, the company even paid delinquent taxes on properties that it might purchase in the future.[71]

Because the process of buying and selling property in the relocation area was relatively quiet, residents lacked a lot of knowledge about what they were to encounter, aggravating splits already forming in each neighborhood. In Butte, as in other areas, differing perceptions of the urgency and magnitude of the problems often divided residents.[72] One journalist described the problem for residents: "Information about the pit's enlargement is not given out, but rumors are abundant. Exaggerated stories cause panic, forcing each family to look out for itself. Any hopes of unified community action are delicately squashed in this manner."[73]

Negotiations by Anaconda with individual owners aggravated suspicions, but tension existed even when communities set up committees to deal with the company. In Meaderville, one lady thought that the local committee "was selling her down the river." Others took Anaconda's side, while still others castigated the first people who broke ranks, sold their houses, and relocated.[74] In McQueen, residents attempted to form a united front to make sure that everyone was treated well by Anaconda. However, community meetings about Anaconda's actions continued to be quite contentious.[75]

Divisions worsened as relocation became a reality rather than a possibility. When considering the relocation of their families, neighbors would sometimes vie with each other over the same desirable, company-owned house. The stressful process of relocation thus wore on longtime friendships in many ways.[76]

Families moved during the earlier phases of expansion had a slightly better chance than those relocated later of reconstituting a familiar environment. At the start of the relocation from Meaderville, Anaconda owned a number of homes in the McGlone Heights district of northwest Butte; those homes previously housed coal miners from the South who had come to Montana during the good times right after World War II. The company also had property in the Walnut Street area and the Drives, projects operated by its Kenwood Realty company; both sat on the suburban Flats below Meaderville. Many people relocated during the first decade of open-pit mining ended up in these areas.[77]

Some in these neighborhoods still held out hope that, as one journalist put it, "if they stuck together, they could force the company to relocate the entire neighborhood in one spot and provide money for schools, parks, and everything else."[78] After it had used up much of its housing stock, the company did not have enough available homes in any one area to agree to such a scheme and Anaconda did not want to invest in more building, especially when Butte itself was not growing. Therefore, despite early efforts by both the company and community leaders to keep neighbors together, large-scale dispersion ensued, far beyond the starting points of McGlone Heights to the north and the Drives to the south.

The reluctance or inability of many families to move to property owned by Anaconda turned much of Butte's real estate into a seller's market, raising costs for now-displaced locals.[79] A number of people from Meaderville moved to McQueen; ironically, they had to relocate once more when mining operations overtook them again ten years later. By the time pit expansion reached the McQueen and East Butte area in the 1970s, families usually relocated to the suburban Flats, or even all the way out to Nine Mile, an area south of Butte. Many ended up in St. Anne's Parish and others in Immaculate Conception, but, for the most part, people who had long gone to the same church for decades, and whose children had attended the same Parish or neighborhood schools, ended up scattered.[80]

For many residents, no place that the company offered them looked as good as their threatened neighborhood. The father of former Meaderville resident Pauline deBarathy was offered a home above Meaderville, for instance, but he refused it. He found it too sad to look out over a beloved neighborhood that was quickly filling with waste dumps.[81]

### Deterioration and Demoralization

Anaconda's success at relocation therefore meant real social costs to the pit's industrial refugees. As each family moved, those who held out found life harder. Demoralized people saw friends leave and their streets turn into an industrial zone. Each neighborhood's experience followed a similar path. The first to leave were often local businesses, because their owners depended on a viable population of locals or travelers. When the Berkeley Pit forced the closing of the highway that ran by East Butte, for example, nearby businesses almost immediately moved or closed. As East Butte resident Carl Gergurich put it, "When they shut down the highway, it was only a matter of time before everything else went also. My father's grocery store lost its main customer base in the travelers. The three motels quickly went out of business and the A&W Root Beer Stand moved to Continental Drive."[82]

As more families left, community services went with them. Since these were mostly Catholic areas, the departure of parishioners forced the diocese to close both the churches and grade schools that served them. Sacred Heart School and Sacred Heart Catholic Church on the East Side closed in 1967 and 1970, respectively. After a down payment of $160,000 from Anaconda to the Catholic Church, Holy Savior Church and School in McQueen/Meaderville followed right behind.[83] Local public schools usually closed after Catholic institutions had disappeared. Washington School and Grant School on the East Side and Harrison School in East Butte had been demolished by the mid-1970s.[84]

Handling the students left behind and those relocated became a problem. Discussions began about busing the remaining students to new schools until their families relocated. That option angered parents who still lived near Harrison School in East Butte; they wanted the school to remain open, even at a low capacity. Meanwhile, as they received now-displaced students, Butte's remaining elementary schools experienced a related problem: severe overcrowding.[85]

Other services deteriorated. When McQueen's population dwindled, Butte Bus Lines decided to cancel its operations there. Elderly people who depended on the service were especially upset. The Montana Public Service Commission made sure that the bus line continued for a few additional months, but the typical lack of an official company announcement about expansion plans confused the issue.[86] When the McQueen Addition's demise drew closer, county officials debated how much Anaconda should pay it for the streets, sewerage facilities, and more.[87] As more buildings disappeared from each area, Anaconda also would often petition to close streets. In 1965, for example, Anaconda had been buying private property throughout the East Side. To relocate its ore haulage road, the company then asked to shut down many sections of streets.[88]

Holdouts also suffered visual affronts and occasional damage to their property. They especially worried when Anaconda left its recently purchased houses empty for long periods.[89] The company eventually gave out contracts to remove salvageable materials and sent in crews to clear unoccupied properties that owners chose not to move. Tearing down structures sometimes caused damage to nearby occupied houses, which created additional tension between the company and residents.[90] Most landmarks were removed or flattened. Holy Savior Church in McQueen became a dump when, instead of moving or salvaging the building, the company instructed truck drivers like Bill Long to dump waste rock from the pit on top of it. Many in town pointed to this sacrilegious act as a final insult.[91]

Fire became a constant problem in emptying neighborhoods. As one journalist wrote, firemen were "left to answer the frequent calls that meant another remnant of the fading neighborhood was in flames."[92] In Meaderville repeated fires hastened residents' departures between 1961 and 1965. Empty homes, major clubs, the Meaderville Mercantile, the Franklin School, and even the old Meaderville firehouse went up in flames. Meaderville's experience was mirrored on the East Side, where "vacant houses became towering infernos."[93] Volunteer fire departments were busier than during any previous time, with fires often putting members in grave danger. Arsonists set many of the blazes, and theories abounded as to their motives. Many alleged the Anaconda Company used the fires to hasten locals' exit and that arsons seemed to follow the path of the pit's expanding boundaries. Others thought that homeowners and business owners set the fires in order to gain insurance money, while still others laid responsibility on youthful pyromaniacs.[94]

Aggravating the arson problem was a decline for each neighborhood's volunteer fire department. In McQueen, for example, the department's budget decreased rapidly from the late 1960s into the 1970s. As Anaconda purchased more property, tax evaluations for the fire district fell, despite the fact that the company was not yet using any of its purchased ground there. As a result, the department had little money to serve what was still functionally a residential area. From 1972 to 1975 the department's annual budget fell from $4,539 to $2,556.[95] To improve its outlook, the department won county approval in 1975 to get rid of some street lighting and fire hydrants in the more abandoned sections. The Anaconda Company protested that the department's action would only "aggravate the problems of fire and vandalism." Anaconda funded a few of the lights and fire hydrants, but the fire department's budget problems continued.[96] The department went from thirty-two firemen at its height in 1955 to twenty-one in 1971. Only five members remained when the department finally shut down in July 1978.[97]

After land clearances and fires, only barren spaces, strewn about with refuse and dumps, were left. Those who stayed in each community saw that pit development "tore the whole town apart," in former McQueen resident Jim Matteuchi's words. Scavenging through abandoned homes became common; people wanted to make a buck on scrap metal.[98] A former Meaderville resident was a child when the pit started. She later recalled, "You went out and you played with all the friends. And then every couple of months somebody'd be gone. Another person'd be gone. Another house would be bulldozed over. Another house would be bulldozed over and it'd just be bigger spaces and bigger spaces and less people and bigger space." The last residents to see Meaderville thought that it looked "like a bomb hit."[99] Journalists used similar terms to describe McQueen's final days: rubble, isolated, deserted, fogged, eyesore, mess. This image stood in contrast to what it had been before. One resident recalled it had been a nice, clean town, while another loved its "deer on the mountainside, all different kinds of wildlife."[100]

On the East Side, a similar picture emerged. On East Mercury Street houses had once extended to the 1200 block, but by the 1980s 527 East Mercury Street was the farthest east address. Only seventy-seven buildings still stood in a once-bustling neighborhood. Analyzing remnants of the East Side, historians Murphy and Shovers described the scene: "Sidewalks [are] marred by gaping holes, and the remnants of stairs and retaining walls are often the only evidence left of the boarding houses, saloons, and retail store fronts that once lined the streets east of Arizona. Some streets have vanished completely under the bulldozer; a sign reading "SCHOOL 15 MPH" tacked to a telephone pole and surrounded by vacant lots is a forlorn reminder of what has happened."[101]

It pained holdouts to see these changes. By the time Louis Govednik left his residence, only two homes remained on his block. He told a reporter, "Chipmunks and mice had invaded their once lush garden. The dumps had come so close that the family once returned from vacation to find a huge boulder in the road in front of their house."[102] The empty, lonely spaces were often hard to take. Those left on the East Side missed their neighbors, but eventually learned to cope or left. If holdouts did not immediately leave and pit expansion plans changed, the company might withdraw its offer, leaving a person stranded without neighbors, as was sometimes the case on the East Side.[103]

A few holdouts remained until the bitter end. Some refused to leave because they were waiting for a better price. George Vidrich, a former resident of the East Butte/Columbia Gardens area, said that some people were simply seeking an outrageously high price for their homes. He was one of the last people left, however, because he had a successful business on his property. To make up for what he would lose, he wanted the company to pay for the value of his radiator shop as well as his land. After his lawyer became involved, the company settled, and he was able to move out.[104]

Other holdouts refused to leave because they had deep personal ties to these places. Vincent Crnich, a resident for seventy years, told one reporter that he remained because "it was just a matter of being deeply rooted in the neighborhood. I lived here from the horse and buggy days to its present form." More poetically, Mitzi Sherick, who lived in McQueen for seventy-six years, told the reporter, "This walk, this dirt, this hill. This is all a part of me."[105]

Frances Spehar, a "strong-willed, outspoken 98-pounder" who refused to take the company offer of $65,000, became locally famous. To the company, reporters, and her customers at Spehar's Bar and Hotel, she complained bitterly about the hazards of the Berkeley Pit. During one interview, she "stooped to a cupboard and produced a rock about the size of a large cantaloupe. 'This is what fell right at the door just after 12 one day. If someone was comin' in the door, it would have killed 'im.'" She was not simply obstinate; she had deep ties to East Butte and her customers. Her father had dug the neighborhood's first well and built the first home. Finally accepting that relocation was inevitable, Spehar sold her property and moved out of town in 1974.[106]

By the 1970s expansion of the Berkeley Pit had become a normalized part of life. Uncertainty about where and when Anaconda would next seek property led to what one local called "a basic resignation"—a real sense of powerlessness.[107] Once Meaderville disappeared, people in other neighborhoods were less likely to see the advantage of a fight. As residents make clear in interviews, pit expansion was an inevitable fact of life. They assumed their neighborhood would disappear one day, even if the company made no declarations about its intentions.[108] Risk researchers suggest that familiar risks are typically seen as less serious than new ones, rare risks as less worrisome than commonplace ones.[109] An expanding pit had become a familiar, commonplace risk to people in Butte's eastern neighborhoods.

Many residents came to look on the Berkeley Pit as a living organism. Miners had long talked about the world of underground mining in the language of living bodies, perhaps partly because it helped them to make sense of an unfamiliar environment. Historian Liza Piper notes that concentrations of ore had long been called "bodies" and strips of ore were "veins" that a miner could "open."[110] The pit likewise took on

Berkeley Pit. This photo is looking across the central business district. Photo courtesy of Butte-Silver Bow Public Archives, HAER (Historic American Engineering Record) Buildings and Archives Photograph Collection, PH002.136a.

its own life. One out-of-town journalist was flabbergasted that residents saw "the pit as a kind of omnipotent creature, as a corpus with needs that simply needs to be fed.... There is no point in fighting The Pit because The Pit will do what The Pit needs to do."[111] The most common metaphor used by insiders and outsiders who described the pit was eating. Local journalist Andrea McCormick told readers that "little by little, the pit grew in size and appetite, nibbling away at the East Side fringes," while East Coast writer Don Moser more dramatically called the Berkeley Pit a "monster" that was "devouring" Butte.[112]

As the final night in a packed Finntown bar made clear, language often severed the connection between the Pit and its maker. Finntown had been one of the more close-knit areas of the East Side, but Anaconda bought one of its final outposts in 1975. As the polka band known as Johnny's Hot Shots played its last song, someone called out, "Don't blame the company, blame the times."[113] By the 1970s people had become more accepting of open-pit mining, even if they were deeply saddened by its effects. The battle that had opened the open-pit era just twenty years before was all but forgotten. In later oral histories few workers or residents remembered either their intense fascination with or their fight against open-pit mining when it first arrived. Neither a technological marvel nor a corporate abomination, the pit simply was.[114] The intrusion of open-pit work into neighborhoods had radically altered people's social lives, while introducing a negative outlook toward the community's future. People like East Butte resident Vincent Ciabattari and Anaconda truck driver Bill Long recognized how devastating the resulting losses could be.

Butte residents were not alone in such reactions, as many communities across the American West, the United States, and the world have faced relocation due to open-pit mines. The expansive logic of open-pit mining demands a lot of land. Despite a belief that this needed space is uninhabited, sometimes it is not, even if it is not typically as urbanized as Butte. Hence, relocation and its attendant social costs are surprisingly common results of moving industrial work aboveground.

The Anaconda Company generally did not intend to cause social harm, but bonds forged over decades between neighbors sadly increased the chances that mass mining would wrench apart social lives. The loss of neighborhoods to the company's Berkeley Pit in the 1950s, 1960s, and 1970s was therefore a variant of urban renewal, a denuding of old neighborhoods, often with at least tacit government support, to benefit the city's major private industry. In Butte, demoralization over this situation became so heightened that it became central to the community's mid-1970s malaise.

## Notes

1. Ciabattari quoted in Moser, "Monster That's Devouring," 50.

2. Long, *Pit Driver*, 99.

3. "The Pit That Ate Butte," *Borrowed Times* (Missoula, MT), Mar. 7, 1973, 1-2, quote on 2. See also "The Undermining of Butte," *Newsweek* (Nov. 6, 1972), 93; Moser, "Monster That's Devouring," 50.

4. Sara Sparks, interview by Brian Leech, July 29, 2009, Butte, MT, transcript and digital recording, OHC, BSBPA; Daily interview.

5. Long, *Pit Driver*, 99.

6. Johansen, "Butte, MT: Sitting Atop a Fortune."

7. "Forward Butte Commission Moves City Backwards," *Borrowed Times*, July 1974, 5.

8. Butte is featured in Hammett, *Red Harvest*. On the company town image, see Hardy Green, *The Company Town: The Industrial Edens and Satanic Mills That Shaped the American Economy* (New York: Basic, 2010).

9. Malone, *Battle for Butte*, 212–17.

10. Finn, *Tracing the Veins*, 83.

11. Swibold, *Copper Chorus*, 313–38; "Hello Butte and Anaconda," *Mont. Std.*, June 2, 1959.

12. K. Ross Toole, *The Rape of the Great Plains: Northwest America, Cattle and Coal* (Boston: Little, Brown, & Co., 1976), 212–13.

13. Ben A. Franklin, "Inexorably, a Copper Mine Is Eating Away Butte, Mont.," *New York Times*, Feb. 15, 1973. See also "Butte Hospitals to Be Merged," *Mont. Std.*, Oct. 17, 1959.

14. Perry, "Future Copper Possibilities."

15. Shea, "Report on Open Pit Copper Sulphide," Nov. 10, 1955, 12; McWilliams, *Mining Methods and Costs*, 11–15.

16. Anaconda Geological Department Report to E. P. Shea, "Report on the Open Pit Copper Sulphide Ore: Butte District," Jan. 5, 1956, file 29801.12, AGDC, AHC.

17. Corbett interview, by Mercier.

18. Discussion of expansion worries is in Edward P. Shea, "Memorandum no. 28, re: Butte District Special Development," Anaconda Geological Department, Oct. 6, 1955, file #29704.02, Fldr. "Progress Reports, Special Development Memos, nos. 23 to 40 (Sept. 2, 1955 to Dec. 31, 1955)," AGDC, AHC; Miller, "Production History"; Shea, "Report on Open Pit Copper Sulphide"; McWilliams, *Mining Methods and Costs*, 11–17.

19. McWilliams, *Mining Methods and Costs*, 15.

20. Miller, "Production History." See also Vincent D. Perry to Clyde E. Weed, file 29705.11, Fldr., "Butte, Mine Operations, 1937–1979," AGDC, AHC; Wallace A. O'Brien, "Progress Report on Deep Pit Development," Aug. 17, 1960, file 29801.17, Fldr. "Butte, Mine Operations," AGDC, AHC.

21. Edward P. Shea to E. I. Renouard, Sept. 21, 1961, file 29801.11, Fldr. "Butte, Mine Operations," AGDC, AHC.

22. McWilliams, *Mining Methods and Costs*, 12–15.

23. Edward P. Shea to E. I. Renouard, Sept. 21, 1961, file 29801.11, Fldr. "Butte, Mine Operations," Doc. 11, AGDC, AHC.

24. Edward P. Shea to E. I. Renouard, Sept. 21, 1961, file 29801.11, Fldr. "Butte, Mine Operations," Doc. 11, AGDC, AHC; Wallace A. O'Brien to Edward P. Shea, "Re: Revised 'Deep Pit' Plan to the 2050 Level," Sept. 8, 1961, file #29801.11, Fldr. "Butte, Mine Operations," AGDC, AHC; Wallace A. O'Brien to Edward P. Shea, "Re: Summary of Deep Pit Plans," Sept. 8, 1961, file 29801.11, Fldr. "Butte, Mine Operations," AGDC, AHC.

25. Edward P. Shea to Martin K. Hannifan, Feb. 28, 1962, file 29801.14, Fldr. "Butte, Mine Operations," Doc. 14, AGDC, AHC.

26. "Surface Map of the Berkeley Pit-East Extension-Continental, Showing Diamond Drilling and Churn Drilling Results," Oct. 1971, file 120807.19, AGDC, AHC; Vincent D. Perry to Leonard C. Powell, Aug. 14, 1975, file 299061A, Fldr. "Butte, Mine Operations, 1961–1985," AGDC, AHC.

27. Vincent D. Perry to James L. Marvin, Apr. 12, 1982, file 30005.02, Fldr. "Butte Mine Operations, Prospect Data and Correspondence," AGDC, AHC. See also Vincent D. Perry to F. M. Monninger, Dec. 5, 1973, file 30005.02, AGDC, AHC; Vincent D. Perry to Clyde E. Weed, May 12, 1958, file 29802.02C, Fldr. "Butte, Mine Operations," AGDC, AHC.

28. "Data Point Set—600 Level," Butte, Aug. 1979, File 121301.26, AGDC, AHC; "Map of the Berkeley Pit," Jan. 1, 1967, File 120807.23 AGDC, AHC; "Surface Map of the Berkeley Pit—East Extension—Continental," Oct. 1971, File 120807.19, AGDC, AHC; McWilliams, *Mining Methods and Costs*, 4–5, 11–12.

29. LeCain, *Mass Destruction*, 108–218.

30. Francaviglia, *Hard Places*, 21–24.

31. McWilliams, *Mining Methods and Costs*, 4–5, 11–15, 25–41.

32. GIS map of Butte using 1950s street map, made by author and geographer Matthew Fockler; Murphy and Shovers, "The Decline," VF 1615, 2.

33. Keeling and Sandlos, "Environmental Justice Goes Underground?," quote on 118.

34. Guy J. Ossello, interview by Brad Spear, Nov. 9, 2001, Butte, Montana, transcript, OHC, BSBPA.

35. Art Kesikitalo quoted in McCormick, "East Park Survivors."

36. Brian Shovers, "Walkerville: Butte's Sister Silver Town," *Report for the Butte Historic Landmark District Project*, June 1986, VF 1637, BSBPA, 2, 5; Malone, *Battle for Butte*, 11–24; Brothers, *Sketches of Walkerville*, 61; Carter, "Walkerville Won't Say Die."

37. On renting, see Peter Zanchi interview; deBarathy interview. On land ownership, see Holter interview, by Leech; Franconi, Maffei, and Petroni, BHS meeting; Petroni interview, by Magnaghi.

38. See sample agreements in Bernard P. Harrington to T. F. Kiely, May 19, 1969, Fldr. "Sherrill, Lewis A.," Box 1B, Dave Piper Collection, MC388, BSBPA; "Bill of Sale Between the Anaconda Company and Orlie Kukus, July 19, 1968," Fldr. "Alto, Sandra," Box 1A, MC388, BSBPA. On the company's later moving process, see Chuck Krause, interview by Brian Leech, Oct. 22, 2009, Butte, MT, transcript and digital recording, OHC, BSBPA; Vidrich interview.

39. Holter interview, by Leech. See also Petroni and Petroni interview; "McQueen: Isolated and Deserted," *Borrowed Times* (Missoula, MT) 3, no. 13 (Mar. 16): 9.

40. Kovacich interview.

41. Carden interview. See also Hugo Kenck, interview by Laurie Mercier, Dec. 8, 1982, Butte, MT, cassette tapes, OHC, MHS.

42. First quote is Helmie Carriger, interview by Thomas Gordon, [no month/day available] 2001, Butte, MT, transcript, OHC, BSBPA. Second is Fran Corcoran in Fisher and Corcoran interview. See also Matteuchi interview; Vidrich interview; "McQueen: Isolated and Deserted," 9.

43. Leeper, "Changing Urban Landscape," 86. See also Holter interview, by Hayward; Holter interview, by Ferriter; Merzlak interview; Kearney, *Butte Voices*, 191, 220–21.

44. Jenks, "An Interview."

45. Jane Jacobs, *The Death and Life of Great American Cities* (New York: Random House, 1961); Herbert J. Gans, *The Urban Villagers: Group and Class in the Life of Italian-Americans* (New York: Macmillan, 1962).

46. Holter interview, by Leech.

47. Kearney, *Butte Voices*, 191.

48. Angelo Petroni in Petroni and Petroni interview.

49. Franklin, "Inexorably."

50. On land and split estates in Butte, see Petroni and Petroni interview; Nicholls interview; Holter interview, by Hayward; Holter interview, by Ferriter; Merzlak interview. On split estates in Montana and the U.S. West, see Eve Byron, "'Split Estates' May Pit Surface-Rights Owners Against Mineral-Rights Owners," *Helena Independent Record*, Aug. 26, 2007; "Do You Really Own the Minerals Under Your Land?," *Bureau of Land Management* (Montana), BLM/MT/GI-06/012, n.d., accessed June 14, 2012 [online version no longer available]; Gary D. Libecap, "The Assignment of Property Rights on the Western Frontier: Lessons for Contemporary Environmental and Resource Policy," *Journal of Economic History* 67, no. 2 (June 2007): 266–71.

51. Maxwell, "Hill City Group"; *Senate Journal for the State of Montana*, 37th Legislative Assembly, Reg. Sess., 1961, p. 71.

52. Alexandra Klass, "The Frontier of Eminent Domain," *University of Colorado Law Review* 79, no. 651 (2008): 654–61. On eminent domain and mining see Curtis H. Lindley, *A Treatise on the American Law Relating to Mines and Mineral Land within the Public Land States and Territories and Governing the Acquisition and Enjoyment of Mining Rights in Lands of Public Domain* (San Francisco, CA: Bancroft-Whitney Company, 1914), 578–82, 594–97.

53. *Laws, Resolutions, and Memorials of the State of Montana*, 37th Legislative Assembly, Reg. Sess., 1961, Chapter 216, 595–98; quote on 595. See also "Support and Objections Are Voiced on Bill Relating to Pit Mining in This Area," *Mont. Std.*, Jan. 21, 1961.

54. Jimmy Shea quoted in "Support and Objections Are Voiced."

55. Maxwell, "Hill City Group."

56. Ibid.

57. Ibid.

58. Petroni and Petroni interview. See also Maxwell, "Hill City Group"; Kearney, *Butte Voices*, 220–21; "Support and Objections Are Voiced"; "Eminent Domain for a Private Mining Company... A Dangerous Precedent," *The People's Voice* (Helena, MT), Jan. 27, 1961.

59. "Eminent Domain for a Private Mining Company."

60. Robert E. Miller, "Mayor James Shea in Bind on Eminent Domain Bill," *Helena Independent Record*, Jan. 31, 1961.

61. Ibid.

62. *Senate Journal*, 288.

63. *House Journal for the State of Montana*, 37th Legislative Assembly, Reg. Sess., 1961, 570–71.

64. *Laws, Resolutions*, 675; *Senate Journal*, 104.

65. *Schara v. Anaconda Co.*, 187 Mont. 377, 610 P.2d 132 (1980); *Anaconda Co. v. Montana Power*, case #56978, vol. 33, Register of Actions, p. 482 (Mont. 2nd Dist. 1970); *Anaconda Co. v. Louise Call*, case #57308, vol. 34, Register of Actions, p. 83 (Mont. 2nd. Dist. 1971). Anaconda also filed for condemnation at least one other time, but it settled out of court.

66. Moser, "Monster That's Devouring," 48. See similar comments in Franklin, "Inexorably."

67. Leeper, "Changing Urban Landscape," 83–87; Petroni and Petroni interview; Matteuchi interview.

68. R. S. Newlin to E. I. Renouard, Dec. 4, 1961, File 29802, Fldr. "Kelley Mine Operations, Butte Mines," AGDC, AHC.

69. Department of Housing and Urban Development, "Analysis of the Butte, Montana Housing Market as of July 1, 1971," Mar. 1972, 7, Fldr. 43, Box 1, Personal Collection 127 "Butte Chamber of Commerce" (hereafter PC127), BSBPA.

70. Leo Berry Jr., acting commissioner of the Department of State Lands, to McQueen Volunteer Fire Department, Oct. 20, 1976, Box 3, Fldr. 15 "McQueen Contracts and Land Agreements," McQueen Fire Department, OC19, BSBPA.

71. Krause interview; R. C. Redlich, John S. Wulf Jr., and Daniel M. Popovich to John Frankovich, Apr. 27, 1966, Fldr. "Kalafatich, Lewis & Francis," Box 1A, MC388, BSBPA.

72. Thomas E. Shriver and Dennis K. Kennedy, "Contested Environmental Hazards and Community Conflict Over Relocation," *Rural Sociology* 70, no. 4 (2005): 491–513; Couch, "Environmental Contamination."

73. "Forward Butte Commission," 5.

74. Angelo Petroni in Petroni and Petroni interview. See also Nicholls interview.

75. Berry Jr. to McQueen Volunteer Fire Department, OC19, BSBPA; Vidrich interview.

76. Petroni and Petroni interview; deBarathy interview; Franconi, Maffei, and Petroni, BHS meeting; Michelotti interview.

77. Leeper, "Changing Urban Landscape," 83–87; Petroni and Petroni interview; Matteuchi interview; deBarathy interview; Peter Zanchi interview; Sparks interview, by Leech.

78. "The Pit That Ate Butte," 1.

79. Petroni and Petroni interview; Nicholls interview.

80. "McQueen: Isolated and Deserted," 9; Bone interview; Holter interview, by Leech; Stepan and Stepan interview; Fisher and Corcoran interview.

81. deBarathy interview. See also Petroni and Petroni interview; Franconi, Maffei, and Petroni, BHS meeting; Holter interview, by Leech; Father Tom Fenlon, interview by Ellen Crain, Oct. 23, 2006, Butte, MT, transcript, OHC, BSBPA.

82. Carl Gergurich quoted in Kearney, *Butte Voices*, 187.

83. "Sacred Heart Church Closes Doors," *Mont. Std.*, July 2, 1970; "Catholic Property Sold," *Mont. Std.*, Dec. 20, 1973; "Old Butte Church Coming Down," *Mont. Std.*, Apr. 11, 1974. See also Kearney, *Butte Voices*, 189.

84. Murphy and Shovers, "The Decline," 7; Kearney, *Butte Voices*, 187.

85. Foote, "Harrison School"; Mike Babcock, "Taxpayers Hostile, *Mont. Std.*, May 15, 1973; "Trustees Tell PTA 'Harrison Open,'" *Mont. Std.*, Mar. 22, 1973. See also W. P. Mufich to Frank M. Monninger, "Memorandum: School Purchase Contract—Trustees' Statement Newspaper Mar. 22, 1973," Mar. 22, 1973, Fldr. "Harrison School," Box 3C, MC388, BSBPA.

86. "Butte Bus Lines to Quit McQueen Run on Friday," *Mont. Std.*, Mar. 22, 1973;

"McQueen Bus Runs," *Mont. Std.*, Apr. 3, 1973; William J. Clark, "McQueen Bus Will Continue," *Mont. Std.*, May 17, 1973.

87. Rick Foote, "Property Compensation Studied in Dying Area," *Mont. Std.*, Mar. 2, 1973.

88. "Anaconda Co. Seeks Closing of Some Streets," *Mont. Std.*, Nov. 11, 1965.

89. Jones to Picarelli, "Blasting Complaint at 26 North Drive."

90. Charles R. Krause to William P. Mufich, "Joseph Cassidy Property," Mar. 22, 1978, Fldr. "Cassidy, Joe," Box "Cabinet 1, Drawer 3 (Box 1)," MC388, BSBPA; P. W. Murray to J. P. Hagerty, June 23, 1972, Fldr. "Filpula, William," Box "Cabinet 1, Drawer 3 (Box 2)," MC388, BSBPA.

91. Long, *Pit Driver*, 99; Bone interview; The *Mont. Std.* and The World Museum of Mining, *Remembering Butte, Montana's Richest City* (Helena, MT: Far Country Press, 2001), 53; "Old Butte Church Coming Down," *Mont. Std.*, Apr. 11, 1974; "Another Landmark," July 25, 1979.

92. McCormick, "Given a Choice."

93. Madison, "Butte's 'East Side,'VF" 6.

94. Holter interview, by Leech; Holayter interview; Peter Zanchi interview; deBarathy interview; Petroni and Petroni interview; Matteuchi interview. See also Frank Quinn, "Years Swallow Memory of Meaderville Merc," *Mont. Std.*, Apr. 2, 1973; Filpula, *Meaderville Revisited*, 94; McHugh, "The Gulch and I."

95. Board of Silver Bow County Commissioners to Harry Shaffer, McQueen Fire Department, Aug. 21, 1972, Box 2, Fldr. 19, file "Correspondence 1970s McQueen Fire Department," OC19, BSBPA; Board of Silver Bow County Commissioners to John W. Jaksha, McQueen Fire Department, Aug. 15, 1975, Box 2, Fldr. 19, file "Correspondence 1970s McQueen Fire Department," OC19, BSBPA. On tax evaluations, see John W. Jaksha and Edward Petrin, McQueen Fire Department, to George McGrath, Silver Bow County Assessor, n.d. (1974), Box 2, Fldr. 19, file "Correspondence 1970s McQueen Fire Department," OC19, BSBPA; Foote, "Property Compensation."

96. Robert P. Corbett to John W. Jaksha and Edward Petrin, McQueen Fire Department, Jan. 23, 1975, Box 2, Fldr. 19, file "Correspondence 1970s McQueen Fire Department," OC19, BSBPA. See also Ray J. Wayrynen, Chairman of State Tax Appeal Board, to Edward Petrin, McQueen Volunteer Fire Department, Feb. 19, 1975, OC19, BSBPA.

97. McQueen Volunteer Fire Department Roll Book, 1954–76, Box 3, Fldr. 6, OC 19, BSBPA.

98. Matteuchi interview. See also Stepan and Stepan interview.

99. First quote is deBarathy interview. Second is Matteuchi interview. See also Franconi, Maffei, and Petroni, BHS meeting.

100. Quote is from "McQueen: Isolated and Deserted," 9. Terms to describe Butte are in that article and in "The Pit That Ate Butte," 1–2.

101. Murphy and Shovers, "The Decline," 2.

102. McCormick, "Given a Choice."

103. McCormick, "East Park Survivors." See also McCormick, "Finn Stories"; McCormick, "Old Drug Store."

104. Vidrich interview.

105. Mitzi Sherick quoted in Ciabattari, "McQueen Spirit."

106. Betty Ann Raymond, "A Long Time Stayin' a Long Time Goin'," *Mont. Std.*, June 3, 1973. On Spehar, see "Frances Spehar Moving Out," *Mont. Std.*, Mar. 24, 1974; Moser, "Monster That's Devouring," 50. For Spehar's complaints to the company, see examples of phone messages to Nelson Jones in Fldr. "Plum St.," Box 3C, MC388, BSBPA: Nov. 8, 1971; Mar. 13, 1973; July 24, 1973; July 27, 1973; Oct. 4, 1973.

107. Frank Gary, interview by Teresa Jordan, Mar. 30, 1986, Butte, MT, partial transcript and notes, OHC, BSBPA. On uncertainty, see Kelling, "Finntown"; Neil Maurer, "Butte, Mont.: The Pit Vs. The City," *City: Magazine of Urban Life and Environment* 6, no. 4 (Fall 1972): 38–43.

108. On residents seeing expansion as inevitable, see McCormick, "Given a Choice"; Ciabattari, "McQueen Spirit"; Holter interview, by Leech; Michelotti interview; Franklin, "Inexorably"; "The Pit That Ate Butte"; "The Undermining of Butte," 93.

109. John Tulloch and Deborah Lupton, *Risk and Everyday Life* (London and Thousand Oaks, CA: SAGE, 2003), 34.

110. Liza Piper, "Subterranean Bodies: Mining the Large Lakes of North-west Canada, 1921–1960," *Environment and History* 13 (2007): 155–86; Piper, *Industrial Transformation*, 113–64.

111. Moser, "Monster That's Devouring," 48.

112. McCormick, "East Park Survivors"; Moser, " Monster That's Devouring," 50. For more examples of consumption language in Butte, see Finn, *Tracing the Veins*, 191–93.

113. Foote, "Mining City Hears."

114. Examples in Darrell C. Reynolds, interview by Cindy Wymer, Nov. 13, 2005, Butte, MT, transcript and cassette tapes, OHC, BSBPA; Robert Melvin, interview by Mary Murphy, Mar. 17, 1980, Butte, MT; Michelotti interview, OHC, BSBPA.

# PART IV

# THE PIT IS DEAD (LONG LIVE THE PIT), 1970–2017

CHAPTER 7

# SALVATION

THE CONTINUAL LOSS of homes, amenities, and social networks to the Berkeley Pit had generated such malaise in Butte that what happened in 1976 and 1977 was truly surprising. Anaconda faced, for the first time, a strong legal and environmental challenge to pit expansion. It came in the form of a complaint by the Hillcrest neighborhood. It was precisely the kind of complaint that blasting consultant Lewis Oriard had warned Anaconda about in 1970, because it alleged pollution and nuisance, topics "not as well defined technically" as those the company had previously faced.[1]

The company sought land near the suburban neighborhood of Hillcrest as a site for dumping overburden from both the Berkeley Pit and a recently started new pit, the East Berkeley. In 1975 Anaconda applied to the Commission of State Lands for a new mining permit in the Hillcrest area. Commissioner Ted Schwinden ordered the drafting of an Environmental Impact Statement (EIS), as required under Montana's Environmental Policy Act. Responding to demands for additional information, Anaconda quickly sent new data to the state on December 19, 1975. Because the Hard Rock Mining Act required the application to be processed within sixty days, state officials did not have time to review the new data or to ask for public comment. They approved a permit without an EIS on December 22, 1975. Without warning, Hillcrest residents learned in a January newspaper report that Anaconda planned a mountainous, 17.4-million-ton dump of overburden only one-quarter mile from their homes.[2]

Hillcrest residents vehemently opposed the dump. Residents contacted the State Environmental Quality Control Council; its staff attorney agreed that violations might have been committed during issuance of the permit. One hundred twenty-five property owners in Hillcrest and the nearby Continental Drive area filed a suit on March 12, 1976. They sought revocation of the permit and an injunction against the company until the state properly reviewed possible environmental damage. In the midst of residents' legal challenge, many became emboldened by a major decision that summer. The Butte City Council had surprisingly rejected efforts to relocate the city's center away from the Berkeley Pit. While the Hillcrest complaint worked its way through the court system, though, the Anaconda Company began the dump in August 1976, building a "mountain of substantial dimensions."[3]

The case went to trial in Silver Bow County District Court on August 27, 1977. Documents submitted revealed greater awareness about mine pollution than in previous cases against Anaconda. The supervisor for the state's air monitoring program, David Maughan, obtained high air pollution readings near the Berkeley Pit.[4] The frequent air inversions over the valley often worsened conditions. Serving as the plaintiffs' expert, University of Montana professor Clarence Gordon found toxic levels of copper in residents' gardens and toxic levels of cadmium in the air. Even at a distance from the vehicles, residents faced a possible sixty-nine decibels of noise from company trucks at the dumps. In addition, the company had given no reclamation plan for areas that suffered water pollution as a result of precipitation washing off the new dump.[5]

When the plaintiffs lost the case, they appealed to the Montana Supreme Court. The decision of that body once again tilted toward the position of the company and the state, but it did offer residents some relief. The justices ruled that, although the mining would "significantly affect the human environment," filing an EIS was not necessary because it would make impossible completion of the permit process within sixty days. They also ruled that public notice was not required for mining plans and that the state's role in controlling air pollution was only discretionary. The court did find that Anaconda's permit application was inadequate; for example, the reclamation portion of it covered only ninety of the five hundred acres. The state had issued the permit illegally, and Anaconda would have to submit a new application.[6]

In the end, the already-large dump remained despite the neighborhood's continued concern.[7] Nevertheless, the case represented a real change for Butte. The environmental era had encouraged a different kind of fight. Although he certainly was exaggerating, reporter Rick Foote called the Hillcrest case "one of the first real citizen efforts to stem the move of the Company to do whatever it wanted to do."[8] Coupled with the recent rejection of relocation for Butte's city center, it seemed that acceptance of open-pit mining plans would no longer be automatic.

It was a big change from just a few years before, when mining-induced relocation seemed inevitable. Reasons for this change were complex, involving factors both inside and outside the city. One major factor was the growth of new attitudes about the environment. Butte's Berkeley Pit featured in 1970s environmentalist attacks, helping to inspire new regulations. The Anaconda Company's own struggles, which made residents less likely to toe the industry line, were another important factor. Finally, a perhaps more amorphous shift had happened: after losing their most prized environmental amenity, the Columbia Gardens, and becoming worried about the survival of their historic city center, Butte residents increasingly believed that the costs of mining, both social and environmental, had simply grown too great. In the view of many locals, open-pit mining now threatened the entire community. Hillcrest residents, like many around Butte, had thus gained a new appreciation for landscape preservation, a new set of laws with which to challenge Anaconda, and an increasing willingness to directly confront their city's primary employer.

## The Environmental Movement

Environmentalism offered a completely new way for insiders and outsiders to react to open-pit mining. The movement's rise and the consequent arrival of the Berkeley Pit as a regional icon challenged Anaconda's standard practices. In the late 1960s concerns about the effects of pollution, fueled by influential thinkers like Rachel Carson, merged with worries about resource availability, typified in works like Paul Ehrlich's *The Population Bomb*. Trying to stay ahead of his opposition, President Richard Nixon signed into law the National

Environmental Policy Act of 1969 and founded the Environmental Protection Agency (EPA) in December 1970. Earth Day in April of that year marked a high point in environmental fervor, as millions of participants became involved in demonstrations, speeches, and college teach-ins across the nation.[9]

Young urbanites in Montana responded to the same initiatives. As of the 1960 census, more Montanans lived in towns and cities than in rural areas. Perhaps partly due to their perceived distance from the natural world, many affluent young residents in those urban areas expressed an interest in environmental issues.[10] Their power became magnified by U.S. Supreme Court decisions in the mid-1960s that forced the reapportionment of state legislatures to better reflect the principle of one person, one vote. Doing so increased the power of urban interests in Montana. Unlike conservative rural interests, which had long been backed by the Anaconda Company and Montana Power, the more urban, liberal, and increasingly female legislatures believed that dramatic change was necessary.[11]

This change marked a "major political metamorphosis" in attitudes about Montana's environment.[12] In the preceding decade, according to Jim Posewitz, an administrator at Montana's Commission of Fish and Game, the state's environmental laws had been "pretty bland stuff." They included mine reclamation acts that depended on voluntary compliance by mine owners.[13] The 1970s, however, witnessed a new generation of politically active Montanans, many of whom held a "new concern for preserving the environment, a renewed pride in community, and a new interest in reforming and improving society and government."[14]

One striking component of Montana's 1970s environmental battles is that the Berkeley Pit was often featured; it had become a primary example of what not to do. Montana's new state constitution, approved in 1972, put in place cutting-edge environmental provisions beyond those of any other previous state constitution in the United States. In Article II, its Declaration of Rights, the constitution proclaimed "inalienable [the] right to a clean and healthful environment."[15] During convention discussions, the legacy of Anaconda and its Berkeley Pit became a common talking point, with many representatives hoping coal companies, then descending on the northern Plains, could be stopped from the kind of environmental destruction Anaconda had wrought. George Harper, a clergyman from Helena, hoped to finally put in place safeguards against Montana's and Anaconda's legacy of corporate plunder.[16]

The state legislature continued an anticorporate bent during much of the 1970s, with the Berkeley Pit similarly continuing as a regional icon of environmental destruction. Laws regulating the eastern Montana coal boom eliminated coal companies' use of eminent domain, a power given to them via Anaconda's earlier success. Walkerville's James Shea even testified in favor of the successful anticondemnation bill, citing Anaconda's use of the power in Butte.[17] Legislators similarly worried about Anaconda's exploration of a palladium/platinum deposit, saying that they feared environmental devastation on a similar scale to the Berkeley Pit. They therefore put in place the Metal Mining Reclamation Act of 1971, that, although relatively weak in its requirements, placed further restrictions on Anaconda and other metal mining companies while foregrounding aesthetic and recreational values.[18]

Anaconda also suffered a defeat when it sought 680 acres of state-owned land near Lincoln, Montana. This land, north and west of Butte, was to be the site of a new copper and molybdenum pit along the Blackfoot River, but a coalition of fishermen, hunters, and environmentalists raised objections. Environmentalists feared an impending "ecological disaster" when they learned the proposal called for "an even larger copper surface mine than Butte's renowned Berkeley Pit."[19] The state land board split two–two on Anaconda's easement request. The tie denied Anaconda a permit for the first time.[20]

According to outside journalists, the Berkeley Pit's new role as an environmental icon surprised Butte residents. Butte did not house any formal environmental groups in the early 1970s, which shocked outside journalists like Bruce Johansen, a reporter for the *Seattle Times*. Locals questioned the goals of "environmentalists from 'back East,'" who seemed more interested in nature than people. Some residents even found a "strange beauty" in the Berkeley operation. Johansen jokingly concluded, "While environmentalists point to Butte as an outstanding example of the way America's insatiable demand for mineral resources—including copper—is ravaging the Mountain West, Butte' s high school students take their dates to The Pit's edge to park."[21] Environmentalists' despair over Butte's passivity was exaggerated, as residents had campaigned in the past to beautify Butte, while others had long looked at the pit in a very skeptical, if not downright hostile, way.[22] Negative portrayals of Butte, however, fit with outsiders' impressions of Anaconda as oppressor and Butte as victim.[23]

Although the Berkeley Pit's size made it the most obvious target, the other major component of the mining complex—the smelter—also came under fire. Mounting evidence of health risks and environmental damage wrought by the smelter in the town of Anaconda hurt both the company's reputation and its pocketbook. In the early 1960s the State Board of Health discovered an incredibly high 0.45 micrograms of arsenic per cubic meter in the air surrounding the town. Sulfur dioxide, lead, sulfates, and fluoride were also far above state standards. By the 1970s new studies connected the company's pollution to high rates of respiratory disease, including lung cancer. Members of USWA Local 117, the smelter workers' union in the town of Anaconda, even walked out when they most feared high levels of air pollution inside the plant.[24]

The Anaconda Company tried to head off the growing number of environmental complaints. Rather than celebrate industrial progress and massive technology, as they had in the 1950s and 1960s, publications made available to tourists at the Berkeley Pit observation deck in the 1970s emphasized Anaconda's roles as "the state's largest single industrial employer" and as "a major taxpayer," while calling attention to the company's environmental efforts. Anaconda described its treatment of 100 million gallons of water a day and its practice of minor land reclamation.[25]

Like other U.S. mining companies at the time, Anaconda was on the defensive. The company paper lamented, "Few people relate mining to their everyday living, yet we all are dependent on it."[26] Shareholders complained of "unreasonable environmentalists" and "self-serving politicians," who backed the environmental movement for the votes, not because they believed in any of its ideas.[27] Although the regulations imposed on Anaconda during the decade were neither particularly stringent nor consistently enforced, the company complained that such laws had forced it either to pass up promising prospects or to devote enormous resources toward satisfying new regulations, claims that held a fair amount of truth.[28]

## Anaconda Struggles

In the past, the Anaconda Company would have been able to handle these environmental challenges and their financial costs. Its U.S. mines, including the ones in Butte, were not the way the company made its income. Almost all of its profits derived from overseas operations. By the early 1970s, though, Anaconda was facing major problems abroad. In Chile, socialists, liberals, and conservatives were all converging on the idea that the country should nationalize its mines. Kennecott, another American copper company with enormous investment in Chile, understood that nationalization was inevitable, so in 1964 it stopped investing much in the country, and in 1967 its management worked out a complex financial deal with Chile over the remaining properties. Anaconda's management, on the other hand, thought that they had a better relationship with the Chilean government and therefore they could keep corporate sovereignty. In overconfident exuberance, the company expanded operations at its El Salvador and Chuquicamata Mines, using a large amount of borrowed money.[29]

By the late 1960s, though, pressure for nationalization in Chile had only grown. In 1969, the administration of Chile's Christian Democratic Candidate, Eduardo Frei, negotiated a Chileanization, or majority ownership, of the copper mines with Anaconda. Anaconda maintained day-to-day control of the mines but hated the government's reduction of its profits. Despite CIA and U.S. corporate funding of his opposition, the socialist Salvador Allende won election as president in 1970. Allende immediately introduced legislation to take complete control of Chile's mines. Almost overnight, Anaconda lost over 75 percent of its net income.[30]

The U.S. press lambasted the company. *Forbes* dismissed Anaconda as "one of the great horror stories of corporate history" and "a classic case of mismanagement."[31] The press claimed the company had become top-heavy and complacent. Anaconda faced challenges in Mexico as well because the government there wanted to Mexicanize Anaconda's operations. More Mexicans came into management and the national government took a controlling interest in Anaconda's Cananea Mine. Domestic operations fared no better. Anaconda's decision to heavily invest in the aluminum industry during the 1950s and 1960s was backfiring, because it could not afford the capital-intensive nature of the operations. The start-up of a new copper operation at Twin Buttes in Arizona also proved costly, especially after one side of the pit fell in. In 1973 the oil embargo of the Organization of Arab Petroleum Exporting Countries (OAPEC) in the Middle East caused fuel prices to skyrocket for everyone, including Anaconda.[32]

Butte residents initially hoped that Chilean nationalization, which took place during the 1971 strike, would convince Anaconda to refocus on its birthplace.[33] Workers made significant contract gains that year, but Anaconda also asked unions to consider its overseas situation in making future demands. Al Skinner, leader of the international USWA, reminded participants at an October meeting, "We must understand this Company took a hell of a bath." Local union leaders were not as forgiving. Said Bernard Rask, "For many years they always held over our heads this Chilean operation. Now we are supposed to feel sorry for them because they lost these properties. Well, I think it's time that this Company showed a different attitude—not just give us the same kind of bullshit."[34] While union spokesmen remained bullish, local residents were growing wary of the company's troubles.[35]

Business leaders were also skeptical. The financial institutions on the company's board—Chase, First National City, Metropolitan Life, and Brown Brothers Harriman—had taken control of the company. They put John Place, a Chase banker without any mineral industry experience, in charge. Place saw that the capital-intensive nature of copper production in the United States—which included trucks whose single wheels cost more than a personal car—was hitting the company hard. The falling domestic price of copper was not helping either. The company's value had dropped from $1.4 billion in 1969 to $260 million in 1972.[36]

By 1972 the Anaconda Company had begun a policy of "reorganization and retrenchment."[37] It cut the national headquarters staff by hundreds of people, laid off additional salaried staff around the country, and curtailed wages for those who remained. Anaconda then closed unprofitable and marginal properties. Officials shuttered four brass mills in eastern states and a lead smelter in Utah. They moved on to Montana-based operations. Closing Anaconda's large zinc works in Great Falls eliminated 770 employees. Disposal of a beryllium manufacturing unit and the sale of a zinc flume plant in East Helena followed, as did sale of the company's 650,000 acres of timber in the state for $117 million. In another blow, President Place moved Anaconda's western headquarters from Butte to Tucson, which was closer to the company's more numerous southwestern operations.[38]

Astute Butte employees recognized the implications. They took seriously Place's warnings that problems with "labor productivity" and "distance from market" were sure to handicap Anaconda's success in the region, despite the recent "record-breaking production at the Berkeley Pit."[39] A *Montana Standard* editorial asked locals to have "understanding and cooperation" with Anaconda since it had so little money left in the coffers.[40]

Butte's unions restrained their demands during the 1974 bargaining period. Around the West, 20,000 workers went on strike against others of the Big Four copper companies, but Anaconda's unions reached an agreement before the deadline.[41] While the unions tried to be kind, the company's troubles had another effect on most Butte residents: Anaconda's image as all-powerful and essential to the community had greatly diminished.

## The Continental Pit and the Columbia Gardens

Butte residents' unease with the Anaconda Company increased even more when open-pit mining factored into a community tragedy. Despite residents' intense attachment to the Columbia Gardens, the treasured amusement park and recreation area had, by 1973, become threatened by a new open-pit mine, the Continental Pit. The fight over this mine inspired grassroots community action and the infiltration of environmentalist language into Butte. The subsequent loss of the Columbia Gardens to mining waste dumps generated a sense of betrayal. Many Butte residents therefore became increasingly willing to challenge Anaconda.

The story of the Columbia Gardens' transformation from a recreational landscape into an industrial one begins with the park's early days. Early promotional literature framed the Gardens, whose construction had originally been funded by copper baron William Clark, as a haven from Butte's industrial work and smelter smoke. In one piece, Adolph Heilbronner wrote, "The contrast between the city and the Gardens is striking because there is in it the element of two extremes. In one are the constant hum, activity, heat, dust, impure air, and microbes of nervousness; in the other, pure air, green grass, peace and the unflecked placid serene realm of genuine rest."[42]

Later oral histories suggest that Butte residents continued to accept this portrait of the Gardens as distinct from the noisy, dirty mining operations closer to the city.[43] As onetime Butte–Silver Bow County Attorney Bob McCarthy explained, residents were long willing to sacrifice the environment surrounding their homes to mining refuse as long as they had a beautiful getaway in the Columbia Gardens as a "trade-off."[44]

The Anaconda Company, which had purchased the Columbia Gardens from William Clark's heirs in 1928, operated the Gardens as a form of welfare capitalism—a demonstration of generosity by the community's major employer and a means of discouraging labor unrest.[45] For two residents interviewed years later, the "beautiful" Gardens proved that the company "was going to take care of us as long as they could."[46] Generations of Butte youth had grand memories of the many prom dances held there, and Butte residents still speak glowingly of the grounds, the extensive playground equipment, and the exciting roller coaster, as well as the playground monitors, grounds crew members, and carnival game operators who helped to make childhood summers memorable.[47] In 1962 Anaconda transferred ownership to the Columbia Gardens Foundation, but it continued to be the sole financial support for the enterprise.[48]

COLUMBIA GARDENS. Columbia Gardens featured many popular children's activities, including the roller coaster shown here. Photo courtesy of Butte-Silver Bow Public Archives, Clinton Peck Collection, PH308.12

Welfare capitalism was expensive. William Clark had incurred thousands of dollars in losses each year he operated the Gardens, and the situation did not improve later. Despite a substantial level of investment, Anaconda put fewer resources into the park than Clark had. According to data from 1972, the Foundation operated the Gardens at a large net loss. That year, Anaconda contributed $170,000, but the Gardens brought in only about $20,000, mostly from concessions. The Gardens was still $12,000 short of covering its costs. Robert Corbett, president of the Foundation, estimated the annual loss for the Gardens even higher—at $50,000.[49]

Remembrances of the Columbia Gardens tend not to mention the dwindling numbers that the Gardens attracted in later years—perhaps because actual numbers are hard to come by and perhaps because residents do not want to admit having any role in the Gardens' demise. Like many American amusement parks after World War II, including New York's famous Coney Island, the Gardens suffered a decline due to the proliferation of new forms of entertainment and the growth in car ownership. The removal in 1937 of a streetcar system that had funneled working-class people to the Gardens was a key factor.[50]

Still, the Columbia Gardens remained an important attraction in the early 1970s. Tourists ranked it third, just behind the Berkeley Pit and the World Museum of Mining. Butte's shrinking population, however, hurt the Gardens' long-term outlook. Ted Beech, the Gardens' superintendent from 1948 to 1973, was soon asking groups like the Cristoforo Colombo Club (an Italian-American club) to hold more events to justify the park's existence, but neither such groups nor other locals were using the Gardens as much as they had in previous decades.[51]

Rumors about the fate of the Columbia Gardens spread rapidly after December 1969, when Anaconda's manager of Montana operations confirmed ore exploration in that area.[52] Anaconda hoped to start a new pit to the east of the Berkeley Pit and to the north of the Columbia Gardens. By 1972 the *Montana Standard* had declared a new pit would "eat" the Columbia Gardens. William E. Quigley, the vice chairman of Anaconda's board, tried to calm locals. He admitted the company was planning a Continental Pit, but he reassured the public that any possible movement south into the Gardens area would be "quite a way into the future."[53]

Much changed within a year. When company president Robert Weed attended the Butte Rotary Club luncheon in February 1973, he announced the Gardens would not reopen that summer. Weed cited the need for the Continental Pit, which would extract copper and molybdenum from a newly identified ore body. He then added, "We feel our handling of the facility smacks of old-fashioned paternalism."[54] Correspondence from Anaconda executives to politicians and union officials pointed instead to financial factors. L. Ralph Mecham told Senator Mike Mansfield that, as Anaconda leaked money from its Chilean losses, any "areas that don't directly relate to our obligations to the workers and shareholders" needed to go.[55]

While some residents accepted the economic argument, others definitely did not. Letters to Senator Mike Mansfield and letters to the editor in the *Montana Standard* suggest that many felt incredible shock and anger. Blatant anticompany statements appeared in the *Montana*

*Standard*, a newspaper long friendly to Anaconda's interests. Most letters mentioned how important the Columbia Gardens park was, not only to Butte now, but also to "future generations." Some letters were heart-wrenching pleas from this future generation—the children themselves—while others pointed out that the Columbia Gardens was not just an amusement park, but also a neighborhood full of beautiful homes.[56]

Butte's citizens began to use language from the environmental movement as a way to attack the transformation of a "clean, non-polluted area" into "a big hole."[57] Citizens who had long accepted mine pollution and expansion as signs of inexorable progress proclaimed the new pit would be "the ultimate in pollution [that] will leave a scab on the fair brow of nature."[58] Another letter-writer claimed that the Gardens was "the last gem in an area scarred by mining, and exploited without regard to the environment or humanity." One more claimed that keeping the park was the only way to save Butte from its label as "the Armpit of Montana."[59]

Other letter-writers viewed loss of the Gardens as a violation of the social contract between the company and the larger community. The state had given mining companies "many breaks," such as the power of eminent domain; in return, the Gardens served as "an example of the Company putting back something for the miners and their families."[60] Although mine expansion into primarily working-class neighborhoods had elicited little communitywide response in the 1960s, the threat to the Gardens, an area dedicated to all of Butte, pulled many more people together.[61]

Public attempts to save the Gardens took a number of forms. Some people questioned the legality of the Foundation's closure of the Gardens. A common belief among Butte locals was that William Clark's heirs had included a clause in the deed from his estate to the Anaconda Company that stipulated that Anaconda had to maintain the park for the people of Butte in perpetuity. Research by the local newspaper's reporters failed to substantiate these rumors.[62] State Representative Bob Harper and County Commissioner Si Holman even sought to exercise the county's own powers of eminent domain to save the park. The idea never came to fruition, partly because Montana's counties held little legal power, but mostly because the approximate $200,000 cost of operating the park each year was too high.[63] Others asked if the place could become a state or federal park.[64]

Grassroots organizers formed the People's Committee to Save the Gardens. They spread bumper stickers around town so that locals could express solidarity, put together a petition drive, promoted a letter-writing campaign, and held popular meetings. One member of the People's Committee connected their fight to an international story of anticorporatism: "We can't import Salvador Allende [the socialist president of Chile who nationalized Anaconda's holdings] but we can show the company we're sick and tired of being pushed around."[65] Their efforts centered around one immediate goal: to keep the Gardens open that summer, which would save the rides and pavilion from deterioration and vandalism, while giving community members more time to formulate a plan for the park's future. The group gained support from the Silver Bow Trades and Labor Council, the Butte Boilermaker's Union, and some government officials for both the county and city. Under pressure, the company opened the Gardens that summer.[66]

Having achieved its first major goal, the People's Committee was unable to sustain its intensity. Members believed the next step was one of two options—either to keep the Gardens open permanently or to convince the company to pay for the Gardens to be relocated. No one in the group could find a solid legal way to attack the decision to close the Gardens, which made the preferred option, number one, difficult. Residents interested in a counteroffensive continued to circulate bumper stickers, to write letters, and to attend city council meetings, but the path forward on option one did not become clear.[67] Neither national nor state politicians were helpful. They did not want to step into a "matter under the jurisdiction of the City of Butte and the Anaconda Company," as Senator Mike Mansfield wrote.[68] The Gardens therefore closed its gates for the last time on Labor Day 1972, in an event that employee Frank Panisco described as "a wake, with the whole town turning out." Even former Butte residents traveled from afar to attend the sad event.[69]

The second option—to relocate the Gardens in toto—did not go much better. Local leaders developed plans to relocate the closed park's

CONTINENTAL PIT DEVELOPMENT. This photograph shows the development of the Continental Pit as its operations enveloped the Columbia Gardens. Courtesy of Butte-Silver Bow Public Archives, Clinton Peck Collection, PH308.13.

equipment. Anaconda appeared willing to help.[70] Those plans literally went up in smoke when a fire struck the Columbia Gardens arcade in November 1973, devouring the beloved carousel and other playground equipment. Although the deputy state fire marshal attributed the fire to electrical problems, many residents believed then (and still believe) that a vindictive company deliberately set the blaze as a way to end debates over a future Gardens.[71] Whether or not it was purposeful, the fire was effective. As one newspaper article put it, "The carousel was the high point" of hopes to relocate the Gardens.[72]

If the fire's aftermath appeared to be the final blow, what happened to the old Gardens site made residents almost as upset. Very little shoveling and drilling took place before the company finished a lucrative, but small, mining operation at the Continental Pit a year and a half later. While the short-lived nature of the mine surprised many locals, Anaconda had planned the pit as a short-term operation from the beginning.[73] With the end of the pit, the old Columbia Gardens first became a dumping ground for overburden and then simply was abandoned.

For locals, this alleged misuse of the land was the ultimate offense; now the beloved playground was no longer even a working landscape. Oral histories make it clear that, for a community that so valued industrial work, the crime was not simply the closure of the Gardens, but the fact that the company "never used anything that was there anyway."[74] One locally famous song about the park, "The Ballad of Willie and Millie," ends with the wistful lines, "65 years all gone down in dust and they haven't found that lode yet."[75] Members of mining communities have long thought of their landscapes as vibrant places, not the derelict, dying areas that others make them out to be.[76] The lack of sustained work at this site therefore was incredibly hurtful to many. Some local historians similarly express disgust. They argue that Anaconda simply wanted to stop paying for the amusement park and came up with the excuse of mining to justify the decision.[77]

Tensions rose. Neither the company nor locals could resolve bad feelings generated by the controversy. The problem for Anaconda was that the industrial intrusion of open-pit mining had now affected the greater Butte population. Almost all residents who commented on the Gardens' fate suggest a growing distrust of the company and its plans for mine expansion.[78] As one Anaconda employee put it, "The Gardens was [*sic*] someplace you could get away from the Pit.... When the Company took the Gardens, that turned a lot of people against them."[79]

Many Butte residents determined that they did not need to heed Anaconda anymore. The company was not holding up its end of the deal. A letter to the editor a week after the Gardens fire, for instance, made a call to "save Butte from demolition."[80] According to one historian, locals' later comments often completely forgot the welfare capitalist role of the park and instead remembered the Gardens as belonging to the people of Butte. Locals explained the removal as a tale of crime against the people of Butte and not as a narrative about a struggling company that had to cut costs.[81]

## Butte Battles Bust

The mining unions were the one group in Butte that had served as a counterbalance to the Anaconda Company, but they were not of much help in the Columbia Gardens fight. Montana's union leadership was rightly worried about the future of underground mining in Butte. James W. Murry, executive secretary of the Montana AFL-CIO, explained to a hearing at the U.S. Congress that Butte's unions failed to fight the new Continental Pit, despite the fact that it was bound to destroy the Columbia Gardens. As he put it, "Perhaps the unions should have opposed the opening of the Continental Pit, because of the land-scarring effects it had and the damage to the environment. They chose not to do that. They felt—at least it was tacitly agreed—that, if anything, it would stabilize the underground operation in Butte."[82]

Given this sacrifice, union leaders were particularly taken aback with an October 31, 1974, letter to employees that announced the complete stoppage of all underground mining in Butte. Since 1960 underground mines in Butte had made a profit only during the period from 1968 to 1971. As Anaconda president Place testified, "There was a gradual decline in the grade of underground ore. We encountered high ground temperatures, water difficulties, and incompetent rock. Good veins became a thing of the past."[83] Labor costs were too high to offset such problems. Following the mines' closure, of the approximately 1,000 employees who worked underground, only 165 kept their jobs. The idling of 1,200 to 1,500 workers was disastrous for unions, particularly the BMU, which lost 40 percent of its members. Butte itself suffered a direct payroll loss of $11 million, and 16 percent of the population felt the immediate effects of closure.[84]

As had been true for nearly a century, as mining went, so went Butte. By the late 1960s, Butte, like more-publicized industrial cities in the Northeast and Midwest, appeared in dire need of federal help, local attention, and business investment. Butte's population had been falling since 1920. Butte County held fewer than 45,000 people, down from as many as 80,000 during World War I, and the city had only 23,368 within its small boundaries. Population loss had accelerated in the 1960s, with the long 1968 strike encouraging more outmigration. That three hundred former Butte residents could gather for fun annually in San Francisco was quite a feat, but such events represented a great loss for Butte itself.[85] Falling employment in mining had driven population loss, but opportunities outside that industry had declined as well. Hence, many of those who stayed in Butte became reliant on public assistance. Periodic strikes made the economic situation particularly volatile. During the 1968 strike, Butte experienced an unemployment rate of over 23 percent.[86]

Outmigration and neglect had diluted the tax base and given rise to social problems. Impoverished and immobile senior citizens were living in substandard facilities. Health and social services were inadequate for both the old and the young, while most neighborhoods lacked recreational facilities. Many Catholic grade schools had closed due to both overall population decline and the disruption of ethnic neighborhoods. The high school dropout rate was climbing.[87]

Increasingly concerned about Butte's deteriorating social, economic, and physical condition, local officials commissioned studies. The data demonstrated a big split. According to one report, "Butte may be said to be divided into two cities; an old one and a new one. . . . Half of Butte's population lives on the urban 'hill.' The other half lives in the suburban 'flat.' The hill-dwellers include the basic mass of the deprived. This group occupies the decaying older buildings of the city. It includes the low income and the aged who cling to the past or are unable to alter their status."[88]

The Butte hill, whose copper veins once made it the "richest hill on earth," had become a sorry city center. In uptown Butte, 86 percent of the buildings showed significant deterioration.[89] According to one report, its neighborhoods were "badly blighted by the remains of mining activity, decaying and abandoned commercial and industrial buildings, and vacant lots that could easily pass for junk yards." Everyone and everything appeared to be fleeing the hill. Between 1960 and 1970 Butte proper lost 44 percent of its individual businesses, including 77 percent of its food retailers.[90]

Many people who left the hill arrived on the suburban Flats, often moving beyond the city's boundaries. As Tom Powers, Butte's mayor in the late 1960s, lamented, "The rush to suburbia has further staggered the City's heroic effort to survive and remain solvent." Investment brought new housing and retail to the Flats, and hence the spot remained attractive, but even the new suburbs had problems. Powers noted that they often lacked "sanitary, educational, and recreational

UPTOWN BUTTE, MAIN STREET AT NIGHT, 1939. This photo was taken before uptown's central business district suffered from economic decline and multiple fires during the 1960s and 1970s. Photo courtesy of Farm Security Administration, Office of War Information Photograph Collection, Library of Congress, Digital Id fsa 8a11332.

facilities," and their "streets, curbs, and sidewalks are lacking or need rebuilding."[91]

In the context of 1960s urban America, Butte's residents were not alone. Old, centrally located neighborhoods experiencing blight was an issue observers commonly perceived in American cities. Like pornography, however, blight was easier to see than to define. Leaders described it as a disease requiring removal to save the rest of the city. Through private or public investment, they pursued various types of urban renewal as the fix. Renewal typically involved dramatic changes to the cityscape. As historian Robert Self explains, "In practice, renewal often meant *redevelopment [emphasis in original]*: denuding neighborhoods of low-income housing and small businesses for the benefit of industry and middle-class homes."[92] Although critics like Jane Jacobs had begun to fight renewal plans, the negative aspects of renewal remained largely hidden in the early 1960s.[93]

Enticed by the Kennedy administration's urban funding, Butte's leaders looked to Washington. In 1962 a citizens' committee, formed by Mayor Vern Griffith, hosted meetings to gather resident feedback about the possibility for urban renewal. Despite enthusiastic approval of a plan by Butte's voters, the Federal Housing Authority decided not to engage with Butte. The agency deemed investment uncertain, given the grave effects of mining on the two areas targeted by Mayor Griffith: the East Side, which faced company buyouts, and central Butte, which had experienced much subsidence.[94]

Griffith's successor as mayor, Tom Powers, re-formed an Urban Renewal Committee in 1965, but his administration came under attack for being too permissive toward prostitution and gambling, long-standing pillars of the mining town's social life.[95] Powers' successor, Mayor Mario (Mike) Micone, successfully worked to clean up this vice, which allowed his administration to turn to broader economic woes and physical deterioration.[96]

The Micone administration's first step was to obtain federal money through the Model Cities Demonstration Program. One of President Johnson's urban initiatives, Model Cities required each of the 150 cities awarded grants to create a five-year plan for antipoverty campaigns and new forms of governmental organization. The Department of Housing and Urban Development (HUD) administered the program, but each community was supposed to carry out its own plans through committees that actively involved local residents. Model Cities encountered criticism alleging political favoritism, Congress proved reluctant to commit funds, and most of the program's implementation came during the Nixon administration, whose New Federalism slowly dismantled many federal antipoverty and urban initiatives. In many places, however, Model Cities was successful at giving a real voice to disadvantaged communities and in starting to tackle deep-seated problems through both so-called soft social service programs, and, to a lesser degree, hard, brick-and-mortar solutions.[97]

Butte's Model Cities program, which began in 1969, was one of the more successful ones. Mayor Micone's "headstrong, demanding" manner helped push through a number of new policies.[98] The program operated until 1975 under the direction of Jim Murphy, an emerging mover and shaker who had previously worked in plumbing, and Don Peoples, who became director of Public Works. Murphy, who would often work twelve-hour days, hoped to make Butte more than an old mining town. As he told one public forum, "I am confident we can drag this community, kicking and screaming, into the Twentieth Century."[99]

Over the duration of Model Cities (and a continuation called Planned Variation), Murphy, and new bureaucrats trained through the program, gained $22 million in grants. They helped to fund a local development corporation, an industrial park, a public safety building, a senior citizens activity center, a nursing home, and an early childhood education program. Infrastructure needs, like repairs to streets and the sewerage system, also gained.[100]

Not all of Butte's citizens were happy about Model Cities and its brand of urban renewal. Aldermen voted down renewal plans for the Front Street area, at the bottom of the hill, and the South Butte area, on the Flats. Residents objected to new home requirements and other intrusions into their lives. One resident argued, "I've lived 53 years in that little house and I'll be darned if I give it up for anybody."[101] As writer Teresa Jordan noted, a "basic contradiction" existed between the citizens' individualism, borne of the very independent work of underground miners, and Butte's increasing reliance on federal assistance.[102]

Micone admitted that administration of the program had seen "many small areas of conflict emerge," but both he and Model Cities continued to be popular enough that he gained another term in office.[103] Minutes from neighborhood council meetings show a fair amount of give and take in budget decisions among the many groups of residents, getting more people involved in city government and the community's future.[104]

Despite its successes, Model Cities proved unable to counteract the structural deterioration of the uptown hill; indeed, some thought it actually contributed. Model Cities' ongoing demolition and neighborhood beautification project targeted more than six hundred buildings beyond the 1,166 housing units already demolished by the Anaconda Company by 1971.[105]

The program reflected the postwar culture of clearance. Nearly 6 million U.S. housing units, particularly in inner cities, were demolished between 1950 and 1970.[106] Butte's Model Cities program could neither beautify every vacant lot nor demolish every building that lacked enough inhabitants. Nor, of course, could it satisfy those who objected to this kind of renewal.

Uptown Butte's high vacancy rate was perhaps the major problem that Model Cities failed to solve. Empty structures in uptown became prime targets for devastating fires, and many major historic buildings constructed between 1890 and 1910 fell prey to burning and subsequent demolition.[107] The city experienced fire after fire across the CBD between 1968 and 1978. In just the first five years of that period, the forty-five fires that occurred almost equaled the total number in the previous two decades. In 1968 fire claimed most of the Beaver Block on North Main Street. In 1969 the Heidelberg Bar fire took down four thriving businesses. In 1972 the vacant Lenox Hotel on West Granite Street became the next victim, and the massive J. C. Penney's fire subsequently damaged two downtown blocks. In June 1973 the recently renovated Medical Arts Building became the center of a nighttime blaze that attracted 10,000 or more onlookers.[108]

Financially draining, the fires posed an even greater cost to the psyche of residents.[109] Just as fires represented economic decline and environmental issues in many deindustrializing U.S. cities, these fires signaled a devastating time for Butte.[110] Like fires in neighborhoods targeted by mine expansion, the uptown fires ignited suspicion. Many locals were convinced that all, or most, of the fires had to be the result of arson and quietly alleged that the Anaconda Company and its friends in city government would do whatever they had to do to get people out of uptown, thus enabling Berkeley Pit expansion.[111] Of the ten most significant fires in the district, the official cause listed one as arson, five suspicious, and four undetermined (but not arson). A "Tardy Arson Report," created for the Butte–Silver Bow Urban Revitalization Agency, suggested that the question of cause "was confined to rumors and supposition," primarily because of "the community's poor sense of efficacy and the lack of trained law enforcement personnel to investigate fires of suspicious origin."[112] Some angry residents asked

MEDICAL ARTS BUILDING, 1973. The Medical Arts Building in Butte was damaged by fire in 1973. This fire was one of many that ravaged areas depopulated due to open-pit mining. Photograph by Cleveland S. Bishop. Photo courtesy of Butte–Silver Bow Public Archives, Postle-Bishop Photograph Collection, PH406.05.

for indictments, while state and local authorities argued over a lack of cooperation in investigating what amounted to fourteen unsolved arson cases by 1976.[113]

Other residents simply sank into depression. A letter to the editor about the J. C. Penney's fire, which took out eleven businesses on Dakota and Park Streets in 1972, claimed that fire had dealt a final blow to uptown Butte. "It rends the heart," the writer stated, "to see a city, once considered a crown princess in western America, wither and die like a decrepit woman of the streets.... The explosions and fire that tore at the heart of uptown Butte tore also at the hearts of her citizens."[114] Local newspaper editor Rick Foote later agreed: "There was the sense that the town was literally falling down or burning down.... The community would be psychologically shaken for months after each fire."[115]

Butte differed from many cities facing similar problems at the time in that residents saw an identifiable villain: the Berkeley Pit. At the start of the 1970s, the pit's perimeter sat about five hundred yards away from the CBD. As a 1971 report put it, long-standing "distrust, suspicion, and animosity" toward the Anaconda Company coupled with the company's refusal to make a clear statement about where it would be mining in ten, twenty, or thirty years convinced most Butte residents "that the big bite is imminent." Having experienced industrial intrusion for years, locals now were certain that the pit would swallow uptown in a matter of years. The constant fires seemed to prove the area was already undergoing pit-inspired destruction. Therefore, instead of investing in uptown Butte, residents either fled the area or accepted their fate, "awaiting the imminent apocalypse."[116]

The Department of Community Redevelopment interviewed company officials and some noncompany geologists and found that perhaps residents were mistaken. Unpredictable copper prices and technology would determine the path of future mining. Expansion westward, farther into town, would not happen for at least twenty years, and the horizon was more likely thirty or forty years, if it ever occurred. In fact, conversations with geologists convinced members of the department that people in Butte should "be speculating that the existing Central Business District would never be devoured by a pit rather than speculating on when it will be."[117] Most residents did not absorb this hopeful outlook—the pit had taken on a life of its own, generating as many problems from imagined hazards as from real ones.

Strangers were even more pessimistic. The expanding pit had become the defining feature of Butte's cityscape—big, dramatic, and unlike anything most reporters had ever seen. Their tales therefore emphasized intrusive industry, citizens' malaise, and uptown's fires, all of which seemed to prove the Berkeley Pit's destructive qualities. During the period, outside journalists began a pessimistic tradition that would continue for decades. A reporter for *Newsweek,* for example, believed that Anaconda's pit would soon devour the town. Its growth had created a "general civic malaise" due to the effects of "living on the edge of an abyss."[118] The *New Yorker* agreed. Butte was a ghost town in waiting that "becomes quieter every year except for the never-ending stream of trucks lugging earth up the steep inclines at the Berkeley Pit."[119]

## An Urban Removal Problem

Although outside journalists and some locals foresaw the beginning of the end for Butte, city leaders were eager to create a more economically and psychologically healthy city—one that could move beyond the dark days of the early 1970s. Local officials' primary desire was to craft a long-term strategic plan for the city, but Anaconda consistently failed to discuss its own long-range intentions. Rumors continued to fill the gap left by the absence of accurate information. Often these rumors made it into the local news, thus legitimizing them.[120] Anaconda continued to defend its reticence to explain itself under the grounds that its long-term plans for the Berkeley Pit were unfixed: "The rate The Pit will grow and the direction in which it will move depend on mining technology and the world price of copper." Skeptics believed that Anaconda's silence was a purposeful effort to encourage the continuing decay of the CBD and the fall of land prices there.[121]

State politicians were quick to join skeptics in rumor mongering. In 1974, when the U.S. Senate Subcommittee on Minerals, Materials, and Fuels of the Committee on Interior and Insular Affairs called for a hearing about the underground's closure, discussion quickly turned to the Berkeley Pit. Anaconda's accusers at the hearing assumed Anaconda was performing an underhanded, sly trick on Butte. As Montana governor Thomas Judge told the committee, "This idea may seem farfetched in Washington, D.C., but it's close to reality in Butte. Is the closure of the underground mine setting the groundwork for the ultimate destruction of the city of Butte?" Having just visited Butte, Senator Lee Metcalf, operating as chair of the hearing, gave some credence to claims that the layoff "is just to drive property values down so you can acquire property" in uptown Butte. Anaconda's Montana manager Frank Monninger strongly disagreed, but the situation shows that rumors were quickly becoming as powerful as the Berkeley Pit's actual effects.[122]

The situation facing Butte's uptown both resembled and differed from situations faced by other neighborhoods by Berkeley Pit expansion. As in those sites, the elderly and poor living in the "numerous older apartment houses, rooming houses, and transient hotels that are remnants of Butte's past" made up the bulk of the most vulnerable residents.[123] The district, however, had an advantage in the stake of the business community in the CBD. Butte's power brokers proved to have more interest in uptown than they had displayed about places like Meaderville and Finn Town. Butte's wealthier residents and city officials were now less passive, even if their initial concern about pit expansion was, as an outside journalist put it, "tardy and rather limited," a situation that today's environmental justice advocates would recognize.[124]

Businesspeople, property owners, and government leaders concerned about uptown divided into two camps: those who sought to save the city, including the blighted CBD, and those who wanted to relocate the city, abandoning the hill for the Flats.[125] Butte's City-County Planning Board suggested relocation in 1969, and Model Cities, the federal antipoverty program, prompted further discussion. Jim Murphy, the director of Butte's Model Cities, became convinced that Anaconda's silence was impeding the city's efforts to get federal funding to revitalize uptown. Frustrated, Murphy came to believe that relocating the business district was the only way forward.[126] As Mayor Mike Micone put it, the Model Cities experience had convinced the city's young leadership that "we could do anything with Federal Grants." Perhaps relocation was next.[127]

Concern was high partly because breakup of the CBD seemed imminent, especially after the 1966 opening of Butte Plaza, a new shopping mall on the Flats, and its 1969 expansion. Several businesses in the Model Neighborhood, a section of uptown Butte that served as the focus for Model Cities funding, were relocating to nearby unincorporated areas with lower taxes. Owners of other businesses, however, were eager to rebuild the old city center. They assumed that increased investment in the city would either dissuade the company from taking uptown or allow the CBD to function well until the company was ready to pay fair market value.[128]

County and city authorities were willing to consider relocation, and they put pressure on Anaconda to help fund the necessary preliminary studies. Between 1969 and 1971, Anaconda retained the Urban Services and Housing Development Corporation of San Francisco to consult with county commissioners, the mayor's office, and the Model Cities board of directors on the relative merits of on-site urban renewal and relocation. Out-of-town interests contributed as well. Butte and the American Institute of Architects invited a regional urban design assistance team, which included planners, architects, a geologist, and a botanist, to the city in 1972. Despite Mayor Micone's warning to the group that it would be "political suicide" to do so, the design assistance team recommended an orderly abandonment of the hill. Pit intrusion was "inevitable."[129] By 1974, perhaps $600,000 had gone into investigating the project by several study groups.[130] Often not recognizing city leaders' role, outside journalists labeled the possible relocation of uptown as another in a long list of company crimes.[131]

What city leaders needed was a better idea of Anaconda's plans. On November 3, 1972, Anaconda finally stated a preference for "a long-range orderly relocation of Butte's central business district to a new area."[132] The declaration was clearer than that offered two years earlier, when P. Largey MacDonald, the business manager for Montana operations, had been willing to say only that "those who like the idea of a new Butte business district off the Butte Hill have a supporter in MacDonald; those who do not will get no strong opposition from him, or Anaconda."[133] Nevertheless, confusion continued because Anaconda failed to predict when, if ever, it would need uptown's property.

Local leaders grew impatient. Shag Miller, the owner of KBOW radio, became particularly concerned, directing editorials to city leaders and to the Anaconda Company. Under pressure from uptown businessmen, the joint board of directors of the Model City Neighborhood Council passed a resolution in December 1973, asking Anaconda to halt westward expansion of the Berkeley Pit until a full relocation plan was in place.[134]

In early 1974 the Forward Butte Commission—a nonprofit coalition of the city government, the Anaconda Company, and uptown's major businessmen—jumped into the leadership vacuum. More popularly known as Butte Forward, the commission was a thirteen-member nonprofit body. Shag Miller led the board. Representatives of big business dominated, but activists from the city bureaucracy also played a role. The former director of Model Cities, Jim Murphy, became executive director. They all took seriously the advice of outside planners and architects that the city needed to start coordinating a move. Inaction had led to lethargy in the community and an inability to raise private or federal investment in uptown.[135]

Butte Forward and the local newspaper issued frequent warnings that the advance of the Berkeley Pit into the CBD was inevitable.[136] Despite having two Anaconda Company executives on its board, though, almost two years passed before Butte Forward was able to cajole Anaconda into an announcement about mine expansion. Finally, in July of 1975, Leonard Powell, head of Anaconda's Montana Mining Division, declared that a decision about uptown might not come for twenty years, and then it would depend on the price of ore, cost of gas, and environmental considerations.[137] Moreover, if the company mined uptown, it would use underground rather than open-pit methods. The former raised the risk of dramatic subsidence but, unlike the latter, would not force complete relocation.[138]

Despite continuing uncertainty, Butte Forward increasingly considered CBD relocation a necessity. The people who worked for and with Butte Forward certainly had mixed emotions at the thought of razing uptown Butte, but they maintained that the action was necessary. As board member Gus Kiely proclaimed, this "was not an Urban Renewal Problem, but an Urban Removal problem." Like other industrial cities, another proponent stated, Butte "was dying."[139]

A grand plan emerged. In closed meetings, the board developed a plan to move businesses off the hill, to demolish old buildings in the business district, to create a landscaped park as a buffer zone, and then to sell most of the land to Anaconda. Businesses and residents would subsequently move to a sleekly designed location on the Flats—a development that urban planners at the time called a new town.[140]

The idea of new towns dated to the beginning of the twentieth century when the English reformer Ebenezer Howard argued that it was more difficult to fix current cities than to create new garden cities, with carefully planned residential, commercial, and industrial sectors, all of which would be surrounded by greenbelts. Believing that population decentralization might cure urban ills, President Franklin D. Roosevelt overrode congressional opposition to involve the federal government in the creation of three so-called green towns in the 1930s. Their subsequent struggles could have killed the movement, but the destruction of cities during World War II revived the concept of planned communities. In England, Parliament's New Towns Act of 1946 put the national government in charge of planning, financing, and building new towns—twelve during the next four years. Following Europe's lead, developers in the United States created a few wholly new communities, like Reston, Virginia; Columbia, Maryland; and Park Forest, Illinois, but those ventures were privately financed.[141]

The crises of inner cities in the 1960s and that decade's willingness to look to Washington for answers revived the idea of federal sponsorship. Private developers, the AFL-CIO, and President Johnson all argued for more government assistance in community development. The New Communities Act of 1968 encouraged such programs, and the Housing and Urban Development Act of 1970 provided for direct loans and grants to develop new towns. Initially skeptical, the mayors of large cities became supporters when the programs encouraged the development of what they called new towns in town, or communities built within an existing city.[142]

New towns were supposed to solve the ills of haphazard postwar suburbanization, which had allegedly increased social isolation through sprawl and left the poor behind in decaying city centers. New towns were to contain their own industrial, commercial, and residential zones; they would not be dependent on central metropolitan areas like the suburbs. They would include housing for the disadvantaged, who would join the retreat of the more affluent from old city centers, increasing community health and even racial integration.[143]

Efficiency was the byword. New towns would "conserve land and energy by situating jobs and residences closer together." The physical form of a new town could "form the building blocks of a community consciousness ordinarily lost in suburban sprawl." New town residents would be able to perform their daily activities "within an environmentally sound, economically viable, and socially interactive framework."[144]

For architects and planners hired to work with Butte Forward, Butte was a textbook example of poor planning. Community services, like reliable fire protection and quality schools, would not be possible if Butte continued along the path of "one and two acre plots that are sprawling south and east."[145] Inaction would result in what many urban planners considered the horror of uncontrolled suburban development, represented by Los Angeles. The primary consultant, Mel Levine, was excited that he might be able to rectify so many wrongs in one swoop. Relocation would instead pull the "two distinct Buttes" of uptown and the Flats together.[146]

The new Butte CBD would re-create the vitality of old Butte; it would include a commercial core with shopping and offices and a surrounding residential area large enough not only to replace lost housing but also to provide for anticipated population growth. Symbiosis between the two would have aesthetic, social, and economic benefits, giving businesses a nearby consumer base, while strengthening the tax base. The result would be a lively community center, with "more rational, greater density development." Several hundred people would now live "within walking distance from work—a desirable objective in energy conscious times," the plan stated. As one Butte Forward staff member put it, in this "beautiful plan...you could have gotten anywhere in the CBD without crossing a street."[147]

The vision of Butte Forward ironically looked backward, to a time in the recent past, when most Butte residents lived near their workplaces, including the mines. Butte's old philosophy of living near work, a local reporter agreed, could now be seen as "sound, both physically and ecologically." It, however, needed modification to compensate for the mines' increasing threat to the good life. The goal of a "new Butte" was therefore to separate the "basic ugliness" of industrial areas from a combined commercial and residential development and thus to restore "a degree of harmony between man and his environment."[148]

City Hall welcomed Butte Forward's report. Mayor Micone believed that its vision could avert the problem of scattered relocation all over the Flats, while Public Works director Don Peoples thought the plan would push forward "a very exciting chapter in Butte history." By late 1975 city officials joined the board of directors for Butte Forward, Inc.[149]

Enthusiastic Butte representatives hoped to raise an initial $50 million from HUD, but their timing was not auspicious.[150] President Nixon had never been supportive of new towns and he favored the devolution of social welfare initiatives to states and localities. Nixon battled HUD secretary George Romney, who favored new town development, and allocated little money for the Office of New Community Development. Following Nixon's directive, the Office of Management and Budget approved loan guarantees for some new town projects, but it continually denied requests for the grant programs present in the 1970 law. By 1975, when Butte Forward released its preliminary report, HUD was announcing that its Office of New Community Development would no longer accept new applications.[151]

Butte Forward still had options. Senator Mike Mansfield helped the group and city officials to gain interest from the Economic Development Administration (EDA), which put together an independent project review commission to examine the need and feasibility of relocation. During the review commission's meetings, one agency member claimed that Butte had a 50 percent chance at gaining enough federal monies to make relocation a reality. The Special Economic Development and Adjustment Assistance program, which was part of Title IX of the Public Works and Economic Development Act, might offer a way forward. Butte Forward put up $50,000 and the EDA another $200,000 to hire American City Corporation, a private consulting firm from Maryland, to make detailed recommendations about relocation.[152]

The EDA's national Project Review Commission noted that it would not be able to give a positive recommendation until Butte could demonstrate strong public support for relocation and firm financial backing by the Anaconda Company. Neither was quickly available.[153] When Butte Forward held public meetings in early 1976, the city remained divided between those who wanted to relocate uptown and those who wanted to save it.[154] That was not surprising; communities considering relocation tend to divide into contentious factions.[155] Moreover, the

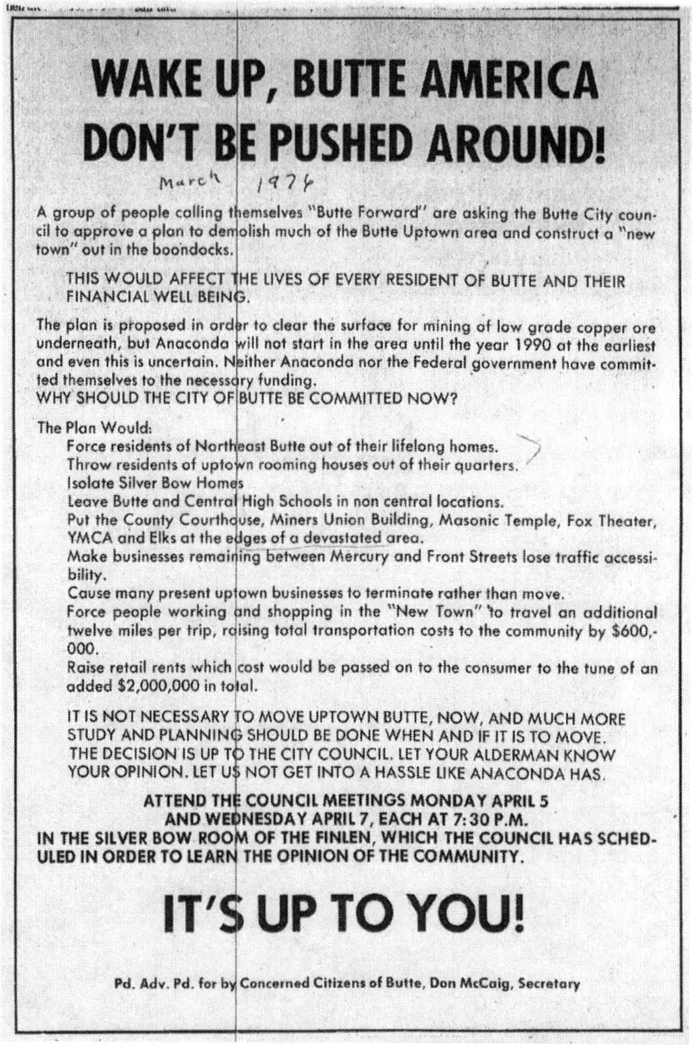

WAKE UP, BUTTE AMERICA
DON'T BE PUSHED AROUND!

A group of people calling themselves "Butte Forward" are asking the Butte City council to approve a plan to demolish much of the Butte Uptown area and construct a "new town" out in the boondocks.

THIS WOULD AFFECT THE LIVES OF EVERY RESIDENT OF BUTTE AND THEIR FINANCIAL WELL BEING.

The plan is proposed in order to clear the surface for mining of low grade copper ore underneath, but Anaconda will not start in the area until the year 1990 at the earliest and even this is uncertain. Neither Anaconda nor the Federal government have committed themselves to the necessary funding.
WHY SHOULD THE CITY OF BUTTE BE COMMITTED NOW?

The Plan Would:

Force residents of Northeast Butte out of their lifelong homes.
Throw residents of uptown rooming houses out of their quarters.
Isolate Silver Bow Homes
Leave Butte and Central High Schools in non central locations.
Put the County Courthouse, Miners Union Building, Masonic Temple, Fox Theater, YMCA and Elks at the edges of a devastated area.
Make businesses remaining between Mercury and Front Streets lose traffic accessibility.
Cause many present uptown businesses to terminate rather than move.
Force people working and shopping in the "New Town" to travel an additional twelve miles per trip, raising total transportation costs to the community by $600,000.
Raise retail rents which cost would be passed on to the consumer to the tune of an added $2,000,000 in total.

IT IS NOT NECESSARY TO MOVE UPTOWN BUTTE, NOW, AND MUCH MORE STUDY AND PLANNING SHOULD BE DONE WHEN AND IF IT IS TO MOVE. THE DECISION IS UP TO THE CITY COUNCIL. LET YOUR ALDERMAN KNOW YOUR OPINION. LET US NOT GET INTO A HASSLE LIKE ANACONDA HAS.

ATTEND THE COUNCIL MEETINGS MONDAY APRIL 5 AND WEDNESDAY APRIL 7, EACH AT 7:30 P.M. IN THE SILVER BOW ROOM OF THE FINLEN, WHICH THE COUNCIL HAS SCHEDULED IN ORDER TO LEARN THE OPINION OF THE COMMUNITY.

IT'S UP TO YOU!

Pd. Adv. Pd. for by Concerned Citizens of Butte, Don McCaig, Secretary

FLYER PROTESTING PLANS TO RELOCATE THE CENTRAL BUSINESS DISTRICT, 1976. Residents protested relocation to make way for mine expansion. Photo courtesy of Butte-Silver Bow Public Archives, Butte Forward Collection, Box 3A, folder "D. J. McCaig," "Wake Up, Butte America Don't Be Pushed Around!"

Anaconda Company did not seem ready to make a full commitment. Nevertheless, on April 21, 1976, the city council voted to further explore relocation.[156]

Anger grew, especially among small business owners; they did not have access to the Forward Butte Commission, which the city's major economic actors enjoyed. Two small business owners, Beverly Hayes and D.J. McCaig, emerged as the opposition's spokespeople. Large ads published in the *Montana Standard* by an organization that called itself the Concerned Citizens of Butte screamed, "Wake Up, Butte America!"[157]

Choosing the site to which the CBD was to be relocated became, in the words of Mayor Micone, a "real circus." Rocker, an area a few miles to the west of uptown, was a contender, as were several sites on the Flats south of uptown, including two near the airport. Many city leaders favored putting the CBD where the airport currently stood. Mayor Mike Micone and others talked about moving the air terminal, because it was inconveniently located right next to the mountains and hence was a relatively poor landing spot. Just as important, the airport was on land that the city already owned and it was far away from Anaconda's mining operations.[158]

The city's bankers, other major uptown business leaders, and a faction in the city council preferred Stodden Park to the airport. In southeast Butte, the park was more centrally located, with great transportation access for most Butte residents. The city's ownership of the park was another advantage, unmatched by competing sites other than the airport. A nine-member, ad hoc committee of major uptown businessmen, put together by Butte Forward, recommended Stodden Park as the best location in May 1976. Jim Murphy said that the city needed to acquire only the sixty-four-acre golf course and seventeen additional vacant acres next to it. A park area containing a new city pool, he claimed, would not be touched.[159]

Endorsement by Butte Forward was enough to stir opposition to the Stodden Park site. The park had quickly become the city's largest recreation area. It featured softball fields, a picnic area, an ice skating rink, tennis courts, and more on its 104 acres. Most of its new features had only been completed in 1974, using almost $1 million of Model Cities federal funds.[160] Butte had lost its treasured amusement park and recreation area, the Columbia Gardens, only two years before the new Stodden opened. "Wasn't it sad when Columbia Gardens was dismantled?," mourned a typical letter to the editor. It continued, "We were helpless to prevent that, as it was company-owned.

So please, good citizens of Butte, since we all own Stodden Park...let your voice be heard."[161]

Butte's residents worried not only about environmental dangers but also about the loss of environmental advantages, such as the park's open space and recreational opportunities. At two public meetings in June 1976, most attendees spoke vigorously against the Stodden Park site. They objected to losing some of the park's grounds and feared that increased traffic would make what remained of the grounds dangerous for children. At the end of the second meeting, Mayor Micone was reduced to pleading for any positive input about possible sites.[162]

In mid-June, Micone set up a special committee of the city council to select an acceptable site. A few members of the body opposed the whole idea of relocation. Other members requested an alternative site, but Butte Forward, Murphy, and major merchants continued to hold firm for Stodden Park. That option was the first to be rejected. One committee member was convinced that the site was so small that the new CBD would have to take up the rest of the park, including the newly built pool and recreation complex.[163]

By July, a new location, close to Stodden Park, with which it shared similar commercial advantages, joined the competition, much to the dismay of Butte Forward. Known as the Spolar site, it had not been the subject of studies, and Butte Forward had not obtained an option on the property, as it had done for other possible locations.[164] California developer Kenneth Cummings, who held options on fifty acres of the site next to the stockyards and the city dump, intended to develop the area into a shopping center. He, however, preferred to act with the city's help and announced that he was willing to sell for an expensive $15,000 an acre. Another fifty-eight acres next to Cummings' site were needed, available, and expensive; the price quote was $1 million.[165]

With some malice, Beverly Hayes and florist Aggie Jean Golubin decided to put their faction of uptown merchants opposed to relocation behind the Spolar site. Its advantage, in their thinking, was being "thoroughly unattractive, both psychologically and financially." The Spolar site reeked so much that opponents nicknamed it Manureville. It was next to a railroad crossing, it faced dust from the old Clark Mill's tailings, and it cost too much money. In addition, the stockyards would have to be moved, which might mean the loss of another business from the area.[166]

On July 6, after much debate, the special committee recommended the Spolar site. Some, like Alderman Terry Kelly, probably chose it as a protest vote against relocation more broadly. Others acquiesced because they had run out of options; their favorite sites had been eliminated one by one.[167]

Critics of Butte Forward, like Beverly Hayes, held a major advantage in the battle over a site. Many Butte residents suspected that Butte Forward was nothing more than a tool of the Anaconda Corporation. An outside reporter noted that "most of the working class residents on the hill regard the Forward Butte Commission as a sham," with one interviewee claiming that the commission's "businessmen are only figureheads for the company," and another arguing that the commission's job was to put a "smiling face" on the company's "dirty work."[168] Suspicion about Anaconda, which was the product of a one-company-town mindset, had long been a way of life in Butte, and Butte Forward had operated mostly in secret for two years before the full plans for relocation became public.[169] In addition, the Anaconda Company provided 80 percent of Butte Forward's budget, gave the group office space, contributed two members to the board, seemed to be the major reason for uptown's problems, and looked to be the main beneficiary of the land deal.[170] As restaurant owner Bill Stimpson argued at one public meeting, "The Anaconda Co. is instigating this whole program. They'd like to see us all go belly up, then they could walk in and pick it (the business district) up for nothing."[171]

Fears of Anaconda's influence were exaggerated. The company remained reticent when Butte Forward pleaded with it for a financial commitment to relocation. Likewise, it took a pass on the offer, in 1974, of Butte Forward director Jim Murphy to testify on Anaconda's behalf during congressional hearings. Murphy said he would do so if the company was willing to announce expanded commitment to the new town project.[172]

By late 1975 Butte Forward's relationship with Anaconda was still surprisingly weak. As Jim Murphy explained, "Anaconda originally seemed enthusiastic about the project and although local [Butte]

Anaconda officials are still supportive, the project now seems to have low priority in overall company picture because of 1. The financial and other problems facing the Company (including the nationalization of its bountiful Chilean properties)" and "2. The benefit to the company, i.e., the title to the vacated CBD land, has only long range significance."[173] In a closed Butte Forward meeting, Murphy pleaded with the president of Anaconda's Montana Mining Division to "assume the risk regarding the feasibility and uncertainty of the time frame inherent in long term ore recovery" under uptown, instead of the company's hesitant approach.[174]

Suspicions remained, despite Anaconda's lack of outward interest in uptown. The company's offer of money to Butte Forward was contingent on the company gaining full title to the CBD by 1981, which fueled speculation that Anaconda's denials of plans to mine there in the near future were untrue.[175] Moreover, most people dismissed or forgot Anaconda's announcement in 1975 that its only scenario for uptown involved underground, not open-pit, mining. Nothing could dissuade either local or outside skeptics from the expectation that the Berkeley Pit would devour uptown.[176]

Anaconda's penchant for secrecy was the problem. Opposition leader Beverly Hayes argued, "If they had come to us and said flat out that they wanted the land and would have built us a new town, we probably would have agreed. But they didn't. They act like we are all idiots."[177] To local politician Don Ulrich, Anaconda's often-conflicting statements about uptown Butte's future diminished the company's credibility with Butte residents and thus Butte Forward's efficacy.[178]

Further impeding relocation plans was the emergence of a historical preservation movement.[179] Both Butte Forward and the local newspaper received letters from people who were quite emotional about the current CBD structures, stating that they were personally and deeply connected to the space of uptown where they lived, ran longtime family businesses, or shopped. A student at one of the packed city council meetings summed up these feelings, decrying a plan to "destroy 100 years of history."[180] For many residents, the damage that mining could one day cause to the business district seemed distant and much less inevitable than the loss threatened by relocation. The Save the Central

Business District Committee gathered six thousand signatures in support of preservation. The new group contacted congressional representatives and even HUD, asking for revitalization of the current CBD, not relocation.[181]

Plans for relocating the CBD nevertheless moved forward. The EDA's national Project Review Commission had given city officials and Butte Forward permission to seek federal funding. Montana's governor, Thomas Judge, had committed the state to assisting Butte, the Butte Chamber of Commerce was in favor, and the major businesses located in uptown had officially committed to joining a new town development.[182]

The turning point on relocation occurred on July 7, 1976, when the final vote for site selection came before the entire city council. After the council president presented a motion in support of the Spolar site, Mayor Micone read letters from John Place, the president of the Anaconda Company, and from Shag Miller, the president of Butte Forward. Together, the communications derailed the hopes of those supporting relocation.

In his message, John Place made a financial commitment of $11 million to the project. The amount was the minimum requested by Butte Forward, and $2.5 million of it consisted of the value of the company's uptown buildings, which would simply be sold right back to Anaconda when the CBD moved. The company's small contribution fell far short of the minimum expected cost of $50 million for relocation. It also increased the concerns of those who were aware that Mike Mansfield's impending departure from the Senate would decrease the likelihood of major federal contributions to the project.[183] Director Jim Murphy asked the council to study the letter from President Place before it made its decision, but, as a reporter put it, "his suggestion was ignored," while Alderman Bill Robinson proclaimed that the company's offer "don't help us a damn bit."[184]

Shag Miller's letter claimed that its drawbacks would make the Spolar site a "grave mistake."[185] Then Robert Geach, an engineering instructor at Butte Vo-Tech, made the surprise announcement that the Spolar site might overlie a large deposit of porphyry copper, which would make it a potential site for mining. A divided city council still voted eight to five for relocation to the Spolar site, but a procedural move by one of the opposed aldermen shelved resolution of the issue for two weeks. In the meantime, Butte's papers erupted in contentious discussion about the merits of relocation sites, public costs, and company involvement. The editors of the *Montana Standard* wondered why the city council was sabotaging a project into which so much effort had already been put.[186]

On July 21, 1976, the Butte City Council finally made what historian Brian Shovers describes as a decision that was "possibly the most important in its one-hundred-year history."[187] During a "tumultuous meeting repeatedly interrupted," Jim Murphy tried to save the relocation effort by stating that property owners next to the airport were willing to make a grant of seventy-five free acres. Most council members, however, deemed the land as being too far from the city's growth pattern. After the council reaffirmed a vote for the Spolar site, Alderman Phil Paull, chairman of the city council's uptown study committee and a member of Butte Forward's board of directors, commented, "I think it is time to quit this exercise in futility and I move we instruct the mayor to stop seeking federal funds for the relocation site and leave the central business district where it is." His announcement was "greeted by whistles and handclapping." In a nine-to-four vote, the council voted down the plan to move Butte's city center. Alderman Bill Robinson advised, "We better drop the subject while people are still speaking to each other."[188]

The city council's about-face ended the drive for relocation. Two days later, on Friday, July 23, Butte Forward met in Washington with the EDA and HUD. Although the federal directors had figured out a way to get the remaining money needed to fund the project, Butte Forward's leaders had to admit their local defeat. Nevertheless, HUD gave the city $32,000 to examine alternative sites for Butte's airport, in case the current site would be needed for a future CBD. Butte Forward and the American City Corporation finished a final report later that year, in the event that the city's residents would come to their senses, asking for relocation in the future. Then Butte Forward's members cleaned out offices and moved on with their lives.[189] As one board member said, "I know that this is all going to have to go sometime. But I'm kind of glad that the plan was rejected, so I won't have any responsibility for tearing it down."[190]

Historian Bridget Barry argues that the opposition's ability to push Manureville to the top of the site selection list explains the relocation effort's failure; nevertheless, it is important not to minimize the importance that Butte Forward's preference for Stodden Park had in galvanizing opposition. Butte residents were unwilling to lose another green space so soon after their failure to keep the Columbia Gardens.[191] Historian Brian Shovers is probably even more astute in alternatively positing that the looming loss of uptown created "a sense of community" in Butte and hence led to a "psychic" reaction.[192]

Were fears that the loss of uptown was imminent based in reality? Probably not. In the early 1970s, William A. Humphrey, who had managed the Anaconda Company's operation in Cananea, became vice president of Planning in Butte. He proposed creating a massive pit, like the one he had created in Cananea. This pit would have devoured not only Butte's CBD, but also much of the rest of uptown, all the way to the edge of the Montana Tech campus, on the far west side. When company leaders refused even to contemplate such a proposal, Humphrey left the city.[193] Indeed, between 1971 and 1975 westward expansion of the Berkeley Pit had stopped.[194]

Technical and logistical problems limited prospects of mining in uptown. In 1974 Anaconda's Montana manager, Frank Monninger, reported that the presence of much worthless ground between the ore and surface ruled out open-pit mining there. Nothing in the Anaconda Geological Department's records suggests that Anaconda had plans for the area beyond the long-term possibility of underground mining. That underground project, however, was "not feasible at this time" because it would disturb many "surface features," including the company's own warehouse and machine shops. Equipment, fuel, and property costs would have to all be very low, and copper prices would have to be quite high, for such a project to proceed.[195]

In the end, the narrative of inevitable mine expansion seems to have had a greater impact on local progressives, outsiders, and even Anaconda Company managers than it had on ordinary residents of Butte, especially those not living in proximity to the hole. The thought of the Berkeley Pit's inevitable consumption drove local progressives to save their community through the creation of a new Butte. Company managers, many of whom were from Butte, were torn between their corporate responsibility not to rule out mine expansion and their eagerness to help the city. After public leaders put their efforts behind relocation, Anaconda became convinced that locals wanted the change. The vote against relocation blindsided its management. A spokesman exclaimed, "We just assumed that most people wanted the move."[196]

Butte Forward's fatal error, according to deputy director Bob Corbett, was its failure to ask for early input from other community groups, like the Butte–Silver Bow Anti-Poverty Council, the City-County Planning Board, and smaller merchants. After Jim Murphy abandoned it to work for Butte Forward, the Model Cities board of directors protested his new employers' "lack of grassroot citizen representation."[197] Corbett explained that Butte Forward wanted greater community involvement, but the group needed to ensure its own financial support and it faced very tight deadlines. Rumors and misinformation began circulating even before Butte Forward released its preliminary report.[198] Its consulting firm, the American City Corporation, subsequently tried to get citizen feedback, but relied on what an opponent called a "perfunctory" questionnaire sent to only some uptown merchants.[199] An ad hoc staff could not save the project during public meetings. As one resident at a public meeting claimed, Butte Forward had a grand vision, but it seemed like "the general populous has no conception of what this dream is."[200]

Residents' concerns about their built and natural environment shaped their reaction to the relocation scheme. Unlike residents of many towns facing what social scientists call development-induced displacement and resettlement, people in Butte were perhaps less concerned about industrial hazards than they were about environmental and historical resources. Thus, the fate of Stodden Park and the symbolic importance of CBD became salient factors in their attitude.

Butte's longtime relationship to the Anaconda Company also severely damaged residents' faith in the relocation plan. Unlike most large cities facing renewal projects, Butte was a *one-company* town. Butte's residents looked to Anaconda for economic sustenance, but the confrontational union culture there had long made locals suspicious of any action involving the company.[201] The one-company-town mindset also made Butte Forward's supporters too dependent on hesitant company speeches and plodding actions. Anaconda vice president Martin Hannifan was prophetic in his 1972 prediction that company involvement would be "the kiss of death" for a relocation project. After the 1976 vote, the local newspaper and at least one Butte Forward board member agreed.[202]

Butte might have been a one-company town, but it was not a *company-owned town*, which would have made relocation easier. When Butte Forward sought advice from other places that faced open-pit expansion, the Phelps Dodge mining company replied that moving Morenci, Arizona, was simple because the company transported the town from one company-owned plot of land to another.[203] Most companies told variations on that story.

The closest parallel to Butte's situation turned out to be the tiny town of Bingham Canyon, Utah. Kennecott Copper's massive Bingham Pit encroached on town during the 1960s. President C. D. Michaelson of Kennecott's Metal Mining Division explained that moving the town involved a natural process spread out over several years. Kennecott bought privately owned parcels "whenever they became available.... As the purchases were made the buildings were demolished and the population of the town of Bingham decreased accordingly. This action, in turn, caused the merchants to move to other locations outside of Bingham.... [Finally,] when the population of the town became small enough to justify the move, the schoolhouses were sold to the company, as were eventually the town hall, firehouse, post office, etc."[204]

Michaelson overlooked many struggles between Kennecott and residents, but he was right that, by the time the company forced Bingham Canyon to disincorporate, its earlier purchases, coupled with threats to use mining's power of eminent domain, had taken their toll. The remaining residents settled a lawsuit with the company in 1971, and Bingham Canyon became part of the Bingham Pit.[205] This parcel-by-parcel operation seemed much more in line with what Anaconda had been successfully doing all along in Butte. Anaconda had been doing just fine without Butte Forward's help and perhaps could have continued doing so if it wished, which is perhaps a good indication that it did not want to do so.

Conflicting visions of the landscape and Anaconda's relationship to Butte thus stifled the possibility for relocation. Butte Forward saw the urban landscape as facing an impending, pit-induced death, but other residents increasingly saw the landscape as having great value.

## Locals Move On

With the city council's decision joining the later Hillcrest case, the years 1976 and 1977 were a turning point in the relationship between Butte residents and open-pit mining. The general citizenry had become more critical of the Anaconda Company, and many believed the defeat of relocation was a defeat for Anaconda, even if company officials were not of that opinion. The battle over urban removal had turned economic and social crisis into an opportunity, partly because newer ideas about landscape preservation had infiltrated Butte in the guise of historic preservation and environmentalism.

Local politicians determined that mine expansion should no longer stall its efforts at city planning. Reporter Rick Foote explained, "Butte developed a sense of going forward despite the adversities that were in place."[206] Leaders managed to successfully consolidate Butte's city and county governments, an initiative that had failed three times in the past. Historian Brian Shovers believes that consolidation had mostly positive effects: "Local government was streamlined, patronage cut, and government costs reduced."[207] Butte residents, who had mostly thought about lives in their own neighborhoods just two decades earlier, seemed now to be governing and thinking bigger. As Mayor Mike Micone put it, the atmosphere was different in Butte by the late 1970s; no longer apathetic, the populace became more involved in the city's future.[208]

Residual fears for the future of uptown gradually receded. The boundaries of the Berkeley Pit's boundaries never moved farther west into town. Although it warned against too much investment in uptown redevelopment, in 1977 Anaconda's management admitted the company would consider mining below the Berkeley Pit long before it would dig under the CBD.[209] In 1979 the company's Butte manager, Dan Rovig, confirmed that mining would not affect the CBD for at least twenty-six years, if ever. Open-pit mining, in the meantime, would continue only to the east and so therefore "uptown renovations can progress."[210] It was not exactly a ringing endorsement, but uptown's future seemed more assured than it had for decades.

Outsiders had obsessed over the pit during the relocation debate because it engaged the trope of bust in the classic mining town narrative, but by the end of the 1970s, Butte's residents were beginning to feel more positive about the town's future. Their views were growing farther apart from outsiders, who still saw Butte as a dying mining camp.[211]

## Notes

1. Oriard to Robichon, June 19, 1970.

2. James DeWolf, "Public Gets No Say on Anaconda Plan," *Billings Gazette*, Jan. 5, 1976; *Kadillak v. The Anaconda Co.*, 184 M 127, 602 P2d 147 (1979).

3. *Kadillak v. The Anaconda Co.*, 184 M 127, 602 P2d 147 (1979).

4. Deposition of David Maughan, May 13, 1977, *Kadillak v. The Anaconda Co.*, 2nd District Court, Silver Bow County, no. 61740, in Box 97, Series 2, Clarence C. Gordon Papers, Manuscript Series 611 (hereafter MSS 611), K. Ross Toole Archives and Special Collections, Mansfield Library, the University of Montana, Missoula (hereafter UMM), 39.

5. Final argument annotated, *Kadillak v. The Anaconda Co.*, 2nd District Court, Silver Bow County, no. 61740, in Box 25, Series 2, MSS611, UMM, 87–90.

6. *Kadillak v. The Anaconda Co.*, 184 M 127, 602 P2d 147 (1979).

7. Jeri Pullum, "Mine Dump Sites Concern Locals," *Mont. Std.*, June 24, 1980.

8. Foote interview.

9. Paul Ehrlich, *The Population Bomb* (New York: Sierra Club/Ballantine Books, 1968); Hal K. Rothman, *The Greening of a Nation?: Environmentalism in the United States Since 1945* (Fort Worth, TX: Harcourt/Brace, 1998), 83–134; J. Brooks Flippen, *Nixon and the Environment* (Albuquerque: University of New Mexico Press, 2000); "The Genius of Earth Day," *Environmental History* 15 (Apr. 2010): 194–205.

10. Harry W. Fritz, "The 1972 Montana Constitution in a Contemporary Context," *Montana Law Review* 51 (1990): 270–74; Dorothy Bradley, interview by Bob Brown, Sept. 29, 2009, Missoula, MT, Oral History 396-62, ir.lib.umt.edu:8080/xmlui/bitstream/handle/10844/630/OH396-62.pdf?sequence=2>, UMM.

11. Fritz, "1972 Montana Constitution," 272–73; Brian Shovers, "Forrest Anderson, the 1972 Constitution, and the Reshaping of Montana," *Montana: The Magazine of Western History* 64, no. 1 (Spring 2014): 41–61; Kelly Kirk, "State of Change: Gender and the 1972 Montana Constitutional Convention," (master's thesis Montana State University, Bozeman, 2011), 1–9; Kelly Kirk, "State of Change: Women and the 1972 Montana Constitutional Convention," *Montana: The Magazine of Western History* 66, no. 2 (Summer 2016): 3–24.

12. Malone, Roeder, and Lang, *Montana*, 395.

13. Jim Posewitz quoted in Rob Krueger, "Relocating Regulation in Montana's Gold Mining Industry," *Environment and Planning A* 34 (2002): 867–81, quote on 873.

14. Malone, Roeder, and Lang, *Montana*, 394–95. See also Harry W. Fritz, "The Origins of Twenty-First-Century Montana," *Montana: The Magazine of Western History* 42, no. 1 (Winter, 1992): 77–81.

15. Barton H. Thompson, Jr., "Constitutionalizing the Environment: The History and Future of Montana's Environmental Provisions," *Montana Law Review* 64 (2003): 157–198, quote on 165; C. Louise Cross, "The Battle for the Environmental Provisions in Montana's 1972 Constitution," *Montana Law Review* 51 (1990): 449–57.

16. Transcript of Proceedings, Montana Constitutional Convention, vol. 6, Mar. 2, 1972–Mar. 5, 1972 (Helena, MT, 1972), Montana State University Library, internet archive, 3933–40, 3959, 3961; Deborah Beaumont Schmidt and Robert J. Thompson, "The Montana Constitution and the Right to a Clean and Healthful Environment," *Montana Law Review* 51 (1990), 414–19. On the Montana coal boom, see Sally Jacobsen, "The Great Montana Coal Rush," *Bulletin of the Atomic Scientists* 29, no. 4 (Apr. 1973): 37–42; James Conaway, "The Last of the West: Hell, Strip It!" *Atlantic Monthly* 232, no. 2 (Sept. 1973): 91–103;

17. "Eminent Domain Bill Is Modified," *Mont. Std.*, Jan. 27, 1973; Cody Ferguson, *This Is Our Land: Grassroots Environmentalism in the Late Twentieth Century* (New Brunswick, NJ: Rutgers University Press, 2015), 19–70; "Judiciary Committee Minutes," Jan. 26, 1973, 43rd Montana Legislative Assembly, 1973, Fldr. 2 "House Judiciary Committee," Box 2, Montana Legislative Assembly Records LR43, MHS; James R. Shea, draft of comments, Fldr. 1, File 3, Box 1, James Shea Papers, PC104, BSBPA.

18. Krueger, "Relocating Regulation," 872–75; Schmidt and Thompson, "The Montana Constitution," 428–30.

19. "Conservationists Rally to Help a Boycotted Friend," *New York Times*, Dec. 9, 1970. See also Mr. and Mrs. Tony Schoonen to Public Instruction Superintendent Dolores Colburg and Attorney General Robert Woodahl, Mar. 4, 1970, in Fldr. 13 "General Correspondence, 1970," Box 19, Record Series 111 (hereafter RS111), MHS; Jim Huggins to Robert Woodahl, Mar. 3, 1970, RS111, MHS; Bill Ohrmann to Robert Woodahl, Mar. 3, 1970, RS111, MHS.

20. "Anaconda Co. Resumes Lincoln Area Exploration," *Mont. Std.*, Aug. 16, 1973; Heather Abel, "Where One Sister Sees Gain, Another Sees Ruin and Loss," *High Country News*, Dec. 22, 1997.

21. Johansen, "Butte, MT: Sitting Atop a Fortune."

22. On the subject of campaigns to beautify Butte, see Don Ulrich, interview by Teresa Jordan, Jan. 7, 1987, Butte, MT, transcript, OHC, BSBPA; "Common Market Advocated for Magicland," *Mont. Std.*, Mar. 29, 1962.

23. Among many examples of negative portrayals, see "Into the Pit," 46.

24. "Robbing the Grave: Arsenic and Old Sulphur—Anaconda Piles It On," *Borrowed Times* (Missoula, MT), Feb. 7, 1973, 18–19; "An Old Battle with New Allies: The Anaconda Company's Fight Against Pollution Control," *Borrowed Times*, Nov. 1975, 4; "Life on the Hill: The Copper Reduction Plant in Anaconda," *Borrowed Times*, Mar. 1–15, 1973. On the Steelworkers' early 1970s stand against pollution, see

"Pressures Mounting to End Pollution," *Nonferrous Report* (USWA) 7, no. 1 (Jan. 1972): 4, Fldr. "Copper Nonferrous," Office Files General, Bernard Greenberg Papers, Collection 1573, PSHCLA.

25. Quotes in "What Is Anaconda?" *Anaconda Company Trailsman*, July 1973, 3, VF "ACM: General Mining," BSBPA; "What Is Anaconda?" *Anaconda Company Trailsman*, Aug. 1974, 4, VF "Anaconda Company Trailsman," MHS. See also "Site of New Smokeless Copper Plant," *Anaconda Company Trailsman*, July 1973, 4, VF "ACM: General Mining," BSBPA; "Environment Program Expands," *The Anacondan* (Butte, MT) 2, no. 3 (Nov. 1965): 1–2, MHS.

26. "We All Depend on Mining," *Anaconda Company Trailsman*, July 1973, 4, VF "ACM: General Mining," BSBPA. On the mining industry's uncomfortable encounter with 1970s environmentalism, see Duane Smith, *Mining America: The Industry and the Environment, 1800–1980* (Lawrence: University of Kansas Press, 1987), 136–61.

27. "Anaconda's Meeting," *Mont. Std.*, May 23, 1971.

28. "Anaconda Co. Problems Aren't Insurmountable"; "Four Copper Producers Cut Exploration Efforts," *New York Times*, Jan. 1, 1972.

29. Finn, *Tracing the Veins*, 58–59; Navin, *Copper Mining and Management*, 215–21.

30. "Anaconda Head Denies Part in Chilean Plots," *Mont. Std.*, Mar. 28, 1973; Finn, *Tracing the Veins*, 64; "Chile Takes Control of Copper Resources: Kennecott, Anaconda, Cerro Lose Mines," *Wall Street Journal*, July 19, 1971; "Anaconda Says Chile to Pay It $253 Million Sum," *Wall Street Journal*, July 25, 1974.

31. "From Riches to Rags," *Forbes*, Jan. 15, 1972.

32. *Anaconda Company Annual Report, 1971*, VF 1432, "General Mining: Annual Reports," BSBPA, 3–9; Navin, *Copper Mining and Management*, 215–18, 222; William A. Humphrey, "Mining Operations and Engineering Executive for Anaconda, Newmont, Homestake, 1950 to 1995," interview by Eleanor Swent, 1994–1995, Western Mining in the Twentieth Century Oral History Series, ROHO, 104–6; Perry, "A Half Century," 113; "Haulage Costs," *Mont. Std.*, Nov. 30, 1973.

33. Finn, *Tracing the Veins*, 64–65; "Welcome Decision," *Mont. Std.*, Jan. 9, 1972.

34. Oct. 27, 1971, Minutes of Anaconda Job Evaluation Meetings, Fldr. "Anaconda Co. General Correspondence 1971," Box AX/B40/HCLA/03098, USWA Wage Department, USWA, PSHCLA.

35. Robert Walker, "Presidential Choice Surprises Butte," *New York Times*, May 23, 1971.

36. Ibid.; "From Riches to Rags," 24.

37. *Anaconda Company Annual Report, 1971*, BSBPA, 8.

38. Testimony of John B. Place before the Subcommittee on Minerals, Materials, and Fuels, Committee on Interior and Insular Affairs, U.S. Senate, Nov. 23, 1974, U.S. Senate Subcommittee on Minerals, Materials, and Fuels of the Committee on Interior and Insular Affairs, *Closure of Underground Copper Mines: Hearing to Review the Anaconda Company's Plans to Close Its Underground Copper Mines in Butte, Montana*, 93rd Cong., 2nd sess., Nov. 25, 1974, 9; "Anaconda Salaried Layoff Under Way," *Mont. Std.*, Oct. 31, 1971; "Anaconda Slices Staff," *Mont. Std.*, Nov. 10, 1971; Bert Gaskill, "Anaconda Sells Lumber Division to U.S. Plywood," *Wall Street Journal*, May 9, 1972.

39. Bert Gaskill, "18.6 Million Is Butte's Share," *Mont. Std.*, May 17, 1972.

40. D. R. "Scotty" Campbell, "Anaconda and Montana: a Partnership," *Mont. Std.*, Feb. 6, 1972.

41. "Contract Miners Settle with Company," *Mont. Std.*, July 12, 1973; U.S. Department of Labor, *Wage Chronology*, 10–11; "Anaconda Gets Tentative Accord with Eight Unions," *Wall Street Journal*, June 20, 1974.

42. Heilbronner, *Sights and Scenes*, VF 1041, 6.

43. Gwen Mitchell, interview by Mary Murphy, May 29, 1987, cassette tapes and written transcript, OHC, BSBPA; Ralph interview.

44. McCarthy interview, by Leech.

45. Matthew Ronald Carl, "The Columbia Gardens Amusement Park: Company Sponsored Community in Butte, Montana" (master's thesis, University of Montana, Missoula, 2011), 1–3. On welfare capitalism, see Andrea Tone, *The Business of Benevolence: Industrial Paternalism in Progressive America* (Ithaca, NY: Cornell University Press, 1997); Sanford M. Jacoby, *Modern Manors: Welfare Capitalism Since the New Deal* (Princeton, NJ: Princeton University Press, 1998).

46. Bert Skinner and Donnetta Skinner, interview by Teresa Jordan, June 22, 1986, written transcript, OHC, BSBPA.

47. Bone interview; Jack Atcheson and Mary Claire Atcheson, interview by Andrew Atcheson, Mar. 6, 2005, Butte, MT, transcript, OHC, BSBPA; Patricia Seccomb, interview by Joanne Seccomb, Mar. 2, 2002, Butte, MT, transcript, OHC, BSBPA; Frank Quinn, *Memories of Columbia Gardens* (Butte, MT: Quinn and Associates, 1973).

48. Walter, "Butte Paradise."

49. Judith Diane Thompson, "Goodbye to the Gardens: A Look at the Controversial Closure of Butte's Columbia Gardens and How the *Montana Standard* Covered the Event" (bachelor's thesis, University of Montana, Missoula, 1987), 19; Pat Kearney, *Butte's Pride: The Columbia Gardens* (Butte, MT: Skyhigh Communications, 1994), 85; Betty Ann Raymond, "Columbia Gardens Trims Sails," *Mont. Std.*, Apr. 22, 1973.

50. Carolyn J. Mattern, "An Era's End: Columbia Gardens Gone," *Montana Post* 11, no. 4 (Nov.-Dec. 1973). On the effects of cars on the Garden's popularity, see Charles C. Godard, interview by Ray F. Calkins, Mar. 21, 1979, written transcript, OHC, BSBPA; Laura Hart, interview by Teresa Jordan, Dec. 9, 1985, Butte, MT, partial transcript and notes, OHC, BSBPA. On Coney Island's decline, see Gary S. Cross and John K. Walton, *The Playful Crowd: Pleasure Places in the Twentieth Century* (New York: Columbia University Press, 2005), 147–62.

51. On the ranking and the Garden's diminished use, see Gruen Gruen and Associates, "The Targets and the Structure of an Economic Development Program for Butte," Oct. 7, 1970, 35, Fldr. "Butte Local Development Corporation," Box 2A, BFC, BSBPA; Ulrich interview; Gardiner interview; Walter, "Butte Paradise," 27. On attempts to get locals to use the gardens more often, see Franconi, Maffei, and Petroni, BHS meeting; Petroni interview, by Magnaghi.

52. Jenks, "An Interview."

53. "Columbia Gardens Area Pit Said Not Imminent," *Mont. Std.*, May 19, 1972; "Butte's Future Is Looking Brighter," *Mont. Std.*, May 22, 1972.

54. William J. Clark, "Gardens Era Over, New Pit Expected," *Mont. Std.*, Feb. 23, 1973.

55. L. Ralph Mecham to Senator Mike Mansfield, Jan. 30, 1973, Fldr. 11, Box 34, series 17, Mike Mansfield Papers MSS65 (hereafter MSS65), UMM.

56. Examples of letters include Dolores H. Collier, Wendell M. Martinell, Corinne C. Martinell, Addison C. Collier, Charles J. Speake, Carole C. Speake, Claire C. Sloan, to Senator Mike Mansfield, Feb. 16, 1973, Fldr. 11, Box 34, Series 17, MSS65, UMM; Alana La Rock to Senator Mike Mansfield, Feb. 13, 1973, MSS65, UMM. For children's messages, see various articles in *Mont. Std.*: Tami Williams, "Don't Tear Down Columbia Gardens," Apr. 2, 1973; Michele Vercella, Tony Gomez, and other McKinley School third graders, "School Children Plead for Gardens," Mar. 7, 1973; Kathy Kain, Kathy Foley, and sixty-four additional West Junior High School students, "Gardens Closure Is Protested," Apr. 8, 1973.

57. Margy Harrington to Senator Mike Mansfield, n.d., Fldr. 11, Box 34, series 17, MSS65, UMM.

58. Martin J. Flaherty, "End of Gardens Saddens Reader," *Mont. Std.*, Mar. 27, 1973. See also John and Peggy Dunstan, "Company Should Replace Gardens," *Mont. Std.*, Mar. 9, 1973; Carol Lowney Wood, "Columbia Gardens Action 'Shocking,'" *Mont. Std.*, Apr. 22, 1973.

59. First quote is John Astle Sr. to Senator Mike Mansfield, Dec. 11, 1972, Fldr. 11, Box 34, Series 17, MSS65, UMM. Second is Rocco Robins to Senator Mike Mansfield, n.d. See also Alana La Rock to Senator Mike Mansfield, Feb. 13, 1973, Fldr. 11, Box 34, Series 17, MSS65, UMM; Marie Campbell to Senator Mike Mansfield, Mar. 2, 1973, Fldr. 11, Box 34, Series 17, MSS65, UMM.

60. Dorothy McBride Stetson to Senator Mike Mansfield, Feb. 26, 1973, Fldr. 11, Box 34, Series 17, MSS65, UMM.

61. McCarthy interview, by Leech.

62. Rick Foote, "No Perpetuity Clause for Columbia Gardens," *Mont. Std.*, Mar. 15, 1973; McCarthy interview, by Leech; Alana La Rock to Senator Mike Mansfield, Feb. 13, 1973, Fldr. 11, Box 34, Series 17, MSS65, UMM; Esme La Breeke to Senator Mike Mansfield, Feb. 23, 1973, MSS65, UMM.

63. "Bob Harper Has Plans for Gardens," *Mont. Std.*, Apr. 27, 1973; "Simatz Says County Can Claim Gardens," *Mont. Std.*, May 4, 1973; "Si Holman Is Wondering If Gardens Worth Saving," *Mont. Std.*, June 15, 1973; "Gardens Acquisition Raises Hard Questions," *Mont. Std.*, May 7, 1973; McCarthy interview, by Leech.

64. Mrs. E. A. Stroup to Senator Mike Mansfield, Mar. 25, 1973, Fldr. 11, Box 34, series 17, MSS65, UMM.

65. "Save the Gardens: The Company Wants to Make a Hole," *Borrowed Times*, Apr. 4, 1973. On the petition drive, see "Save the Gardens Committee Meets," *Mont. Std.*, Mar. 26, 1973; "Gardens Meeting Thursday," *Mont. Std.*, Mar. 28, 1973; "Gardens Meeting Draws 150," *Mont. Std.*, Mar. 30, 1973; "Bob Harper Plan Backed," *Mont. Std.*, Apr. 28, 1973; "Columbia Gardens Meeting Scheduled," *Mont. Std.*, June 1, 1973.

66. "Save the Gardens"; "Mayor Makes Plea to Save Columbia Gardens," *Mont. Std.*, Mar. 3, 1973; Rick Foote, "Gardens to Open in May," *Mont. Std.*, Apr. 5, 1973.

67. "Save the Gardens"; Mattern, "An Era's End"; "ACM: Again," *Borrowed Times*, Apr. 18, 1973.

68. Senator Mike Mansfield to Rocco Robins, May 25, 1973, Fldr. 11, Box 34, series 17, MSS65, UMM.

69. Panisco quoted in "Butte Sheds Tear as Gardens Close," *Mont. Std.*, Sept. 4, 1973. See also "Columbia Gardens: The Company Taketh Away," *Borrowed Times*, Sept. 15–30.

70. "Continental Pit Work Starts," *Mont. Std.*, Aug. 12, 1973; Don Peoples, interview by Brian Leech, Oct. 26, 2009, Butte, Montana, transcript and digital recording, OHC, BSBPA

71. Pam Swiger, "Priceless Equipment Lost in Columbia Gardens Fire," *Mont. Std.*, Nov. 13, 1973; "No Arson in Columbia Gardens," *Mont. Std.*, Nov. 14, 1973; Holter interview, by Leech; Gardiner interview.

72. "Charcoal Replaces Carrousel," *Mont. Std.*, Nov. 19, 1973.

73. Gardiner interview.

74. Robert Osier, interview by Ed Brown, Mar. 30, 2002, Butte, MT, transcript, OHC, BSBPA. See also Eli Platisha, interview by Lisa Speegle, Nov. 23, 2001, transcript, OHC, BSBPA; Shirley McCloskey, interview by Sarah Nyland, Apr. 9, 2002, Butte, MT, transcript, OHC, BSBPA; Reverend Joe Warren, interview by Teresa Jordan, Apr. 7, 1986, Butte, MT, transcript, OHC, BSBPA.

75. The song's lyrics, written by Walkin' Jim Stoltz, are in Karen Sullivan, "Gardens Gone, Not Forgotten," *Mont. Std.*, Aug. 29, 1988.

76. Robertson, *Hard as the Rock Itself.*

77. Kearney, *Butte's Pride*; Teresa Jordan to Judi Thompson, Jan. 25, 1987, Fldr. "Columbia Gardens: Columbia Gardens Mine," VF1040, BSBPA; Walter, "Butte Paradise."

78. Betty Ann Raymond, "Spark from Gardens Fire May Ignite Community Spirit," *Mont. Std.*, Dec. 2, 1973; John Hughes, interview by Teresa Jordan, Feb. 7, 1985, cassette tapes and transcript notes, OHC, BSBPA; Duncan Perdita, interview by Mary Murphy, Mar. 18, 1980, cassette tapes, OHC, BSBPA; Daniel R. Ramirez and Wilma J. Ramirez, interview by Teresa Jordan, June 20, 1986, Grand Junction, CO, partial transcript and notes, OHC, BSBPA; Warren interview.

79. Mike Shutey and Kathy Shutey, interview by Teresa Jordan, Apr. 12, 1988, cassette tapes and written notes, OHC, BSBPA. See also John Kriskovich, interview by Teresa Jordan, June 10, 1986, Butte, MT, partial transcript and notes, OHC, BSBPA.

80. Becky Taylor, "Let's Save Butte from Demolition," *Mont. Std.*, Nov. 20, 1973. See also McCarthy interview, by Leech.

81. Carl, "The Columbia Gardens Amusement Park."

82. Testimony of James Murry, *Closure of Underground Copper Mines*, 61.

83. Testimony of John B. Place, *Closure of Underground Copper Mines*, 17.

84. Testimony of John B. Place, John D. Morgan Jr., and James Murry, *Closure of Underground Copper Mines*, 12, 17, 52, 59, 75, 83, 84.

85. Department of Housing and Urban Development, "Analysis of the Butte, Montana Housing Market as of July 1, 1971," Mar. 1972, 9, Fldr. 43, Box 1, PC127, BSBPA. For figures, see Gruen Gruen and Associates, "Targets and the Structure," 23; Howard, Needles, Tammen & Bergendoff, "Butte–Silver Bow, Montana: Population and Economic Inventory-Analysis," June 11, 1971, 3-1-3-2, Box 14, BMCC, BSBPA.

86. Kaplan, Gans, and Kahn, *Butte Community Renewal Program Report*; Department of Housing and Urban Development, 7; *Butte Model Cities Comprehensive Demonstration*

*Plan, Part I: Problem Analysis* (Butte, MT: Standard Printing, 1969), 9–16, Box 8, BMCC, BSBPA.

87. *Butte Model Cities Comprehensive Demonstration Plan*; Foote interview.

88. City of Butte Department of Community Redevelopment, "Goals and Premises for Renewal," 3–6, Fldr. 47, Box 1, PC127, BSBPA.

89. Ibid., 10.

90. Kaplan, Gans, and Kahn, *Butte Community Renewal Program Report*, vi, 2–4; quote on vi.

91. Thomas F. Powers, *Butte Urban Renewal Report*, Apr. 28, 1967, Box 14 "Permanent Record," BMCC, BSBPA; quotes on 2.

92. Robert O. Self, *American Babylon: Race and the Struggle for Postwar Oakland* (Princeton, NJ: Princeton University Press, 2003), 139–40.

93. Jacobs, *Death and Life*.

94. James M. McCoy to Mayor Tom Powers, Apr. 27, 1967, Box 14 "Permanent Record," BMCC, BSBPA; A. J. Massman, Federal Housing Authority, to Vern Griffith, May 14, 1963, Fldr. "Urban Renewal up to Dec. 1963," Box 55, Micone/Peoples/Fisher Collection, Butte-Silver Bow Chief Executive's Office, GR.CE.BSB.002 (hereafter BSBCE), BSBPA; Nov. 28, 1962 meeting notes, BSBCE, BSBPA; Press Release, n.d., BSBCE, BSBPA.

95. Foote interview; Tim Babcock to Forrest Anderson, Oct. 21, 1968, Fldr. 29 "Law Enforcement 1967–1968," Box 13, RS111, MHS; Mrs. Don Percival to Forrest Anderson, June 14, 1968, RS111, MHS; Mark P. Sullivan to Forrest H. Anderson, Aug. 12, 1968, Fldr. 1 "Complaints 1967–71," Box 5, RS111, MHS; Ellen Baumler, "Devil's Perch: Prostitution from Suite to Cellar in Butte, Montana," *Montana: The Magazine of Western History* 28, no. 3 (Autumn 1998): 4–21.

96. Foote interview; Peoples interview, by Leech; Ulrich interview; Mike Micone, interview by Teresa Jordan, Jan. 5, 1987, Helena, MT, transcript, OHC, BSBPA; William J. Clark, "Gambling Must Go, Mayor Tells Police," *Mont. Std.*, Dec. 22, 1971.

97. Bret A. Weber and Amanda Wallace, "Revealing the Empowerment Revolution: A Literature Review of the Model Cities Program," *Journal of Urban History* 38, no. 1 (2012): 173–92; Bret A. Weber, "Denver's Model Cities Program" (PhD diss., University of Utah, Salt Lake, 2007); Jeffrey Craig Sanders, *Seattle and the Roots of Urban Sustainability: Inventing Ecotopia* (Pittsburgh: University of Pittsburgh Press, 2010), 65–98; Roger Biles, *The Fate of Cities: Urban America and the Federal Government, 1945–2000* (Lawrence: University Press of Kansas, 2011), 134–41.

98. Foote interview.

99. Jim Murphy quoted in Lewis T. Poole, "Mike Micone Election Top News," *Mont. Std.*, Jan. 4, 1970. On Micone and Peoples in Model Cities, see Foote interview; Peoples interview, by Leech; Micone interview.

100. Brian Shovers, "Remaking the Wide Open Town: Butte at the End of the Twentieth Century," *Montana: The Magazine of Western History* 48, no. 3 (Autumn 1998): 40–53; *Butte Model City Program: Fifth Year Plan*, May 15, 1973, Action Year: July 1, 1973–June 30, 1974 (Butte, MT: City of Butte Office of Programming and Management, 1973), Box 14, BMCC, BSBPA, II-A-8-II-A-8-19; "Model Cities Fund Allocation Fifth Year Simple Priority Distribution," Apr. 14, 1973, Fldr. "CDA Material (City Demonstration

Agency)," Box 2, BMCC, BSBPA, II-A-8-II-A-8-19; Don Peoples, interview by Teresa Jordan, Jan. 14, 1987, Butte, MT, transcript, OHC, BSBPA.

101. Pam Swiger, "South Butte Rejects Urban Renewal Proposal," *Mont. Std.*, Feb. 2, 1973. See also Pam Swiger, "Residents Critical of Urban Renewal," *Mont. Std.*, Jan. 19, 1973; Pam Swiger, "City Aldermen Close Door on South Butte Urban Renewal," *Mont. Std.*, Feb. 6, 1973.

102. Teresa Jordan quoted in Foote interview.

103. "Mayor Micone Deserves a Second Term," *Mont. Std.*, Mar. 19, 1971. See also "Model City Extension Pushed by Butte Mayor," *Mont. Std.*, Apr. 16, 1974; Foote interview; Micone interview; "The Butte Primary and Model Cities," *Mont. Std.*, Mar. 22, 1973.

104. Neighborhood Council Joint Board of Directors Minutes, Feb. 6, 1973, Fldr. "CDA Material (City Demonstration Agency)," Box 2, BMCC, BSBPA; Butte Model Cities Program, By-Laws of City Demonstration Agency Board, Feb. 2, 1972, Fldr. "CDA Material (City Demonstration Agency)," BMCC, BSBPA.

105. "Evaluation Project PE-3a, Demolition," Mar. 1972, Fldr. "Model Cities—Demolition Program," BMCC, BSBPA; Department of Housing and Urban Development, 10, BSBPA.

106. Francesca Russello Ammon, "Unearthing Benny the Bulldozer: The Culture of Clearance in Postwar Children's Books," *Technology and Culture* 53, no. 2 (Apr. 2012): 306–36.

107. Janet Cornish to Joy Brandon, Jan. 20, 1981, "Tardy Arson Report," VF 1111, BSBPA.

108. Ibid.; Tracy Thornton, "Mining City Fires Take Their Toll," *Mont. Std.*, Dec. 1, 1991.

109. On money, see Pam Swiger, "1972 Area Fires Costly in Lives, Dollars," *Mont. Std.*, Mar. 12, 1973.

110. On fire as a symbol for deindustrialization, see David Stradling and Richard Stradling, "Perceptions of the Burning River: Deindustrialization and Cleveland's Cuyahoga River," *Environmental History* 13 (July 2008): 515–35.

111. This interpretation continues today. See Stacie Lynn Barry, "Coming to the Surface: The Environment, Health, and Culture in Butte, Montana, 1950–2010" (PhD diss., University of Montana, Missoula, 2012), 64–66; Morin, *Legacy of American Copper Smelting*, 121; Peoples interview, by Leech.

112. "Tardy Arson Report."

113. "Federal Officials Probe Store Fire," *Mont. Std.*, Mar. 2, 1972; "Three Questioned about Butte Fire," *Mont. Std.*, Mar. 4, 1972; William J. Clark, "After Five Months Still No Arrests in Penney's Fire," *Mont. Std.*, July 28, 1972; Jonathan Krim, "Arson Work in Butte Ignites Hot State-Local Debate," *Mont. Std.*, July 1, 1976; Jim Tracy, "Arson Squad Targets Six Butte Blazes," *Mont. Std.*, Nov. 5, 1981.

114. Thomas C. Satterthwaite, "Letter to the Editor: Fire Signalled End of the Hill," *Mont. Std.*, Mar. 3, 1972.

115. Foote interview.

116. City of Butte Department of Community Redevelopment, 3–6.

117. Ibid., 3–6.

118. "Undermining of Butte," 93. See also "Open Pit Mine Chews at Butte," *Star Citi-*

zen (Tucson, AZ), Aug. 23, 1970; "Pit Presses Butte to Move Off Richest Hill on Earth," *Salt Lake Tribune*, Aug. 16, 1970; Thomas O'Hanlon, "The Perilous Prosperity of Anaconda," *Fortune*, May 1966, 117–21, 235–36.

119. Phillip Hamburger, "Notes for a Gazetter," *New Yorker*, Sept. 15, 1962, 106–15, quotes on 107, 115.

120. Rumors, news, and planning in Rick Foote, "Anaconda Co. Plans Southeast Expansion," *Mont. Std.*, Apr. 2, 1971; "Community Can't Gain Under Anaconda Cloud," *Mont. Std.*, Apr. 6, 1971; "Anaconda and Community Must Plan Together," *Mont. Std.*, Apr. 12, 1971; "School Site Offered by Anaconda Co.," *Mont. Std.*, Apr. 20, 1971; Peoples interview, by Leech.

121. Quote in Johansen, "Butte, MT: Sitting Atop a Fortune." See also "Finlen Safe From Pit Until 1991," *Mont. Std.*, Apr. 10, 1971; "Letter from Anaconda," *Mont. Std.*, Apr. 13, 1971; Maurer, "Butte, Mont.," quote on 42.

122. Quotes in testimony of Thomas Judge, Lee Metcalf, and Frank Monninger, *Closure of Underground Copper Mines*, 2, 12, 45–52, 59.

123. Leeper, "Changing Urban Landscape," 96. See also McCormick, "East Park Survivors"; City of Butte Department of Community Redevelopment, 9; *Butte Model Cities Comprehensive Demonstration Plan*, 9.

124. Quote in "The Pit That Ate Butte," 2.

125. Leeper, "Changing Urban Landscape," 113.

126. Mike Mansfield to Mary J. O'Neill, Sept. 16, 1970, Fldr. 1, Box 34, Series 17, MSS65; Maurer, "Butte, Mont.," 38–43; "Montana Says Company Deters Urban Renewal," *Mont. Std.*, Oct. 16, 1970; Kathleen Cook, "Council Gets Ready for Relocation Vote," *Mont. Std.*, Apr. 20, 1976.

127. Micone interview.

128. Leeper, "Changing Urban Landscape," 107, 125; Maurer, "Butte, Mont.," 41; William Wyckoff, "Postindustrial Butte," *Geographical Review* 85, no. 4 (Oct. 1995): 478–96.

129. Maurer, "Butte, Mont.," 38–43; quotes on 41. See also "Anaconda Answers News Article," *Mont. Std.*, Mar. 13, 1973.

130. Leeper, "Changing Urban Landscape," 141.

131. Franklin, "Inexorably"; "Uptown Butte, Mont. Hit By Fatal Case of Urban Blight," *Greely Daily Tribune* (Greely, CO), Nov. 16, 1973; "Anaconda Answers News Article," *Billings Gazette* (Billings, MT), Mar. 15, 1973; "Forward Butte Commission," 5; "The Pit That Ate Butte," 1-2.

132. MacDonald quoted in "Fjare Opinion Answered, *Mont. Std.*, Nov. 15, 1972.

133. "Anaconda Co.'s Largey MacDonald on Butte," *Mont. Std.*, Nov. 8, 1970.

134. "Forward Butte Commission," 5; "The Pit That Ate Butte," 1-2.

135. Butte Forward Minutes, Jan. 3, 1974; American City Corporation, *Central Business District Development Program Butte, Montana: Final Report, Phase I* (Feb. 13, 1976), pages 2, 7, BFC, BSBPA; Butte Forward, Inc., *CBD Relocation and New Town Project, Butte, Montana: Preliminary Feasibility Report*, Sept. 1, 1975, Fldr. 46, Box 1, BMCC, PC 127, BSBPA.

136. Richard Kaudy, "Some Uptown Firms Already Have Moved," *Mont. Std.*, July 19, 1976; "New Town Plan for Butte Ready for Drawing Board," *Mont. Std.*, Sept. 13, 1974; Rick Foote, "Butte Forward's Staff Report Details Proposals for Uptown," *Mont. Std.*,

Nov. 7, 1975; City Council Hearing #1, Apr. 5, 1976, Fldr. "Public Meetings-Questions," Box 3A, BFC, BSBPA.

137. Brian Mertz, "Leonard Powell Sees Bright Copper Glitter in Mining City's Future," *Mont. Std.*, July 16, 1976. See also Butte Forward Minutes, Sept. 9, 1975.

138. L. C. Powell to Mike Micone, Dec. 18, 1975, Fldr. "The Anaconda Company 1977–78, Box 5, Micone/Peoples/Fisher Collection, BSBCE, BSBPA; "Anaconda Undercuts Any Hope for Uptown," *Mont. Std.*, Oct. 10, 1975; Butte Forward Minutes, Dec. 11, 1974.

139. For first quote, see Butte Forward Minutes, May 15, 1974. For second, see City Council Hearing Minutes, Apr. 5, 1976, Fldr. "Public Meetings—Questions," Box 3A, BFC, BSBPA. See also Micone interview; Ulrich interview.

140. Butte Forward, Inc., *CBD Relocation*; American City Corporation, *Central Business District Development Program*, 27–32; Pam Swiger, "Butte Forward Unveils 'New Town' Plan," *Mont. Std.*, May 23, 1975.

141. Roger Biles, "New Towns for the Great Society: A Case Study in Politics and Planning," *Planning Perspectives* 13 (1998): 113–32; William Petersen, "The Ideological Origins of Britain's New Towns," *Journal of the American Institute of Planners* 34, no. 3 (May 1968): 160–70; Carlos C. Campbell, *New Towns: Another Way to Live* (Reston, VA: Reston Publishing, 1976), 17–25; Lynne C. Burkhart, *Old Values in a New Town: The Politics of Race and Class in Columbia, Maryland* (New York: Praeger, 1981).

142. Biles, "New Towns," 116–21.

143. Herman G. Berkman, "The New Town and Urban Change Form," *Land Economics* 48, no. 2 (May 1972): 93-103; Margaret Mead, "New Towns to Set New Lifestyles," in *New Towns—Why and For Whom?*, H. S. Perloff & N. C. Sandberg, ed. (New York: Praeger, 1973), 117–29; Biles, "New Towns," 113–32.

144. First quote is in Biles, "New Towns," 114; second is in Burkhart, *Old Values in a New Town*, v; third is Campbell, *New Towns*, 17. See also Irving Lewis Allen, ed., *New Towns and the Suburban Dream: Ideology and Utopia in Planning and Development* (Port Washington, NY: Kennikat Press, 1977).

145. Butte Forward, Inc., *CBD Relocation*.

146. Leeper, "Changing Urban Landscape," 105. See also Maurer, "Butte, Mont.," 42; Robert J. Corbett, interview by Teresa Jordan, Jan. 21, 1986, Helena, MT, transcript and notes, OHC, BSBPA.

147. First quote is "Residential Land Needs," Butte Forward, Inc., *CBD Relocation*. Second is American City Corporation, *Central Business District Development Program*, 42. Third is Corbett interview, by Jordan.

148. Rick Foote, "New Butte Could Rekindle Old Lifestyle," *Mont. Std.*, Nov. 6, 1975.

149. Peoples quote is in Kathleen Cook, "Merchants, Citizens Still Divided on Uptown Future," *Mont. Std.*, Apr. 8, 1976. See also "Butte Forward Expands Board," *Mont. Std.*, Sept. 12, 1975.

150. American City Corporation, *Central Business District Development Program*; Micone interview.

151. On new town troubles, see Biles, "New Towns," 121–28; Letitia C. Langford and Gwen Bell, "Federally Sponsored New Towns of the Seventies," *Growth and Change* (Oct. 1975): 24–31; John V. Ard, "Economic Crunch Shakes Financial Footing on New Town Developments," *Industrial Development* (Jan./Feb. 1975): 11–14.

152. American City Corporation, *Central Business District Development Program;* Micone interview. On the EDA, see Eugene Boyd, "Economic Development Administration: A Review of Elements of Its Statutory History," *Congressional Research Service Report for Congress*, June 3, 2011, Butte, MT, 1–18, R41241, www.fas.org/sgp/crs/misc/R41241.pdf.

153. Butte Forward Minutes, May 12, 1975; "Consultants See Little Chance of Refurbishing Uptown Butte," *Mont. Std.*, Mar. 30, 1976; Kathleen Cook, "Council Votes Move, New Site Proposed," *Mont. Std.*, May 6, 1976; "Baucus Expands on Funding for Uptown," *Mont. Std.*, May 11, 1976.

154. "Butte, Forward, Publicly," *Mont. Std.*, Sept. 30, 1975.

155. Thomas E. Shriver and Dennis K. Kennedy, "Contested Environmental Hazards and Community Conflict over Relocation," *Rural Sociology* 70, no. 4 (2005): 491–513.

156. Council Resolution 5427, Apr. 21, 1976, Butte City Council Minutes, 1973–Apr. 1976, Clerk and Recorder's Office, Butte–Silver Bow County Courthouse, Butte, MT; "Max Doubts Funding for Butte Relocation," *Mont. Std.*, Apr. 24, 1976.

157. "Wake Up, Butte, America, Don't Be Pushed Around!" Fldr. "D. J. McCaig," Box 3A, BFC, BSBPA. On big business support, see "Forward Butte Commission," 5; "Utility Pledges Relocation Move," *Mont. Std.*, Apr. 8, 1976; "Bank Joins Move List," *Mont. Std.*, Apr. 20, 1976. On growing protest, see Rick Foote, "Little Interest Shown in Butte Relocation Project," *Mont. Std.*, Mar. 11, 1976; "Quake Shakes Could Imperil Possible New Town Locations," *Mont. Std.*, Mar. 12, 1976; D. J. McCaig, "Comments of D. J. McCaig on the American City Corporation Report Re: 'The Central Business District Development Program for Butte," Mar. 5, 1976, Fldr. "D. J. McCaig," Box 3A, BFC, BSBPA.

158. Micone interview; American City Corporation, *Central Business District Development Program*, 93.

159. Ad Hoc Site Selection Committee to City Council, Fldr. "Ad Hoc Site Selection Committee," Box 5B, BFC, BSBPA; Special Committee Meeting on Relocation Sites, minutes, City Council Chambers, Butte, MT, July 6, 1976, Fldr. "Council Committee Meeting Minutes," Box 2B, BFC, BSBPA. See also Kathleen Cook, "Aldermen Have Close Shave After Protest During Meeting," *Mont. Std.*, May 20, 1976; "Hearings Planned on Relocation Sites," *Mont. Std.*, May 21, 1976; "Hearings on Relocating Uptown Are This Week," *Mont. Std.*, May 31, 1976.

160. "Stodden Park Grows," *Mont. Std.*, May 18, 1973; "Stodden Park in Phase 3," *Mont. Std.*, July 22, 1973; "Outdoor Playtime," *Mont. Std.*, Dec. 6, 1973.

161. Laverne Koprivka, letter to the editor, *Mont. Std.*, July 15, 1976. See also City Council Meeting minutes, June 3, 1977, Fldr. "Public Meetings—Questions," Box 3A, BFC, BSBPA.

162. Andrea Ciabattari, "Micone Encourages Vote Next Week on Relocation," *Mont. Std.*, June 9, 1976. On the meetings, see City Council Hearing #2 Site Selection notes, June 8, 1976, Fldr. "Public Meetings-Questions," Box 3A, BFC, BSBPA; "Debate Continues Over New-Town Site," *Mont. Std.*, June 25, 1976; Andrea Ciabattari, "Vote on Butte Relocation Site Deferred to July 7," *Mont. Std.*, June 17, 1976.

163. Andrea Ciabattari, "Developer Outlines Hopes for a New Business District," *Mont. Std.*, June 18, 1976; "Debate Continues Over New-Town Site," *Mont. Std.*, June 25, 1976; Special Committee Meeting on Relocation Sites Minutes, June 22, 1976, June 30,

1976, July 1, 1976, July 6, 1976, Fldr. "Council Committee Meeting Minutes," Box 2B, BFC, BSBPA.

164. On options, see Keith P. Johnson to Joe Picarelli, June 4, 1975, Fldr. "Butte Forward Site #2, Joe Picarelli File," Box 2A, BFC, BSBPA; "Butte Forward Takes Option on Third Townsite," *Mont. Std.*, Oct. 7, 1975.

165. Ciabattari, "Developer Outlines Hopes"; Bridget R. Barry, "Toxic Tourism: Promoting the Berkeley Pit and Industrial Heritage in Butte, Montana" (master's thesis, University of Nebraska Lincoln, 2012), 34–36.

166. Nancy Cardwell, "How Two Women Outwitted Big Firm, At Least for a While," *Wall Street Journal*, Dec. 8, 1976. See also Richard Kaudy, "A New Pitch for a New Site for a New Butte," *Mont. Std.*, July 2, 1976; Barry, "Toxic Tourism," 34–36.

167. Special Committee Meeting Minutes, July 6, 1976, Fldr. "Council Committee Meeting Minutes," Box 2B, BFC, BSBPA; Ciabattari, "Developer Outlines Hopes"; "Aldermen Continue Butte Move Study," *Mont. Std.*, June 23, 1976; Richard Kaudy, "Spolar Site Is Recommended," *Mont. Std.*, July 7, 1976.

168. "Forward Butte Commission," 5.

169. "Some Minor Matters," *Mont. Std.*, Oct. 20, 1975; "Butte Forward Meetings Opened at Council Urging," *Mont. Std.*, Oct. 16, 1975.

170. Tim Clark to Shag Miller, June 11, 1974, Fldr. "Correspondence to July 1, 1974," Box 2A, BFC, BSBPA; Butte Forward Meeting Minutes, Oct. 3, 1974, Sept. 9, 1975, Fldr. "Meeting Minutes," Box 3A, BFC, BSBPA.

171. City Council Hearing #2, Apr. 7, 1976, City Council Hearing #1, Apr. 5, 1976, BFC, Box 3A, Fldr. "Public Meetings-Questions," BSBPA.

172. Jim Murphy to Frank Monninger, Memo, Nov. 8, 1974, Fldr. "1973 Hearings on New Communities," Box 2B, BFC, BSBPA.

173. James J. Murphy to Stan Kimmit, Fldr. "Correspondence to Nov. 1, 1975," Box 2A, BFC, BSBPA. See also Jim Murphy to Gene Tidball, Oct. 16, 1975, Fldr. "Correspondence to Nov. 1, 1975," Box 2A, BFC, BSBPA.

174. "Minutes of the Sept. 9, 1975 Meeting, Butte Forward, Inc.," BFC, Box 3A, Fldr. "Meeting Minutes," BFC, BSBPA.

175. Cardwell, "How Two Women"; City Council Hearing #2 Minutes, Apr. 7, 1976, Fldr. "Public Meetings—Questions," Box 3A, BFC, BSBPA.

176. "More Hole Predicted for Butte," *Billings Gazette*, Oct. 11, 1975; Carrie Johnson, "Future Haunts Battered, Worn Butte," *Salt Lake City Tribune*, Nov. 25, 1976; Cardwell, "How Two Women."

177. Cardwell, "How Two Women." See also City Council Hearing #2 Minutes, Apr. 7, 1976, Fldr. "Public Meetings—Questions," Box 3A, BFC, BSBPA.

178. Ulrich interview. See also Cook, "Council Gets Ready."

179. McCarthy interview, by Leech; Peoples interview, by Leech; William Langewiesche, "The Profits of Doom: One of the Most Polluted Cities in America Learns to Capitalize on its Pollution," *The Atlantic* (Apr. 2001), www.theatlantic.com/magazine/archive/2001/04/the-profits-of-doom/302177/

180. "Relocation Vote Faces New Test," *Mont. Std.*, Apr. 22, 1976. See also Hazel Carter, "Let's Get Together and Save Our City," *Mont. Std.*, July 21, 1976; Corbett inter-

view, by Jordan; Hazel E. Carter to Jim Murphy, n.d., Fldr. "Readers Speak," Box 3A, BFC, BSBPA.

181. Beverly Hayes and the Save the Central Business District Citizens Committee to Carla Hills, Department of Housing and Urban Development, Apr. 24, 1976, Fldr. "Mields, Hugh," Box 3A, BFC, BSBPA; Mike Mansfield to D. J. McCaig, Mar. 29, 1976, Fldr. "D. J. McCaig," BFC, BSBPA. See also David O. Meaker Jr. to Beverly Hayes, May 11, 1976, Fldr. "Mields, Hugh," BFC, BSBPA; McCaig, "Comments of D. J. McCaig"; Richard Kaudy, "Group of Uptown Merchants Fight Move Off Historic 'Butte Hill,'" *Mont. Std.*, July 12, 1976.

182. "National Review Group Endorses Plan for Butte to Pursue Relocation Funding," *Mont. Std.*, May 14, 1976; "Anaconda Co. Backs Butte Forward Plan," *Mont. Std.*, Apr. 15, 1976. Also see Fldr. "Commitments," Box 2A, BFC, BSBPA.

183. John B. Place to Mike Micone, July 1, 1976, Fldr. "Commitments," Box 2A, BFC, BSBPA. See also Richard Kaudy, "Spolar Site Approved, 8–5," *Mont. Std.*, July 8, 1976; "Butte Relocation Effort Gets $11 Million Boost," *Mont. Std.*, July 3, 1976; "Mansfield's Decision Casts Shadow on Butte Forward," *Mont. Std.*, Mar. 5, 1976.

184. Kaudy and Robinson quoted in Kaudy, "Spolar Site Approved."

185. Kaudy, "Spolar Site Approved."

186. "An Odd Choice," *Mont. Std.*, July 12, 1976; "Murphy Says Spolar Site Bad Choice," *Mont. Std.*, July 12, 1976; "Timber Butte Profit Margin Varies," *Mont. Std.*, July 16, 1976; Richard Kaudy, "Relocation Only Solution for Some Uptown Stores," *Mont. Std.*, July 17, 1976; Cardwell, "How Two Women."

187. Shovers, "Remaking the Wide Open Town," 47.

188. First three quotes in Richard Kaudy, "Council Rejects Relocation and Free Land," *Mont. Std.*, July 22, 1976. Robinson quoted in Richard Kaudy, "Aldermen Reflect on Relocation," *Mont. Std.*, July 23, 1976. See also Butte CBD Project City Council, votes tally, July 21, 1976, Fldr. "Public Meetings-Questions," Box 3A, BFC, BSBPA.

189. Shag Miller to Murray Paper, Sept. 9, 1976, Fldr. "Correspondence During September 1976"; Robert J. Corbett to Gene Tidball, Oct. 15, 1976, Fldr. "Correspondence During Oct. 1976"; James J. Murphy to George Miller, Economic Development Administration, Aug. 20, 1976, Fldr. "Correspondence During Aug. 1976." All above documents in Box 2B, BFC, BSBPA. See also "Micone Reports Funding Hopes Nil," *Mont. Std.*, July 27, 1976; Corbett interview, by Jordan.

190. Board member quoted in Carrie Johnson, "The Undermining of Butte, Montana," *Washington Post*, Nov. 22, 1976.

191. Barry, "Toxic Tourism," 35–36. On the import of Stodden Park to relocation's failure, see Kaudy, "Aldermen Reflect on Relocation" *Mont. Std.*, July 23, 1976; Bill R. Tait to Shag Miller, Aug. 11, 1976, Fldr. "Correspondence During Aug. 1976," Box 2B, BFC, BSBPA.

192. Shovers, "Remaking the Wide Open Town," 47. See similar views in Morin, *Legacy of American Copper Smelting*, 122; Foote interview.

193. Humphrey interview, 107.

194. "Anaconda Answers News Article," *Mont. Std.*, Mar. 13, 1973.

195. John Place, letter to Lee Metcalf, Jan. 10, 1975. See also testimony of Lee Metcalf, *Closure of Underground Copper Mines*, 45–52, 116.

196. Spokesman quoted in Cardwell, "How Two Women."

197. Tim Clark to Shag Miller, June 11, 1974, Fldr. "Correspondence to July 1, 1974," Box 2A, BFC, BSBPA. See also Mike Doto to Shag Miller, June 28, 1974, Fldr. "Correspondence to July 1, 1974," Box 2A, BFC, BSBPA; Kathleen Cook, "Alderman Says Citizens Not Included in Planning," *Mont. Std.*, Feb. 19, 1976.

198. Corbett interview, by Jordan.

199. McCaig, "Comments of D. J. McCaig." See also "Merchants Cool to Relocation," *Mont. Std.*, Mar. 11, 1976.

200. Kathleen Cook, "Few Oppose Plans to Relocate Uptown," *Mont. Std.*, Mar. 3, 1976.

201. "The CBD Problem: Background and Strategy for Redevelopment," Nov. 7, 1977, Fldr. CBD Background and Strategy Redevelopment, Box 12, BSBCE, BSBPA. On urban renewal in Flint, another one-company town, see Andrew R. Highsmith, "Demolition Means Progress: Urban Renewal, Local Politics, and State-Sanctioned Ghetto Formation in Flint, Michigan," *Journal of Urban History* 35, no. 3 (Mar. 2009): 348–68.

202. Quote in Rick Foote, "Anaconda Co. Promises Cooperation in Central Business District Study," *Mont. Std.*, Apr. 6, 1972. See also "Butte's Future Up to the Council," *Mont. Std.*, July 28, 1976; Richard C. Timmerman to Shag Miller, Aug. 5, 1976, Fldr. "Correspondence During Aug. 1976," Box 2B, BFC, BSBPA.

203. John A. Lentz Sr., vice president of Phelps Dodge, Douglas, AZ, to James J. Murphy, executive director of Butte Forward, Aug. 5, 1974, Box 2A, BFC, Fldr. "Analogous Situations," BSBPA.

204. C. D. Michaelson, President of Kennecott Copper Corporation's Metal Mining Division, New York, to James J. Murphy, Aug. 2, 1974, BFC, Box 2A, Fldr. "Analogous Situations," BSBPA.

205. On Kennecott's piece-by-piece purchases and eminent domain threats, see "Property Owners Mailed Contract Offers this Week," *Bingham Bulletin* (Utah), Sept. 9, 1960; A. M. Ferro to City of Bingham, Oct. 17, 1961, in Correspondence and Subject Files, 1904–1972, Series 4590, reel 4, Utah State Archives (hereafter USS). On fights between Bingham Canyon and Kennecott, see "Public Hearing Regarding Zoning Ordinance," Apr. 22, 1959, in Correspondence and Subject Files, 1904–1972, Series 4590; "Public Hearing Regarding Amendment to Zoning Ordinance," July 30, 1960, in Correspondence and Subject Files, 1904–1972, Series 4590; "Special Meeting of the City Council," July 21, 1960, in Bingham Canyon City Council Minutes, Series 4587; all on reel 4, USS. See also Marion Dunn, *Bingham Canyon* (Salt Lake City, UT: Publisher's Press, 1973), 139–49.

206. Foote interview.

207. Shovers, "Remaking the Wide Open Town," 46. See also Peoples interview, by Jordan; Peoples interview, by Leech; Ulrich interview; McCarthy interview, by Leech.

208. Micone interview.

209. Donald R. Peoples to Uptown Redevelopment Committee, Dec. 16, 1977, Fldr. "CBD Background & Strategy Redevelopment," Box 12, BSBCE, BSBPA.

210. Rick Foote, "Butte in 1979: Rise in Copper Price Boosts Chances of a Stable Mining Economy," *Mont. Std.*, Feb. 10, 1979. See also Ralph F. Cox to Mario Micone, June 1, 1978, Fldr. "The Anaconda Company, 1977–1978," Box 5, BSBCE, BSBPA.

211. For an example, see Francis J. Moriarty, "Tough, But Crumbling," *San Diego Union*, Mar. 30, 1980, D3.

## CHAPTER 8

# RECLAMATION

THE ANACONDA COMPANY'S financial struggles deprived Butte residents of the peace of mind that saving the uptown area might have earned them. The company could not make up for past mistakes in Chile and Mexico, nor had it devoted enough energy toward locating new domestic resources. Labor problems and low metal prices challenged it. By 1977 the company had been purchased, and then, just a few years later, in 1983, all open-pit operations in Butte shut down, devastating the region. Residents' subsequent attempts to rebuild and reshape Butte have sometimes been successful, but at other times have been frustrated.

Sometimes Butte's attempts came under fire. On June 22, 2006, correspondent Jason Jones filed a report with the popular topical comedy television program, *The Daily Show with Jon Stewart*. As producer Miles Kahn said, this report had a "naturally funny premise"—the decision by the Butte Chamber of Commerce to increase the admission price to see a toxic lake.[1] The lake was actually an enormous open-pit copper mine called the Berkeley Pit. When major mining operations ended in the early 1980s, workers turned off the pumps that kept water out of the mine—a mine that was now a mile wide, a mile and a half long, and 1,200 feet deep. The result was an artificial lake filled with 40 billion gallons of what the *High Country News* labeled "a cocktail of heavy metals," including elevated levels of arsenic, copper, cadmium, and lead.[2]

In *The Daily Show*'s take on the pit, Jason Jones sarcastically suggested that "the city has taken lemons and turned them into something that if you drank could kill you." The fake news team interviewed city leaders, who argued that treating the pit as a bona fide tourist attraction could help out the area's economy. On the other side of the issue stood former state representative Fritz Daily, who expressed disgust with the state of the Berkeley Pit. He asserted, "The Berkeley Pit is not an amusement park, it is not our Dollywood." In typical *Daily Show* fashion, Jones narrated an amusing commentary comparing the Berkeley Pit to an amusement park: "While Dollywood has the Rampage Ride and Dolly's Rockin' Rodeo Way," he joked, "the soon to open Toxi-Land will wow you with the exciting Pesti-Slide or hop on the Contamination Station and head for No-Tomorrow World. It's a day that will possibly alter your genetic structure."[3]

That the comedy show's producers, based in New York City, should fail to comprehend why a pit filled with water can hold economic and cultural value is not surprising. Outsiders and insiders to Butte have often disagreed about the meaning of the Berkeley Pit specifically and the mining landscape in general. Even scholars argue about the relative balance of pessimism and optimism in people's understanding of Butte and other postindustrial areas. In *Hard as the Rock Itself*, David Robertson argues that longtime residents of mining communities feel strong, warm attachments to landscapes that outsiders often deride as physically (and even morally) derelict. Smokestack nostalgia—a vague, collective longing for the industrial past among the established—lives in tension with the mining imaginary, a popular negative view often adopted by newcomers and outsiders. For authors like Robertson, the two views are typically consistent and monolithic—positive for insiders, negative for outsiders.[4]

Other scholars, including John Sandlos and Arn Keeling, worry that upbeat local stories ignore the obvious fact that many postmining communities have largely experienced economic decline and environmental challenges. Technology cannot fix all of a closed mine's lasting problems, nor can all old mining landscapes transform into productive, treasured places. Instead, so-called "zombie mines" live on—continuing to poison the landscape and, all too often, the people.[5]

Neither Robertson nor Sandlos and Keeling fully capture the situation in Butte. Scholars are right to emphasize a disconnect between insiders and outsiders, but they are wrong that the trend is consistent and monolithic.[6] Both groups had agreed on Butte's big bust during the Depression and on its postwar boom. The Berkeley Pit, however, played a key role in separating outsiders from insiders. After it shut down, the Berkeley Pit became the centerpiece of the nation's largest Superfund environmental cleanup site, as designated by the EPA. For outsiders, the postindustrial pit became a symbol of all that was wrong with resource extraction and its communities in the American West. Its visual presence made it an easy target.

Among Butte insiders, views were more mixed. Right after the Berkeley Pit's closure, not every resident treasured the mining landscape. Newcomers and the young were often more upset about mining waste, more invested in historic preservation, and increasingly ready to celebrate an industrial past that they had never known. Industrial work, especially in the open pit, was so powerfully affecting that it took a whole new generation of people to see Butte in a new light and to start it down a new path.

Butte residents have therefore spent the past three decades engaged in a complex attempt at what environmental cleanup experts call reclamation. Their reclamation involved not only land and water, but also economics, social life, and historic buildings. They have had many successes in confronting the Berkeley Pit's lasting legacies. Former residents of lost neighborhoods even founded reunions on the edge of the Berkeley Pit, meetings that many found to be healing. Yet the Berkeley Pit's presence as a massive toxic lake, never to be fully decontaminated, has sometimes stifled city leaders' efforts to move beyond Butte's mining past. The permanency of an open-pit metal mine means that anyone who drives through Butte will continue to encounter the legacy of industrial work; Butte's leaders thus decided to try transforming that encounter into a community revival.

## Shutdown

The Berkeley Pit's eventual closure was one major result of a fateful series of company decisions, beginning in the late 1970s. In 1977 the Atlantic Richfield Corporation (ARCO), a wealthy oil company that had purchased Sinclair Oil a few years earlier, took over a struggling Anaconda Company. Anaconda became one of nine different American copper producers bought by oil companies between 1963 and 1982.

ARCO hoped to continue metal mining operations as usual but, despite initial optimism in Butte, it struggled with the same problems that had plagued Anaconda. It battled regulators over pollution and failed to improve labor relations.[7] Fundamental to the problem was that ARCO lacked experience in metal mining. Longtime Anaconda geologist Vin Perry opined that ARCO failed to understand underground exploration. Nor did ARCO do well with mining itself: used to hitting oil and pumping it out immediately, oil companies tended not to recognize that mining properties with large deposits require substantial, ongoing capital and energy.[8]

Other problems followed. ARCO originally bought Anaconda to use its expertise in large-scale shale oil exploration, but an international oil glut and lower petroleum prices thwarted such plans. The oil glut damaged the company's bottom line, causing it to sell off refineries and gas stations in the eastern United States. Then a recession hit, which meant skyrocketing corporate debt. For Butte-area residents, historian Laurie Mercier argues, ARCO "symbolized the new multinational order"—a diversifying company with no ties to the community.[9] As Robert Driscoll, a laid-off Berkeley garage employee put it, "The people who count are no longer here in Butte."[10]

With little to stop it and a lot to compel it, ARCO began to unload its least-profitable operations. It targeted Montana. Citing the costs of environmental regulations and, to a lesser extent, union demands, ARCO announced a smelter shutdown in the town of Anaconda during the 1980 strike. ARCO also closed the Great Falls refinery in 1980. In Butte it shut down underground exploration and trimmed the workforce at both the Berkeley Pit and Weed Concentrator. By 1981 ARCO had eliminated two thousand Montana employees.[11]

With open-pit mining the only going concern, the Berkeley's declining ore-to-waste ratio became particularly troubling. By the early 1980s it cost $1.30 to produce an ounce of Butte copper, but that ounce brought only 70 cents on the open market.[12] *Businessweek* described Butte's story as representing the "sad state" of U.S. copper companies. Others agreed; the western United States could not compete globally, especially with the "Saudi Arabia of Copper," Chile, which enjoyed both low labor costs and massive copper reserves.[13]

ARCO's local management tried to put a positive spin on the situation, but the Berkeley Pit was doomed. Early in 1981, ARCO reported that it planned to spend $25 billion on Montana improvements and that the Berkeley Pit could last until 2010. With the closure of its Anaconda plant, however, ARCO's metals division sent copper concentrates to Japan for smelting. Butte ore had more arsenic than the Japanese smelter could handle, and so the Japanese converted it into inert arsenic trioxide to be returned to the United States. The expense of shipping and storing fell to ARCO. With copper prices tumbling, the metals'division cost ARCO $265 million in 1981.[14]

At the end of April 1982, company leaders announced they would shut the Berkeley Pit. Management laid off some employees, then shifted the remaining eight hundred to the new East Berkeley Pit, which contained a widely disseminated deposit of copper and molybdenum.[15]

Area residents denied the company's complaints that union demands and environmental regulation led to the closing of the Berkeley Pit. Unions claimed that they were willing to bargain on the first issue, while the local, state, and federal governments said they would be happy to work with the company on the second.[16] Environmental concerns by ARCO employees and other Butte residents had already faded, primarily because ARCO's financial situation so worried them.[17] Locals instead charged that a top-heavy pay structure, inefficient management, overuse of outside contractors, and a lack of capital investment were the main contributors to the pit's shutdown.[18]

With both the Berkeley Pit and the underground closed, executives at ARCO's Denver headquarters concluded they should save the $10 million spent on pumping costs each year. On Earth Day, April 22, 1981, ARCO announced it would discontinue pumping water out of Butte's thousands of miles of underground mine workings (and hence up into the bottom of the Berkeley Pit). That decision alarmed residents almost as much as the shutdown. Learning about the pumps one day after ARCO's decision, elected officials had no time to change the minds of company officials.[19]

Millions of dollars of equipment underground was abandoned as the waters rose. The company claimed it was too expensive and labor intensive to salvage this equipment, but the loss truly angered local miners, who believed the company was receiving big tax write-offs. They also knew that a flooded underground would be very expensive to reopen.[20] Although ARCO's leaders insisted that they could dewater the Berkeley Pit for a future restart, their own newsletter had stated, just two years before, that "months, even years" would pass before start-up operations could take place, if water seeped into the underground and the Berkeley Pit.[21]

On January 6, 1983, after a further drop in copper prices, ARCO's Frank Gardner announced the "tough [and] traumatic" decision to end all Butte mining operations later that year.[22] Although made public only in 1983, plans to close had been in the works for years. Only the smaller East Berkeley Pit operation had been profitable in Butte. Management again cited inflexible work practices, high labor costs, and, especially, low ore prices. Labor unions reacted with anger. ARCO was able to relocate some miners elsewhere in the country, but others thought that ARCO had failed to protect the jobs and seniority of most former employees. Despite some verbal protest, workers trucked out the last loads from the East Berkeley Pit on June 30, 1983. By then, mining employment in the county had dropped from 6,006 people in 1950 to 70.[23]

Following in ARCO's footsteps, other multinational companies began to sell off or close their U.S. metal mines. Fabricators seemed more interested in low-cost imports than in U.S. copper. The U.S. Senate attempted to restrict copper imports, but President Ronald Reagan refused to consider the action.[24] From 1981 to 1986 more than 20,600 Americans in the copper industry became part of the 11.5 million U.S. workers who lost industrial jobs due to shutdowns and relocations during that period. The contraction of metal mining, according to a December 1984 issue of *Businessweek*, "was creating modern-day ghost towns in mining regions of the country."[25]

To many, it seemed like Butte would become the next ghost town. Unemployment exceeded 20 percent. Some younger families left town. Students followed these families out. Some out-of-work men struggled with becoming househusbands, no longer able to fulfill the blue-collar ideal of the male breadwinner. Unemployment worsened preexisting problems with alcoholism. The county's chief executive personally knew three mining men who committed suicide.[26]

Outsiders connected Butte and its now-empty open pits to broader national and regional stories. In one reporter's eyes, Butte was part of a 1970s and 1980s trend toward deindustrialization, in which "the production and processing of all basic materials" had shifted "from the industrial countries to the Third World."[27] For journalists and scholars, Butte's quiet minescape symbolized the end of an economy in the American West based on natural resource extraction. Butte was an Old West "relic town [undergoing] death throes." This economy would soon be replaced by a "New West" of urban growth and outdoor adventure.[28]

While a town with working open-pit mines had often been a target for outsiders, a place with empty holes in the ground seemed worse. Butte residents faced the stigma of bust, which aggravated a long-standing perception of the city as ugly, dirty, dangerous, and rough. One young woman interviewed in the 1980s went to college in nearby Bozeman, Montana. She explained that people from Butte stuck together there because "we are the butt of everybody's jokes.... When I was in Bozeman, I stopped telling people I was from Butte because some people would just walk away from me."[29] Newspaper editor and Butte local Rick Foote said, "There's a sense of us versus them," and it was only hardening.[30]

Despite the hard times, a few reporters saw hope. Paraphrasing Mark Twain, one remarked, "The reports of Butte's death are greatly exaggerated."[31] The local newspaper reported that residents actually had a "remarkably good attitude about the future," likely because they had already been through many "tough times."[32] Interviews with residents suggest that most people thought highly of the town's resilient attitude.[33]

## Recovery

The immediate concern after the ARCO shutdown was economic. Don Peoples, who had served as both the city's director of Public Works and director for the Model Cities program, became a key figure in recovery efforts. Succeeding Mike Micone as executive of the city-county government, Peoples was aggressive, and not paralyzed by fear, during the shutdown. His administration first took emergency measures, expanding the Butte Alcohol and Drug Program as well as the city's food bank.[34]

Having taken care of some basic services, Peoples pulled locals together to work on diversification projects, including an Economic Futures Program, a small-business incubator, and a high-altitude speed skating center. The presence of a major employer, the St. James Community Hospital, was an asset, as were a small number of high-tech companies and temporary union programs that sought to retrain former mine workers. The state-run college in Butte also played a key role. Known first as the Montana School of Mines, then as the Montana College of Mineral Sciences and Technology, it rebranded itself in 1994 as Montana Tech. Student enrollment rose from a thousand students in 1971 to almost three thousand students by the 2000s.[35]

Citizens' groups assisted the people most affected by the shutdown and the areas most affected by mining. The Butte Community Union came into existence in 1982 to assist locals hurt by economic bust. It organized a number of demonstrations with lower-income people, trying to build a sense of empowerment. It then engaged in a scrappy legal battle with the state, successfully halting the dismantling of a general assistance program for the unemployed. The group later took up the cause of the central Butte neighborhood, which had experienced subsidence due to underground mining and now hosted both mining hazards and many unemployed residents.[36]

Economic advancement also came for places that had suffered from open-pit mining. In an area of the East Side abandoned for pit development, Community, Counseling, and Correctional Services, Inc. was incorporated in 1983 as a private, not-for-profit organization to provide services for local, state, and federal agencies, including correctional and other human service programs designed for adults and juveniles. Years later, a senior center was constructed and the Maroon Activity Center emerged as an athletic facility for Butte Central High School and as a place for a wide variety of popular activities, including fitness training, athletic tournaments, and receptions.[37] Uptown, once troubled by Berkeley Pit fears, also gained investment, though it suffered setbacks too, such as the downfall of Montana Power, which had transformed itself from an important power company into a high-tech telecommunications company, but then filed for bankruptcy. Once an employer of three thousand, perhaps half of whom worked in uptown Butte, it laid off much of the workforce, leaving only about four hundred employees by March 2003.[38]

By that time, though, Butte's situation had been bolstered by the comeback of open-pit mining. Faced with a number of lawsuits, including one by the state for unpaid taxes, ARCO by 1985 was aiming to scrap or sell off everything it owned in Butte.[39] The men left in charge of ARCO's Butte operations, Frank Gardiner and Ray Tilman, looked for a way to save what remained. They contacted the city–county government, which gathered locals together in a Mining Development Committee. The committee sought to make a new mining start-up easier in terms of finance, labor, taxes, energy, freight, and environment. In 1985 the committee identified Montana businessman Dennis Washington as a possible buyer. Washington was a construction contractor who came to buy equipment from ARCO. Local ARCO people instead convinced Washington of Butte's possibilities for future mining. Taking advantage of ARCO's desperation, Washington bought everything it owned in Montana. He even obtained from the EPA an exclusion of the active mining area from a hazardous waste cleanup area. Washington's Montana Resources Incorporated (MRI) hired Gardiner as president and then reopened the East Berkeley as the East Continental Pit.[40]

MRI's official literature loudly proclaimed it was union free, which made it part of a new era in western hard-rock mining. Many in labor felt betrayed; union leaders had supported the project with the understanding that MRI management would not oppose efforts to organize. Union advocates accused MRI of screening its new hires to eliminate any of Anaconda's old union-friendly workers. The company disagreed; it had very few positions open, so hiring all former Anaconda employees back was impossible. MRI noted that more than 90 percent of its workforce came from the local area, and many of them had worked for Anaconda at some point.[41] The BMU started picketing actions, but state and local officials sided with MRI, heeding an economist's advice that people "in desperation" must learn to do things differently from the way they had done them in the past.[42]

Picketers also did not have youth on their side; Butte's young people were increasingly less interested in unions than the older generation had been.[43] Meanwhile, white-collar, often technical, workers began to arrive in Butte under Peoples' reign, bolstering the economy even as they challenged the existing culture.[44] Butte's unions thus faced the same kinds of pressures that diminished union power across the United States during the 1980s. The large open-pit Clifton-Morenci Mine in Arizona, for instance, similarly became nonunion, although that result followed a far more dramatic fight, one that involved the National Guard and decertification.[45]

Union or not, Washington's MRI did quite well. Using profit-sharing as a significant part of the pay, MRI has operated successfully until the present day, except for a hiatus between 2000 and 2002 when high energy costs and low copper prices forced a temporary shutdown.[46] MRI's success, however, did not solve Butte's early unemployment problem. When MRI opened, more than three thousand people applied for only three hundred positions. Despite retraining efforts, many former copper workers continued to be unemployed into the late 1980s.[47] Open-pit mining made something of a comeback in other parts of the American West, but precious metals and rare earths became the focus of the new boom. As historian Michael Malone noted at the time, an industry that had been key to the "economic, political, and social development of the West" had failed to recover its "former status in the region" or in Butte.[48]

## HERITAGE

Historic preservation, which initially focused on places most affected by open-pit mining, contributed to Butte's economic development, but undertaking it required a dramatic shift in the mindset of locals. Butte had become so used to sacrificing structures for open-pit mining that the idea of preserving them for the sake of tourist dollars and historic importance seemed alien. Preservation also helped to fight the stigma of bust. As one 1980 report stated, "The city has a widely known reputation as a deteriorating mining camp plagued by fire, violence and economic depression."[49]

Nostalgia could hamper the preservation movement, as the case of the Columbia Gardens shows. After the park closed in the early 1970s and a subsequent serious fire destroyed many of its treasures, hopes for relocating it faded but did not disappear. The Butte Ski Club donated land, and the Columbia Gardens Foundation negotiated with the Ski Club to create a new board to oversee the creation of a new park. To be known as Columbia Gardens II, the new facility would include a playground and picnic area. The Foundation, however, failed to notify the public about its board of directors' election, to involve any members of the grassroots People's Committee to Save the Gardens, or to consult the local government about the new site—the old Beef Tree Ski Area, four miles south of Butte. According to Butte historian Pat Kearney, money determined the project's fate. The Anaconda Company had donated $450,000 to the Foundation, but the Foundation spent that money quickly. When the Foundation requested another $23,000 from the county in 1978, angry city-county leaders denied the request. Allegations of financial mismanagement followed.[50]

A lack of community involvement also undercut the project, and that absence was largely due to powerful memories of the Columbia Gardens and the old Butte it represented. Industrial nostalgia—a reverence for the past era of good industrial jobs and a healthy economy—had come into play. With unemployment skyrocketing, residents looked at symbols of the area's past with such reverence that it actually made preservation of the remaining Columbia Gardens equipment more difficult. Many locals' strong childhood memories of the park transformed the loss of the park into the loss of their own childhoods. As one reporter wrote in the *Montana Standard* two years after the Gardens' closure, "In memory, perhaps the Columbia Gardens lives better and brighter than it really was."[51]

A 1983 survey of ninety Butte residents, conducted by students at Montana State University, found that, for many residents, what was being built at Columbia Gardens II "fell too far short" of the original. As one resident lamented, "The memory is so strong of how beautiful it was.... People don't want to see another Gardens because it won't be the same."[52] Many of the written comments included in the report suggest that duplicating the attractive wonder of the Gardens would be impossible. The proposed site was "a depressing and inadequate substitute—a travesty." One respondent even wrote in jagged lines, "It is like being given a plastic flower after being raped." When asked why they had not either gone to the Columbia Gardens II site or been involved in the project, 57 percent of residents simply marked the box that they were not interested. The lack of connection between the community at large and the Gardens II project was evident when the operators went against popular feeling to hire a nonunion landscaping company.[53]

Columbia Gardens II never officially opened. Surviving equipment from the original was abandoned at the foot of the old ski hill until 1985, when community members rediscovered it. They took pictures of the neglected remnants, submitted them to the *Montana Standard*, and, with the resulting public pressure, convinced the government to transfer what remained to a city park.[54]

Another survivor of open-pit fears, Butte's uptown, became even more of a focus for local preservationists. Although the secretary of the U.S. Department of the Interior had designated uptown Butte as a National Historic Landmark District in 1961, partly due to fears that the Berkeley Pit would devour it, the pit never reached the designated area and little was done to make use of the district's official status. Instead, uptown had mostly experienced the decline of its older buildings.[55] A few weeks after the vote against relocating the CBD, the city council voted instead to investigate the possibility of renovation, and the mayor appointed a special committee. Despite concern about future mining, the committee pushed back against ARCO's recommendation that the city not invest too much into redevelopment. One city leader argued, "Butte should decide its future—not ARCO." Shortly thereafter, the Butte Uptown Association announced support for uptown merchants willing to make improvements to their buildings.[56]

The city-county government and the state historic preservation office together invited the Historic American Engineering Record, a department of the National Park Service, to conduct a rehabilitation survey of uptown. This project led to the formation of the Urban Revitalization Agency in 1980, which provided guidance and monies from a tax increment district to fix uptown's dilapidated buildings and their facades. A new group, Butte Citizens for Preservation and Revitalization, supported those actions. After mining ended, the city-county hired preservation professionals, while residents began to make more use of the National Historic Landmark District designation for purposes of tourism.[57] Some interested citizens focused on preserving uptown Butte's headframes, which had held the elevators at old mining sites. Other locals founded a county historical society in 1981, which collected company records, labor materials, and oral histories, all of which it housed in uptown Butte. Even some of the old mining equipment was saved; it found a home in Butte's World Museum of Mining.[58] A "streetcar" (bus) tour of Butte, began in 1971, and the Mai Wah Museum, which showcases materials from Butte's Chinatown, have similarly enhanced interest in Butte's industrial past, as did an early 1980s project to build Our Lady of the Rockies, a statue of the Virgin Mary now overlooking town, thanks in part to many miners' technical contributions.[59]

Annual summer festivals, held in historic uptown, have become another means to highlight Butte's industrial heritage. Since the early 2000s, these have included one celebrating its most famous native son, daredevil Evel Knievel, one celebrating its ethnic heritage, the An Ri Ra Montana Irish Festival, and one, the Montana Folk Festival, created after three years' success hosting the National Folk Festival. People invested in Butte suddenly had an opportunity to explain the mining landscape to hundreds, even thousands, of visitors. Through preservation, Butte's city leaders have sought to refocus outside attention from the story of a failed industry to the story of the city's cultural mining heritage.[60]

Longtime Butte residents were not initially driving the movement. Preservation advocates were often neo-locals—younger newcomers who saw value in the built environment. In the late 1970s, many young, progressive- and tech-minded people had moved to Butte to work at the new National Center for Appropriate Technology and the new lab for Mangetohydrodynamics, which was a high-tech coal conversion process. Both of these facilities were sponsored by Montana's congressional representatives as a way both to deal with the declining mining industry and to develop environmentally friendly technologies. When the Reagan administration cut funding for both programs, some young activists who had lost their jobs stayed in Butte and became active in the preservation movement. As industrial archaeologist Bode Morin recalls, "New people who moved into the region in the late 1970s and 1980s, outsiders at first, did not necessarily see the loss of the previous decades, but rather saw the considerable wealth of cultural resources remaining, and organized, along with some long-time locals, to protect those enduring resources."[61]

As newcomers, the activists were "captivated by the region's rich heritage, but not 'affected' as much by its decline."[62] The collective memory of people who had not lived through the mining past as adults—who "didn't grow up with the pollution and the noise and the constant rumble," in one resident's words—had a "romantic idea" of that era. Older, longterm residents instead had more critical memories of the past, so they were less likely, at first, to participate in preserving symbols of it.[63] Now that neo-locals had arrived, industrial nostalgia was working for preservation, not against it.

## Connecting Neighbors

More romantic ideas of the past also had a strong effect on a different kind of reclamation, as neighbors displaced by mine expansion have reconnected. Scholarship about the movement of white ethnics out of old neighborhoods suggests former residents of Meaderville, McQueen, East Butte, the East Side, and Finn Town faced two options. Historian David Schuyler argues that urban renewal projects in major cities led to "tears [in] the fabric of community, the network of institutions and patterns of human interaction." Scholars dealing more explicitly with white ethnics have added that immigrant traditions disappeared with a move out of the former enclave.[64]

Kathleen Neils Conzen, in her study of ethnic suburbanization, claims, however, that ethnic identity cannot be reduced to common neighborhood residence. Even after moving out of old neighborhoods, ethnics continue to be connected through cultural organizations and ties. Margaret Brennan Lee further argues that ethnic identity continued to hold traditional white ethnic communities together during and after suburbanization.[65] Butte's situation does not neatly fit either the urban renewal or the suburbanization pathway. At first former residents experienced the tears common in urban renewal, but later the most drastic effects were ameliorated and many social connections were rebuilt.

Broader changes in Butte and the United States combined with pit expansion to break up the neighborhoods. During the 1940s and 1950s, young men in each neighborhood had access to well-paying jobs for Anaconda. As jobs in the mines declined and more men as well as women had the opportunity to go to college in the 1950s, 1960s, and 1970s, they became more likely to move out of town, while increased car ownership made living farther from work possible.[66] Expansion of the Berkeley Pit sped up the process by undercutting prized neighborhood institutions, like volunteer fire departments, churches, and grade schools. Former residents became disconnected from all but relatives in their old neighborhoods. They spread across Butte, with adults ending up in new parishes and their children in new schools.[67] Most people who were children or young adults in the 1960s and 1970s, when mine expansion forced them to relocate, claim in oral histories to have coped quite well. They were readily accepted into new neighborhoods.[68]

The older generations bore the brunt of relocation's social effects, as recent cases involving industrial development and relocation elsewhere have documented.[69] Helen Nicholls, whose family home was purchased by Anaconda, explains the great difference between the reaction of her younger generation and that of the elderly: "I can't say that for us or for my family, there were any great emotional ties to the house—it wasn't like we had lived in the house for a hundred years.... Times were changing.... [But for] some of the older people, it really did hurt: these were people that had lived side by side for 50 years, they had grown up as children, their parents had owned the house and then they inherited when their parents died, and these people had lived together all their lives.... For many of them, it was terribly, terribly hard." Indeed, she thinks that relocation had dramatic effects for residents of these old areas: "When the Anaconda Co. started the pit and they moved out Meaderville and McQueen and

East Butte, that was definitely the end of the old ties and the old bonds. People that had been neighbors for a lifetime were maybe living ten miles apart."[70]

Some older people managed to stay in touch with friends from the old neighborhood, but most struggled with isolation. One former resident of Finn Town remembered that her neighbors had been "kind of clannish, they were in their own little section, and they did a lot of visiting back and forth" before the move. Some moved nearby, into the Silver Bow Homes, but most of the other "Finnish people" were sadly "scattered to every direction."[71] In 1983 former Meaderville resident Angelo Petroni told an interviewer that dispersion meant that "the only time we see each other anymore is at weddings, wakes, and funerals."[72]

Unwelcome change occurred even when groups of residents moved from old neighborhoods to developments on the Flats. Houses in the suburbs were more dispersed than in high-density older districts like the hill and the East Side.[73] Greater physical distance seemed to create greater social distance as well. The housing situation did not change as much for people in Meaderville or McQueen, where residents often had large yards and space to themselves, but the problems of social distance remained. One former resident remembers having to gather all of his relatives to convince his weeping mother to leave her Meaderville house. Unlike many of the displaced people, his mother was able to find housing near her former neighbors. She lived next to six of her closest friends from Meaderville in northwest Butte's McGlone Heights, but her formerly frequent visitations with friends disappeared. He explained, "She still stayed friendly, but I guess it just breaks up a thing, breaks up a community."[74] Similar stories of resistance and heartbreak come from former residents of McQueen and East Butte.[75]

Older residents who could not speak English when they left their neighborhoods found it hard to connect with new neighbors. The mother of Bernice Favilla Maki had communicated almost solely in Italian with her neighbors and never felt compelled to learn English until her relocation.[76] People tell tales of older Italian American men from Meaderville who moved away from friends and committed suicide during the 1960s and 1970s. County records do not readily yield evidence of such events, but the existence of such stories hints at just how deeply relocation affected the older generation.[77]

Living in areas where car ownership was the norm also affected the older generation. Many had never learned to drive, which made reconnecting with old neighbors difficult. People who had been closely connected through their churches now saw that the idea of a neighborhood parish was declining.[78] Scholarship suggests that such spatial changes can often be very disorienting, and Butte's relocation certainly was.[79]

Food traditions became difficult to maintain. Although housing was fairly dense in communities like Meaderville, McQueen, and East Butte, many families kept large gardens and even farm animals. Keeping a large garden after a move into the city or suburbs, however, was difficult. Former Meaderville resident Marlene Holayter related the story of an older Italian American lady who "had cows. She had her chickens. You know, so she had milk and she had her eggs. And she was extremely generous. And so, yeah, this was her whole little world and she never spoke English. And then she's gone. That's gone.... [After the move] she can't have her chickens, she can't have her cows, she can't have her, you know, she could have a little garden. But, she's in a place where she doesn't speak English, and the neighbors, I mean, they don't like that."[80] Her whole environment, from language to land, had changed. One relocated family from East Butte who similarly could not re-create their garden in Butte proper eventually left the area altogether.[81]

For Angelo Petroni, one of the biggest problems that Meaderville's displaced Italians faced also revolved around food. In Meaderville, families had earthen cellars where they cured salamis and aged wine. The dry ground of the suburbs, however, contrasted greatly with the moisture of Meaderville's earth. His dad built a cellar, but the wine barrels would start to dry out and crack because "the Flats—it's all sand."[82]

Other ethnic traditions maintained by the older generation also suffered without the neighborhoods' nurturing support. The Cristoforo Colombo Club, which had been an important fraternal order for

Italian migrants and their ancestors, survived the neighborhood, but the high membership numbers that it easily maintained during Meaderville's existence dropped sharply.[83] Similarly, polka music and saunas gradually disappeared, along with Finn Town's bars.[84]

Attempts at neighborhood reunions in the 1970s and 1980s proved too emotionally draining for that generation of retirees. As former Meaderville resident Angelo Petroni recalls, "It was more than they wanted to get involved in. It was upsetting to them."[85] Hurt feelings about the relocation process remained. During the later 1980s, however, Butte residents began to incorporate "increased historic awareness" into their daily lives, with ethnic meals, murals, and other activities becoming part of the community fabric.[86] As historian Matthew Frye Jacobsen has shown, an overall growth in white ethnic awareness happened nationwide during this period.[87] Remembering lost communities soon became part of the trend. Early efforts came through the churches. There were reunion services for former members of Sacred Heart Parish into the 1990s.[88]

Meaderville serves as a good example of the reemergence of neighborhood traditions. Members of most of the Italian families from Meaderville did not move far and were still well represented in Butte twenty to thirty years later. Despite their struggles, many kept up or returned to Italian traditions surrounding food, such as making wine and salamis and cooking other family favorites.[89] Although Meaderville's once-beloved nightclubs are long gone, the relocated Lydia's supper club maintains the traditions of a Meaderville-style meal. Other neighborhood traditions have been resurrected. Butte High School teacher Chris Fisk initiated a 2003 project that brought together his students and community members (including former Meaderville fire department members) to re-create one of the famous, moving, brightly lit Meaderville Christmas displays.[90]

Although the saunas and polka music mostly disappeared from Butte's Finn Town, the area's cultural life is resurrected for special events, including an invented tradition called St. Urho's Day, that originated in Minnesota during the 1950s as a parody of St. Patrick's Day. St. Urho supposedly drove the grasshoppers from Finland. In Butte today St. Urho's Day serves as a way to start the city's massive

St. Patrick's Day celebrations early, although the party usually includes many elements, including the polka music, of earlier Finn Town celebrations.[91]

Following this trend of community celebration, members of younger generations, who had been children or young adults when the neighborhoods disappeared, became involved in activities related to their old homes. In 1985 Esther Francone began a Meaderville Italian luncheon, at which up to two hundred women could "visit and reconnect with their Italian roots." The tradition restarted in 2005, with a younger generation of Meaderville women reinstituting a luncheon. Instead of speaking Italian, as many women at the original luncheons had done, the younger organizers "knew but a few phrases." They would have known Meaderville only as children in the 1950s and 1960s.[92]

The idea to have big neighborhood reunions in the 2000s emerged out of smaller, less formal gatherings like these luncheons. While an older generation had eschewed reunions with former neighbors as too painful to endure, younger residents did not have such qualms. Most organizers were in their fifties and sixties in the early 2000s. Some had already retired, giving them both a sense of the neighborhoods' importance to their later lives as well as more time to devote to the project, but they were also young enough to be quite energetic about reconnecting with old friends.[93]

Organizers first spent months gathering contact information for former residents. Pauline deBarathy, who was just six years old when the Berkeley Pit began, became a lead organizer for the Meaderville reunion. Starting with a 1956 city directory, deBarathy began a database with the names of people who lived in Meaderville. She sent a copy of her current directory to people she could easily find, requesting that they add names and addresses. She soon had a database with 363 entries. Through this social chain, deBarathy and other reunion organizers were able to start rebuilding connections between people who had lost touch.[94] One of the McQueen/East Butte reunion organizers, Rose Marie Ralph, said that additional methods, including "advertising on the radio, on the T.V. and just word of mouth" were helpful, as was the internet, which greatly improved organizers' ability to find people who had moved far away. Organizers also put together committees to plan for food and entertainment, as well as to gather old photographs.[95]

The first reunion, held by the neighborhood of East Butte, generated a lot of interest among former residents of other neighborhoods. That reunion's organizers decided it would be difficult to host other old neighborhoods, but they were willing to help new groups set off on their own. The East Side hosted a reunion in 2005. Meaderville then hosted a 2006 reunion, while McQueen residents joined with East Butte for their next reunion.[96] Each reunion has been successful at gathering people from around the country. At the East Side/Finn Town reunion in 2005, for instance, organizers expected only a hundred attendees. To their surprise, as word spread the number ballooned to 450.[97] Reunions proved so popular that neighborhoods like Meaderville and McQueen/East Butte held repeat reunions every two or three years. Recent reunions have taken place in the Maroon Activity Center, which sits near the edge of the Berkeley Pit, about as close in space to the lost neighborhoods as a person physically can be.[98]

Reunions have become important places to reconnect neighbors, celebrate ethnic culture, and to think warmly about the now-lost landscape. Each of the reunions for the largely Italian American neighborhood of Meaderville has featured catered Italian food. Other reunions showcase the once ubiquitous accordion music. Former residents at each reunion share maps, old photos, and newspaper clippings by posting them on boards or by giving them to a person who projects them via computer. Many also bring memorabilia from scout troops and Copper League baseball teams.[99] Pauline deBarathy reported that some of the older men in particular were very touched to see people with whom they closely worked in the mines or bonded during neighborhood events.[100]

Butte's reunions are not the only ones of their kind. People relocated by megaprojects, like mines and dams, in other regions have also held reunions.[101] Former residents of towns in central Massachusetts, now submerged under the Quabbin Reservoir, have gone on pilgrimages to the edge of the dam and have held a Remembrance Ball.[102] Former residents of Santa Rita, a town in New Mexico devoured by an open-pit mine, have also held events. Because the town is now nothing but a hole—nothing but space—reunion attendees call themselves the Society for People Born in Space.[103]

The lack of the actual place has made former residents more devoted to the memory than they might otherwise have been. It might explain why a younger generation is now "bringing a new sense of energy" to the work of remembering lost Butte neighborhoods.[104] Former residents can stand on the ridge above the area that once housed McQueen and East Butte. They can see the eerie markings where streets and homes once stood. They can also drive around Butte to find homes that had been relocated from old neighborhoods, but otherwise they can no longer experience the old landscape. Local Janie Payne told one interviewer that the absence of these places "affects your roots in some ways, because I can't take my son to my old grade school and say this is where I went to school, this is the neighborhood that I grew up in, because it's not there. So to maintain our family history, it's all pictures."[105] Former Meaderville resident Marlene Holayter explains the collective memory of her former neighborhood: "People keep it alive… because it's not there, you carry the traditions with you."[106]

Efforts to remember these neighborhoods long after their demise can also be attributed to a feeling of nostalgia about the heyday of mining. Some have tried to see positives in the loss of the old neighborhoods—they represent an old Butte that was too parochial and neighborhood-centric for some.[107] It has been more common, however, for Butte inhabitants to hold very warm feeling about ethnic enclaves, seeing their bulldozing as representing the destruction of Butte's "ethnic sense."[108] It no longer seemed like a vibrant industrial town made up of diverse peoples. All events but St. Patrick's Day seemed to have disappeared, partly because, when the neighborhoods disappeared, "the ethnic activities connected with those places were just gone," as one local put it in 1986.[109] The reemergence of ethnic-related activities in the late 1980s and 1990s, followed by reunions in the 2000s, partly came from this impetus to celebrate the mining era.

Childhood nostalgia joined smokestack nostalgia as a cause for the reunions. For many interviewees, a place like Meaderville or McQueen was not just a better place—it was also a better time. For some, the lost neighborhoods represent a time when Butte was more friendly—when saying hello in the post office did not elicit suspicion or fear. For others, it represents a time when there was less on television (or no television at all), so friends would instead spend their time visiting with each other. Perhaps this warm view of the lost neighborhoods comes from the fact that reunion organizers and attendees were children or young adults when the neighborhoods disappeared. Reunion organizer Rose Marie Ralph believes that her pleasant time growing up in East Butte encouraged her to plan a reunion: "Everything seemed simpler, but maybe that's because you were a kid at that time too."[110]

Jim Michelotti, who has helped to plan McQueen reunions, hypothesizes that perhaps his younger generation never knew the real difficulties of the industrial era—the strikes and hardships: "In a lot of ways for my father and mother's generation—those were hard times." His dad worked seventy hours a week and a number of family members died while he lived in that neighborhood. Michelotti believes that his father would not have wanted reunions because the neighborhood did not represent everything that was good about the past, as it does for the generation now in their fifties and sixties. Now that Michelotti's generation is getting older, reconnecting with childhood friends seems like an excellent idea: "There were good and bad times when you're adults. But when you're kids, there's always good times."[111] Members of the younger generation who recognize their parents' hardships tend to believe such problems served a greater purpose—teaching grit, work ethic, and the importance of quality friendships.[112]

For former residents of these neighborhoods, the reunions therefore represent a chance to go back in time and in space. Jim Michelotti put it best: "If you asked somebody, you know, 'Would I go back to McQueen?,' I think you're asking me, 'Do you want to go back to your youth?'"[113] Former Meaderville resident Marlene Holayter shares this idea. She believes that her neighborhood "lives because it has never—it hasn't lost time, you know? It hasn't gotten old, it hasn't—I mean, if Meaderville was there today, and you look at all the old buildings and somebody came in with a bulldozer and put it down, you might say to yourself, 'Oh, yeah, it got old.' But, Meaderville will never get old. Meaderville will always stay as it was."[114] Geographer Doreen Massey has warned that "we can no more go back in space, return to whence we came, than we are able to go back in time." For her, just as time has moved on, space has changed as well.[115] This generation of residents is fighting against the material truth of this concept in order to revive their neighborhoods.

Residents at reunions are not void of critical opinions about the past. A reporter who attended a McQueen/East Butte reunion in 2008 wrote a news article that claimed residents showed a surprising lack of resentment toward relocation. One former resident sent a rebuttal to the paper the next day. The resident said that the reunion actually brought up "bitter" feelings. As she put it, "Some of us cannot bear looking at some photos (particularly the one showing Holy Savior being buried by massive amounts of dirt) without shedding tears." To her, some of the destruction was "unnecessary and malicious. . . . We were once in a better place," she argued, "and some of us just can't forget that."[116] Despite the presence of a few still-incensed former residents, most reunion-goers are now more saddened about relocation than they are angry. Some say that the economic benefits of continued mining outweighed the losses, while others, especially those from McQueen and East Butte, claim that the company gave them a good deal during the move, so they harbor little resentment toward it. For former residents, reunions primarily create connections, and do not renew discontent.[117]

Formal and informal community reunions are therefore doing their best to repair and strengthen social bonds, suggesting that ties were not entirely broken by relocation in the first place, but simply strained. Indeed, many of the reunions' attendees are working hard to keep the neighborhoods alive by bringing whole families to the events. Attendees often bring their children and their grandchildren to meet former neighbors and to talk about past traditions. People who had never seen these neighborhoods are therefore coming to reunions to learn about places that might no longer exist physically, but still exist mentally. Even as they lament the passing of their old neighbors each year, former residents hope that the involvement of younger generations means that these vanished places will paradoxically continue to survive.[118]

## Cleanup

Generational issues also played a major role in cleaning up a century of mine waste. In 1981 residents of Milltown on the Clark Fork River discovered that their wells had been contaminated with arsenic from Butte, which lay upstream, 120 miles to the southeast. Mining wastewater dumped in the river by the Anaconda Company and its predecessors had backed up to the Milltown Dam reservoir and was seeping into the groundwater. Mining companies had treated Silver Bow Creek as an open industrial sewer since the nineteenth century, and both Anaconda and state officials had long known that mine waste was causing damage downstream. Winds and runoffs thwarted expectations that mine wastes would stop in settling ponds near Warm Springs, Montana. Believing that acidic mine water neutralized sewage, Butte had even allowed its sewers to empty into Silver Bow Creek.[119]

Starting at the Berkeley Pit and continuing to the Milltown Reservoir, the Clark Fork drainage basin became the EPA's largest Superfund cleanup site, by area, in the nation. The Superfund program had its roots as a response to the contamination of groundwater at Love Canal, New York, as a result of chemical dumping by Hooker Chemicals and Plastics. On December 11, 1980, President Jimmy Carter signed into law the Comprehensive Environmental Response, Compensation and Liability Act (CERCLA or Superfund). This statute granted the EPA authority to clean up sites containing uncontrolled hazardous waste.[120] The year the EPA named the upper Clark Fork to the National Priorities List, 1983, triggered plans for the cleanup of water and soils contaminated by heavy metals. ARCO, the State of Montana, and the local government were all responsible parties.

The EPA arrived in 1985, and designated a number of sites in the Clark Fork River drainage basin for cleanup, including the mine flooding unit, which involved the old mines and the Berkeley Pit. ARCO, which was both funding and often conducting the analysis and cleanup, developed a good working relationship with the EPA. By 1985 essentially the only ARCO employees left in Butte were those involved in Superfund cleanup.[121]

The difference in reactions between insiders, who had experienced Butte mining, and outsiders, who had not, was stark. On one side stood

Butte residents. On the other were people in the college town of Missoula, a white-collar, environmentalist-friendly community downstream. Sitting ten miles from the Milltown dam, Missoulians had become quite concerned with mine waste. One environmentalist group, the Clark Fork Coalition, called for thorough cleanup of the river basin. To this Missoula-based group, mining had no positive qualities—it meant only poisoned water and toxic soil. No one in the group had much experience with mining or mined areas, which partly explains the emphatic tone of its early newsletters as it railed against mining in general and Milltown Dam waste in particular. The Clark Fork Coalition even argued that efforts to save Butte's mining landscapes through historic preservation were flawed; preservation would limit the effectiveness of remediation.[122]

These developments were a shock to many in Butte. As County Attorney Bob McCarthy explained, the city's residents were now asked to look at the landscape as a totally different place—alien to the way they had previously understood their home. With pollution everywhere, theirs was no longer a working landscape, but a poisoned one.[123]

The absence of Butte people in the environmental coalition of the late 1980s implies that Butte residents were more hesitant than Missoulians to become involved in environmental cleanup.[124] Indeed, the Clark Fork Coalition's director during the 1980s, Peter Nielsen, thought, "It's a cultural thing in Butte. They want mining, that is who they are."[125]

Even into the 1990s, longtime Butte residents maintained a more "accommodating attitude" toward mine pollution. According to public policy researcher Douglas Isokait, "Those who inhabit the Upper Basin have a special relationship with the arsenic and cadmium leaching from the millions of tons of mine tailings they live with and made a living from."[126] Oral histories with longtime residents offer one example of this special relationship. When asked about their childhoods, older residents discuss playing on mine dumps and in toxic streams. Such remembrances usually end with some variation on, "I turned out OK," so the waste must not be harmful.[127]

In an interview, David Stiller, an environmental consultant and former state regulatory official, concluded, "They [Butte residents] were split. At least half of them, particularly the older generation, didn't see the damages for what they were. Until very recently [since the collapse of the mining economy] they really saw environmental issues as important to outsiders, not necessarily a big deal to them." In the mid-1980s, residents believed, "This stuff [tailings and mine waste] was always part of the City. It was just part of their lives.... As a whole they minimized the effect of the damages."[128]

City-county officials also worried. They did not want their homes to be labeled as another Love Canal, thus halting economic development. The city was bewildered as to why it was being named a PRP (potentially responsible party) to some of the cleanup by the EPA and the Montana Department of Environmental Quality (DEQ), because its storm sewerage system was a conduit for transporting mine wastes. They warmed only gradually to environmental cleanup by the late-1980s.[129]

Other locals were simply afraid. To them, being another Love Canal meant major health concerns. As the EPA's Butte project manager once stated, the EPA would tell residents, "Not only are you living on a Superfund site, but your kids' health is in danger." The EPA would then go into years' worth of studies before it could arrive on practical solutions, meaning that residents' initial fear combined with prolonged uncertainty to transform into anger.[130]

More Butte residents eventually became involved in the cleanup but, just as with historic preservation and reunions, neo-locals and younger residents drove the increasing environmental interest. New residents did not identify with the mining landscape and its waste in the same way as longterm residents. Instead, they told stories about Butte that emphasized physical ugliness. Waste for these new people was dangerous and worth jettisoning—and was not a nostalgic landmark of the mining era.[131] City-county officials also became more cooperative when they realized that ARCO was to be responsible for paying the cleanup bill. Their role was mostly to watch over the operations. In 1991 the city even took over and tried to clean up Butte's water service, which had previously been provided by Anaconda, then ARCO, then MRI.[132]

## A Permanent Poison

Although the entire watershed was troubled, the Berkeley Pit generated the most fear, complicating Butte's attempts to move forward. The decision to shut down the Kelley Mine's underground pump station in 1982 scared many people in the area. The pump had been removing up to five thousand gallons of water a minute. Before the water level had been artificially lowered, it had once been high enough that the Flats, on which so many residents today live, had been a marshy wetland. After the mine shut down, water began to rise through thousands of miles of tunnels toward its old equilibrium, much more quickly than anyone at ARCO had anticipated. ARCO had predicted that water would not reach the bottom of the pit for 20 to 120 years, but the waters actually reached that point in November of 1983, when more than 66 percent of the underground workings had already flooded.[133]

The possibility of water rising so much that it would eventually fill the pit caused "a great deal of alarm in the citizens," as the deputy county attorney put it.[134] To make matters worse, ARCO did not immediately release its consultants' monitoring reports. Management worried about panicking residents with the worst-case scenarios that the reports laid out. Chief Executive Peoples demanded the reports (and reflected the area's Democratic bias), stating, "There are some things you know. You know you are going to die, you knew that when Reagan was elected president the country was going to hell and you know that this flooding is not good."[135]

Locals quickly developed worst-case scenarios anyway. The most obvious was that water, polluted from traveling through the mines below, would seep through the porous alluvium at the top of the Berkeley Pit and into Silver Bow Creek, thoroughly contaminating the waterway. If water rose that far, the Montana Department of State Lands speculated, it could also flood the basements of homes and businesses on the Flats, necessitating relocation. Another rumor was that rising water would deteriorate wooden supports under Butte, causing subsidence. A geologist worried that catastrophe could come even sooner: he believed a possible earthquake from the nearby Continental Fault would cause the Berkeley Pit and the Yankee Doodle Tailings ponds, which sit above Butte, to overflow.[136] Attempting to get more information, the local government called on officials and legislators at both the state and federal levels. According to them, the EPA was working closely on other parts of the Clark Fork drainage basin Superfund complex, but not on the Berkeley Pit.[137]

Fritz Daily, a Butte representative in the state legislature, tried hard to get faster action on the flooding situation. In one failed bill, he attempted to move all reclamation activities to the Department of State Lands. He also pushed the Berkeley Pit Re-Industrialization and Mineral Recovery Program in 1987, which studied possible ways both to clean and to use the pit's water. The state legislature refused to fund the program.[138]

In the meantime, the level kept rising in the bottom of the pit. By the end of 1987, water had filled the bottom six hundred feet of the pit. Just three years later, almost 16 billion gallons of liquid sat in the basin, whose hard rock acts like a big bathtub.[139] Early in 1990 one local proposed that the city's garbage should simply be dumped into what was increasingly becoming a polluted lake.[140]

More-comprehensive management of the site began to take shape as ARCO shared more information with the EPA and the state. From 1990 to 1994, the EPA and the Montana DEQ oversaw a large feasibility study that installed additional monitoring wells and mapped out a variety of options for dealing with massive underground flooding. Tensions existed among all parties, with the state demanding more cleanup money from ARCO in court. In addition to ARCO, the city-county and state sometimes had trouble dealing with the EPA, which they felt was more interested in finishing work at the pit than in solving the problem.[141]

The final *Record of Decision,* issued by the United States EPA and the Montana DEQ in 1994, ensured that some water would be diverted away from the Berkeley Pit and that both British Petroleum (BP)–ARCO and MRI would be responsible for building a water treatment plant.[142] Although locals preferred this option over less-comprehensive ways of dealing with the water, the EPA and state "received considerable public comment opposing certain aspects of the preferred alternative presented in the Proposed Plan." Many commenters believed that the

Open-Pit Mining Complex in Butte, 2017. The Berkeley Pit, now filled with water, is the large dark hole. The current Continental Pit operation sits to the east (right) of the Berkeley. To the north (above) of the Berkeley sits the Yankee Doodle Tailings pond, which fans out in shades from gray to green. The pond holds waste from mining operations behind a large earthen dam. Butte's hill sits to the west (left) of the Berkeley and the Flats to the south (below). Image from Google Earth Pro.

EPA and DEQ had too easily acquiesced to an inadequate plan, which required major water treatment only when the level reached a certain elevation and not earlier. Reflecting the town's industrial heritage, other commenters were just as concerned with the treatment process itself. They hoped that pit water would not only receive treatment but also serve as a source of recovered metals.[143]

Today, 2 million to 3 million gallons of acidic, metal-laden water a day continue to flow out of the mines and into the Berkeley Pit, adding to the 40-billion-gallon body of contaminated water. The presence of oxidized sulfide minerals means that the water measures between the

Ph level of cola and that of vinegar, 2.5 to 3. The water contains fifteen times the acceptable EPA level of arsenic for drinking water, seven times the lead, and 430 times the cadmium, all pollutants picked up during the liquid's travel through the underground mines. There is no cost-effective way to remove all of the water or to treat the water completely, once and for all. Therefore, only 3 million gallons of water is withdrawn each day (about the same amount of surface and groundwater that flows into the pit). The Horseshoe Bend Water Treatment Plant uses lime and other processes, like aeration, to neutralize this acidic water. Most of the water is used by MRI's modern-day mining operations, which extract the copper, use the water in the concentrator, and then shoot the residual sludge right into the pit. Sometime around 2023, when rising Berkeley water is expected to reach the critical level of 5,410 feet elevation, after which it could disastrously flow out of the pit and into the alluvial aquifer, MRI and BP–ARCO will kick the plant into full gear. More than 7 million gallons of treated water a day will then be discharged into Silver Bow Creek. The companies will be responsible for continually treating the water that enters the pit, in perpetuity.[144]

Fears about the pit's "toxic soup," as many call it, have continued since this plan went into action, but on a lesser scale. Some worries are no more than irresponsible rumors, but others raise reasonable questions about the efficacy of the chosen treatment method and about the continuing availability of money and resources for future cleanup. As resident Fritz Daily says, "Perpetuity's a long time," so the continued responsibility of the mining companies is a concern.[145] In the early 1990s a local citizen's group posed a question: "Must Butte's residents resign themselves to living next to the nation's largest, most contaminated lake—in perpetuity?" The answer has been "yes." During the study period on the site, the EPA hosted many public meetings, but the 1994 *Record of Decision* and the 2002 Consent Decree, which is a judge-approved agreement between the EPA and potentially responsible parties, have essentially ended the debate on remedies, limiting later public participation.[146]

An independent local committee, the Berkeley Pit Education Committee, publishes an update on flooding called "Pit Watch." It does not critically review the authorities' procedures (perhaps partly because it receives funding from BP-ARCO), but it does offer solid answers to many of the public's questions. Reflecting the decreasing local concern, the publication has gone from quarterly to semiannual to annual.[147] The Berkeley has joined a number of open-pit mines around the world that are now filled with water—difficult to regulate and impossible fully to remediate.[148]

Locals have tried to learn how to live with the Berkeley Pit. Some worry that the dense fog that rises out of the Berkeley Pit on cold days contains contamination, despite EPA protestations to the contrary.[149] More typically, residents have found ways to joke about the pit. The Helsinki bar, on the east side of Butte, recently reopened as the "Helsinki Yacht Club," because it sits "on the western shore of the fabulous Berkeley Pit."[150] Others simply try to ignore it.

While locals have become selectively attentive to the pit in their day-to-day lives, this artificial lake has become the focus for outsiders' opinions about Butte. Almost every news story on Butte since the early 1990s has focused on the incredible scale of environmental cleanup and the toxic lake. Nothing but the pit could better symbolize a Butte that was "Poisoned, Ruined and Self-Cannibalized," as one publication put it.[151] To many, the deaths of 342 migrating snow geese who landed on the polluted waters of the Pit in 1995 has become the most important symbol of Butte's toxicity. The story has been repeated by almost every news story about Butte over the past twenty years as well as in documentary films about the area, such as *An Injury to One*.[152] Today, MRI uses blasts and electronic sounds above the pit's waters in order to ward off migratory waterfowl. If any birds land, employees work to move them within four to six hours, but mass bird deaths are still possible. In 2007 ducks, geese, and a swan all were found dead after fog lifted from the pit.[153]

More recently, Butte hit the world's papers for an even more massive geese event. In December 2016 a large storm in the middle of the migratory season meant that thousands of geese landed on the Berkeley Pit, covering its seventy acres. Fireworks, noisemakers, and a remote-controlled boat could not convince the tired geese to leave. The resulting deaths of thousands of geese shocked residents, while bringing renewed national and international attention to the pit's perpetual pollution.[154]

Concerns about remediation have often increased what can be a tense relationship between the EPA and locals. It did help that the longtime EPA remedial project manager for the Silver Bow Creek/ Butte Area Superfund site, Sara Godbout Sparks, was a Butte native whose father had earned a living in Anaconda's mines. She had also lost family members in mining accidents, including a brother in the Berkeley Pit. She was just as concerned as city-county leaders about the effects that environmental worries could have on attracting new businesses and young families. The EPA opened a local office and its employees were willing to go to meetings around town. Starting in 1989, a Citizens Technical Environmental Committee provided a conduit for residents' opinions to the EPA and ARCO, even if its funding and hence output has since decreased.[155]

Since the 1980s local companies have used Butte as a research site and testing ground for cleanup technologies, which has not only meant more jobs for cleanup-minded neo-locals, but also has turned a number of longterm residents into remediation supporters. Butte–Silver Bow's chief executive in the 1980s, Don Peoples, opposed Superfund when it first arrived. "Over my dead body," he remembers thinking. He deeply feared the national stigma attached to the Superfund label. He later supported cleanup efforts after he saw the economic and environmental good that it was doing for the area. After leaving office, he ran a cleanup technology company that employed two hundred people.[156]

Montana politicians, including former representative Pat Williams (U.S. House of Representatives) and recent governors Brian Schweitzer and Steve Bullock, have tried to turn the region's cleanup into a model restoration economy that would put people to work by cleaning up ecological damage. Additional monies have been made available for cleanup in the area through the Montana Natural Resource Damage Program, which provides millions of dollars from a lawsuit with ARCO, separate from the Superfund program. Their plan for a regulation economy has faced challenges; it will be difficult to fund additional projects as ARCO's contribution dwindles and the state legislature loses interest.[157]

Waste disposal from cleanup has also been an issue. The repository for much of the cleaned-up waste has been the Opportunity tailings ponds, near the town of Anaconda and the even smaller town of Opportunity. An idea that additional waste would supposedly only be "a drop in the bucket" when compared to what was already there has engendered some controversy, as have plans to leave other mine waste in Butte, simply covering some of it with clean dirt.[158]

Heritage concerns have also led some Butte locals to contest how the EPA goes about its cleanup efforts. With support from the State Historic Preservation Office, the grassroots heritage movement of locals and neo-locals eventually persuaded the EPA that mining dumps and headframes could be historic resources and tourist draws.[159] Debates have continued, however, as others worry about the EPA doing too little—leaving Butte residents in contact with poisons such as arsenic, mercury, and lead. The tension between preservationists and remediationists waxes and wanes, proving that although locals generally have a more positive view of Butte than outsiders, they often cannot agree on how to move forward. The Butte Restoration Alliance, a community volunteer organization, sought to put more emphasis on preservation in the remediation process, which has been helpful in bridging the divide.[160]

Residents have become relatively positive about Butte's future, but outsiders still see a stigmatized city. They ridicule Butte's residents for living in such an ecologically damaged area, which the regional joke that Butte should really be pronounced "Butt" represents. One Butte resident recalled attending a high school basketball championship game, where Missoula fans chanted "dirty water, dirty water," or "clean your pit, clean your pit" at the Butte girls team.[161] Journalists also have tended to be skeptical about locals' love of a landscape that includes a pit full of toxic water.[162] The Cool Water Hula, a dance event at the pit's edge, performed in 2000 and 2010, often appears in journalists' stories more as an oddity than as an artistic attempt to bring attention to environmental cleanup.[163]

Writers have also focused on the surprise of living organisms near or in the pit. *Mother Nature News* used the tale of a stray dog, nicknamed Auditor, who lived alongside the Berkeley Pit without incident for years, as a metaphor for Butte's residents, because both were still "hanging on" despite the area's troubles. Other articles have expressed bemusement or amazement at the efforts of local biologists to find cures for contagious diseases by studying microorganisms that make toxic pit water their home.[164] Some of these stories provide a nice window into Butte, but others of them trivialize the city, its pit, and its environmental problems.

Despite the negative coverage, the altered minescape has long been home to much life. A 1976 environmental review of the Berkeley Pit complex portrayed a complex ecosystem that, although altered by human activity, still involved grassland, coyote, beaver, marmots, rabbits, chickadees, crow, mule deer, and elk among the dumps.[165] Butte's situation is therefore not as black and white as Berkeley Pit journalism suggests. The pit's prominent place in the news cycle continues to affect both policymaking on multiple levels and the area's ability to attract new employers.[166] No matter how much locals have done to transform their city, Butte still seems to outsiders like an oddity at best, an environmental horror at worst. Long after it closed, the Berkeley Pit has continued to affect Butte.

The attempt to give the Berkeley Pit an even bigger role in the area's industrial heritage tourism suggests a way to bring outsiders and insiders together. The largest tourist destination in the area, the Berkeley Pit had 23,018 visitors during the summer of 2006, with 27,005 visiting in the summer of 2009.[167] Add to those numbers the free admissions offered at other times of the year and the total rises to 40,000 to 50,000 visitors annually. Guests now pay $2 for a pleasant entry into the pit's strange beauty—its colorful walls and glistening waters. The Butte Chamber of Commerce received only $18,606 from visitors during its first year running the stand, but it was making $39,040 by 2009. The spruced-up visitors' center, which now features a gift shop, benches, green grass, interpretive signs, and a spot to take "selfie" photos, increasingly plays a role in tourism.[168]

Despite the pit's relative success as a tourist site, putting it at the center of Butte's heritage industry is controversial, both nationally and locally. The 2006 episode of *The Daily Show* mentioned above showcases this problem, and so do multiple national news outlets, which

BERKELEY PIT VIEWING STAND, JULY 2013. Photo courtesy of Fredric L. Quivik.

similarly label the plan questionable, if not ridiculous.[169] Inside Butte, Matt Vincent, once head of the Clark Fork Watershed Education Program and later the county's chief executive, has worried that overly focusing on the pit gives the city a far worse reputation than it deserves, while allowing the erasure of an important social and economic factor in Butte—the long history of underground mining. Others, like the chamber of commerce, have invested in the pit's promotion as a way to gather tourists who will enter Butte for the pit, but then stay to enjoy its historic charms (or perhaps vice versa).[170] Bridget Barry's master's thesis argues that Butte's promoters struggle with a paradox. On the one hand, people visit the Berkeley Pit because it is a massive, polluted hole in the ground. On the other, talking about the toxicity too much could hurt tourism, so the environmental problems of pit mining are hidden to a degree in promotional materials at the viewing stand.[171]

Berkeley Pit boosters do benefit somewhat from one trend: visiting places that others used to avoid has become a niche tourist market. Two emerging forms of tourism, one labeled disaster tourism, which involves the act of traveling to a disaster area as a matter of curiosity, and the other toxic tourism, which involves going to poisoned places in order to better understand (and perhaps later prevent) such events, have Butte's Berkeley Pit firmly on their radar. Websites that promote these types of tourism devote attention to the Berkeley Pit.[172]

Instead of viewing the Berkeley Pit as a symbol of everything that is wrong with America, as many outsiders still do, or viewing it as a symbol of progress, like so many did during the 1950s, the pit should become a place that people visit to learn more about work and life in one of America's industrial centers. Instead of just presenting the pit as an oddity, an intelligent use of pit tourism could lead to a better understanding of open-pit mining and its effects. The posted boards at the viewing stand are a start, but, as Bridget Barry points out, they not only sanitize the environmental effects but also neglect Butte's social history, including stories of labor and the now-lost neighborhoods.[173] In 2010 journalist Nathaniel Miller complained that the recording at the viewing stand justified the vanishing neighborhoods as economic necessity, without going into what was lost or how.[174] A proposed museum next to the pit, which could have further explored the effects of industrial development, never came to fruition.

If even a fraction of the 150,000 visitors to the Montana Folk Festival stop at the Berkeley Pit, though, perhaps a more positive understanding of Butte's mining legacy can be engendered.[175] Turning additional visits to the landscape into a deeper understanding of it is the next step. What can result from educational efforts is a more positive outlook by visitors to mining communities like Butte, while still encouraging a mindfulness of mining's often-detrimental social and environmental legacies. If developed carefully, these encounters can bridge the divide between outsiders and insiders, bringing them together in discussions over Butte's past and future.

## Moving Forward

When the Berkeley Pit ended its long run, Montana governor Ted Schwinden exclaimed, "One thing is clear, if copper was once king in Montana, that dynasty has ended."[176] Without king copper, Butte's leadership has to find new ways to revive the community, while being mindful of mining's polluted legacy. Although there have been missteps and generational tensions, locals have found ways to heal from pit-induced social tensions and inner-city decline, while combining economic development, historic preservation, and environmental remediation. A good example of such efforts is a 1993 Regional Historic Preservation Plan that tried to more formally bring heritage, cleanup, and development together, and resulted in a proposal for and some major work done on a regional greenway and trail system in the 2000s.[177]

Butte's residents have therefore started to shape a new-town identity. Casualties at the bust end of mining's economic cycle have taken different paths. Some have withered into ghost towns. Others have thrived, developing into cultural havens, tourist centers, and ski meccas. Lacking the right kind of hills, Butte can never be a ski town like Aspen, Breckenridge, or Park City. Nor is it trying to simply function as a college town, like Platteville, Wisconsin, or as a gambling center, like Deadwood, South Dakota. It is a historic mining community, with all of the environmental cleanup problems and oft-failed, sometimes successful, typically contentious attempts to gather tourists that such an identity entails. It is more akin to Minnesota's iron range than Telluride, Colorado.[178] Citizens hope to make use of mining's legacies, and not just to live with them.

Yet the Berkeley Pit often gets in the way of residents' struggles to move past Butte's past. As Butte-born writer Edwin Dobb notes, Butte's industrial landscape, particularly its pit, "functions as a firewall, thwarting the weak of mind and spirit, the unimaginative, the affluent second-homers, and recreational thrill seekers, who owe no particular allegiance to any particular place or time, the disappointed, disoriented souls who come looking for the Montana they've drooled over in travel brochures and glossy magazines, not realizing that when they enter Butte they leave Montana behind."[179]

The Berkeley Pit's toxic waters will always sit next to the city, pushing away the professionals, businesses, and vacationers who might appreciate Butte, but also repelling those who might radically change Butte into something else. The landscape, particularly the Berkeley Pit, therefore might be what makes Butte, Butte. Industrial heritage in the form of a huge toxic lake is resistant to either Disneyfication or complete omission. Outsiders might look with disdain, but many in Butte like it that way.

## Notes

1. Jason Jones, "Bad Pit," *The Daily Show with Jon Stewart*, Episode #11079, June 22, 2006, www.cc.com/video-clips/iunb4u/the-daily-show-with-jon-stewart-jones---bad-pit; Roberta Forsell Stauffer, "Berkeley Pit as Tourist Attraction? 'Daily Show Does Butte,'" *Mont. Std.*, June 20, 2006.

2. Duncan Adams, "Did Toxic Stew Cook the Geese," *High Country News*, Dec. 11, 1995.

3. Jones, "Bad Pit."

4. Robertson, *Hard as the Rock Itself.* Similar analysis in Gavin Bridge, "Contested Terrain: Mining and the Environment," *Annual Review of Environment and Resources* 29 (2004): 205–29; Ben Marsh, "Continuity and Decline in the Anthracite Towns of Pennsylvania," *Annals of the Association of American Geographers* 77, no. 3 (1987): 337–52; Dublin, *When the Mines Closed*; Peter Goin and Elizabeth Raymond, "Living in Anthracite: Mining Landscape and Sense of Place in Wyoming Valley, Pennsylvania," *The Public Historian* 23, no. 2 (2001): 29–45; Piper, *Industrial Transformation*, 113–64; Francaviglia, *Hard Places.* On smokestack nostalgia, see John Russo and Sherry Lee Linkon, *Steeltown U.S.A.* (Lawrence: University Press of Kansas, 2002); Steven High, "Deindustrializing Youngstown: Memories of Resistance and Loss Following 'Black Monday' 1977–1997," *History Workshop Journal* (Great Britain) 54 (2002): 100–21.

5. John Sandlos and Arn Keeling, "Zombie Mines and the (Over)burden of History," *Solutions* 4, no. 3 (June 2013): 80–83; John Sandlos and Arn Keeling, "Claiming the New North: Development and Colonialism at the Pine Point Mine, Northwest Territories, Canada," *Environment and History* 18 (2012): 5–34; Arn Keeling and John Sandlos, "Introduction: The Complex Legacy of Mining in Northern Canada," in *Mining and Communities in Northern Canada: History, Politics, and Memory*, eds. Arn Keeling and John Sandlos (Calgary: University of Calgary Press, 2015), 1–32. Another take on "zombie mines" is Traci Brynne Voyles, *Wastelanding: Legacies of Uranium Mining in Navajo Country* (Minneapolis: University of Minnesota Press, 2015), 211–18.

6. Research in oral and memory history suggests there can be great differences between personal and collective memory, showing that residents often have divergent understandings of historical events. See Robert Summerby-Murray, "Interpreting Industrial Heritage in the Mining Towns of Cumberland County, Nova Scotia: Landscape Examples from Springhill and River Herbert," *Urban History Review* 35, no. 2 (Spring 2007): 51–59; Alistair Thomson, "Anzac Memories: Putting Popular Memory Theory into Practice in Australia," *Oral History* 18, no. 2 (1990): 25–31.

7. Charles K. Hyde, *Copper for America: The United States Copper Industry from Colonial Times to the 1990s* (Tucson: University of Arizona Press, 1998), 194; Finn, *Tracing the Veins*, 67–68; Mercier, *Anaconda*, 195–97; Gardiner interview; Barbara Miller

and Paul Driscoll, "Company No Longer God as Butte America Rediscovers Its Soul," *Borrowed Times* (Missoula, MT), Spring 1980.

8. Perry, "A Half Century," 115–16; Rick Foote, "'Copper's Friend' Gets Bad News on Pal," *Mont. Std.*, Feb. 18, 1980; Finch interview.

9. Mercier, *Anaconda*, 198. On corporate problems, see Rosalind Klein Berlin, "The U.S. Business Hall of Fame," *Fortune*, Apr. 14, 1986; Lydia Chavez, "The Acquisitions that Haven't Paid Off," *New York Times*, Mar. 28, 1982; "The Death of Mining: America Is Losing One of Its Most Basic Industries," *Businessweek* December 17, 1984, 64–70.

10. Robert Driscoll quoted in Martha H. Hamilton, "But Recovery May Bypass Butte," *Washington Post*, May 1, 1983.

11. Shari Meats, "Anaconda Co. Closes Smelter," *Mont. Std.*, Sept. 30, 1980; Ed Kemmick, "Anacondans Blast Arco for Closure," *Mont. Std.*, Oct. 2, 1980; "Hopes for Smelter Dim in Denver," *Mont. Std.*, Oct. 8, 1980; "Layoffs Increase Budget Pressures," *Mont. Std.*, June 16, 1981; Rick Foote, "Company Trims 400 Workers," *Mont. Std.*, June 16, 1981; "Operations at Two Montana Plants Suspended Indefinitely," *ArcoSpark* 8, no. 39 (Oct. 3, 1980): 1, 6 VF 1376, BSBPA; Mercier, *Anaconda*, 199–203.

12. "Anaconda Closing Open-Pit Copper Mine at Butte," *New York Times*, Apr. 25, 1982. On declining ratio, see "Montana's Copper Land," 1967, 1–2, VF "ACM: General Mining," BSBPA; "Butte One of World's Greatest Mining Areas," July 1973, 3, VF "ACM: General Mining," BSBPA.

13. First quote is "The Death of Mining," 64–67. Second is Paul Taylor, "The Sun Is Setting in the West on the U.S. Copper Industry," *Washington Post*, Apr. 15, 1985. See also Bob McCarthy, "ARCO: The Chilean Connection," *Montana Eagle*, no. 3 (Apr. 15, 1981).

14. Rick Foote, "Anaconda Co. Anticipates Few Changes," *Mont. Std.*, Jan. 25, 1981; Thomas C. Hayes, "Arco Picks an Inside Man," *New York Times*, Mar. 28, 1982; Gardiner interview.

15. "Anaconda Closing Open Pit Copper Mine at Butte," *Mont. Std.*, Apr. 25, 1982.

16. John Leo, "Put the Blame Where It Belongs," *Mont. Std.*, July 16, 1974; Peoples interview, by Leech.

17. Jeri Pullum, "Mine Dump Sites Concern Locals," *Mont. Std.*, June 24, 1980; Foote, "Anaconda Co. Anticipates"; Mercier, *Anaconda*, 194–96.

18. Mercier, *Anaconda*, 199–202; Jim Murry, interview by Brian Leech, Aug. 2, 2010, Montana City, MT, transcript and digital recording, OHC, BSBPA.

19. Patrick Fleming to Frank Gardiner, June 21, 1982, in Fldr. "The Anaconda Company 1982," Box 5, Micone/Peoples/Fisher Collection, BSBCE, BSBPA; Daily interview.

20. Nick Geranios, "It's Still the Richest Hill on Earth," *Mont. Std.*, May 19, 1982.

21. Adele Wolff, "Shutdown in Montana: Industrywide Copper Strike Halts Anaconda Mining, Smelting, Refining," *ArcoSpark* 8, no. 27 (July 11, 1980) 1, 6, VF 1376, BSBPA.

22. Frank Gardner quoted in Lydia Chavez, "When ARCO Left Town," *New York Times*, July 25, 1982.

23. On low profits, see Gardiner interview. On union and community reaction, see Rick Foote, "Union Seeks Benefits for Laid-Off Members," *Mont. Std.*, June 17, 1981;

Rick Foote, "Flexibility and Union Benefits Still Unresolved, Official Says," *Mont. Std.*, Aug. 31, 1982; Jim Hendon, "Anaconda Copper Cuts Off Montana," *Mont. Std.*, Jan. 10, 1983; "Former Miners Sue Arco," *Mont. Std.*, June 24, 1986; Helen Stewart, "Minerals Will Suspend Butte Operations," *ArcoSpark* 2, no. 2 (Feb. 1983): 4, VF 1376, BSBPA; Peoples interview, by Leech.

24. "Oil Companies Try to Dump Their Copper Firms," *Mont. Std.*, Jan. 28, 1985; "Baucus Copper Bills Passed," *Mont. Std.*, Sept. 18, 1984; "Reagan to Act Soon on Copper Imports," *Mont. Std.*, Aug. 30, 1984; "What's Next for Copper Dilemma?," *Mont. Std.*, Sept. 11, 1984; "Baucus Ready to Do Battle with Reagan on Copper Aid," *Mont. Std.*, Sept. 5, 1984; Richard Boudreaux, "Chile Shivers Over U.S. Copper-Limiting Ideas," *Mont. Std.*, Sept. 2, 1984; "Copper Limits Refused," *Mont. Std.*, Sept. 7, 1984.

25. "The Death of Mining," 65. For figures cited in this paragraph, see Nicholas D. Kristoff, "Copper's Plunge Is Pushing Mining Towns to the Brink," *New York Times*, Mar. 23, 1986. On other metal mines and their 1980s problems/shutdowns, see Aiken, *Idaho's Bunker Hill*, 168–204; "Copper Smelter Closes," *Mont. Std.*, Jan. 2, 1985; "Kennecott Copper Division May Be Forced to Shut Down," *Mont. Std.*, Jan. 2, 1985; "Tacoma Smelter to Close," *Mont. Std.*, June 28, 1984; "Smelter Plan Hurts Bunker Hill," *Mont. Std.*, Aug. 1, 1984; "Phelps Dodge to Close Morenci," *Mont. Std.*, Dec. 23, 1984.

26. Finn, *Tracing the Veins*, 201–46; "Mine Closure Revisited," *Mont. Std.*, Jan. 6, 2003; Gary Carle, interview by Teresa Jordan, Feb. 5, 1986, Butte, MT, partial transcript and notes, OHC, BSBPA; Jim Dick, interview by Teresa Jordan, May 28, 1985, Butte, MT, partial transcript and notes, OHC, BSBPA; Ruth Dockins, interview by Teresa Jordan, Feb. 7, 1986, Butte, MT, partial transcript and notes, OHC, BSBPA; Paul Hovan and Judy Hovan, interview by Teresa Jordan, June 22, 1986, Elko, NV, partial transcript and notes, OHC, BSBPA. On the breadwinner ideal among Montana workers, see Mercier, *Anaconda*.

27. "The Death of Mining," 64–65.

28. Quotes in Bryan Di Salvatore, "Et Tu, Butte," *Northern Lights* 1, no. 3 (May/June 1985): 3–4. For outside reporters' views, see also "Anaconda Closing Open-Pit Copper Mine at Butte," *New York Times*, Apr. 25, 1982; Chavez, "When ARCO Left Town"; Norman Sklarewitz, "Diversification Plan Backfires on ARCO as Recession Takes Toll," *Christian Science Monitor*, May 3, 1983; "Misery in the Minefields," *Newsweek*, Dec. 6, 1982. For scholars, see Duane A. Smith, *Rocky Mountain Heartland: Colorado, Montana, and Wyoming in the Twentieth Century* (Tucson: University of Arizona Press, 2008), 235; Michael P. Malone and Richard W. Etulain, *The American West: A Twentieth-Century History* (Lincoln: University of Nebraska Press, 1989), 245–46; William G. Robbins and Katrine Barber, *Nature's Northwest: The North Pacific Slope in the Twentieth Century* (Tucson: University of Arizona Press, 2011), 176.

29. Theresa Tierney, interview by Teresa Jordan, Mar. 12, 1986, Butte, MT, partial transcript and notes, OHC, BSBPA. See also Hovan and Hovan interview; Dick interview; Dockins interview.

30. Foote interview.

31. Jack Kisling, "Butte Hangs Tough," *Denver Post Magazine*, Dec. 8, 1985. See also Albert R. Gini, "A 'Butte' of a Hole in Montana," *Journal of Business Ethics* 5 (1986): 79–83.

32. Jeff Gibson, "Down but Not Out: Without the Company, Butte Seeks Diversification and Hopes for the Pit," *Mont. Std.*, Aug. 18, 1985.

33. Jana Fought, interview by Teresa Jordan, Nov. 23,1985, Butte, MT, partial transcript and notes, OHC, BSBPA; Dick interview; Dockins interview; Peoples interview, by Leech.

34. Celia Gibson, "Don Peoples," *Professional Montanan* 1, no. 1 (Dec. 1, 1993): 5–43; Kisling, "Butte Hangs Tough"; Peoples interview, by Leech; Candy Brown, interview by Teresa Jordan, Jan. 31, 1986, Helena, MT, partial transcript and notes, OHC, BSBPA.

35. Gibson, "Down But Not Out"; Peoples interview, by Jordan; Ulrich interview; "History," Montana Tech, mtech.edu/about/history.htm.

36. Laurie Mercier, "'We're All *Familia*': The Work and Activism of Lula Martinez," in *Motherlode: Legacies of Women's Lives and Labors in Butte, Montana*, eds. Janet L. Finn and Ellen Crain (Livingston, MT: Clark City Press, 2005), 268–77; Butte Community Union, "Preservation of a Neighborhood," 1–11; Bob J. McCarthy, interview by Teresa Jordan, Mar. 29, 1986, Butte, MT, partial transcript and notes, OHC, BSBPA; Robert Rodin, interview by Teresa Jordan, Jan. 16, 1986, Butte, MT, partial transcript and notes, OHC, BSBPA.

37. Peoples interview, by Leech; Mark Reavis, "Our Readers Speak: Preservation of Belmont Spurred Building in East Butte, *Mont. Std.*, Aug. 24, 2012; Don Peoples Jr. "MAC Helped Bring Back East Butte," *Mont. Std.*, July 7, 2011.

38. Charles S. Johnson, "Montana Power Co. Joins Anaconda Co. in Corporate Graveyard," *Helena Independent Record* (MT), Mar. 29, 2000; Dave Reese, "Montana Power Company: Decline and Fall of an Industrial Icon," *Daily Inter Lake* (Kalispell, MT), Mar. 3, 2003; "Company News; Touch America Tiles Chapter 11 Bankruptcy Papers," *New York Times*, June 21, 2003; Deidre McNamer, "How the Big Sky Went Dark," *New York Times*, Apr. 6, 2008; John Grant Emeigh, "Montana Power Co. Shareholders Reach $115M Settlement with Defunct Company," *Missoulian* (Missoula, MT), Oct. 3, 2009.

39. Rick Foote, "Suspension Anniversary in Butte Is All Quiet," *Mont. Std.*, July 1, 1984; Tom Cook, "Firm Loses Tax Appeal," *Mont. Std.*, Feb. 6, 1985; Jim Tracy, "Butte Mines Up for Sale; but no Buyers Yet," *Mont. Std.*, May 29, 1985.

40. In Fldr. "Mining Development Committee 1985," Box 36, BSBCE, BSBPA, see Evan Barrett and Pete Antonioli to Mining Promotion Steering Committee, "Report on Committee Mission, Structure, Organization," July 12, 1985; Butte-Silver Bow Mining Development Committee Minutes, Sept. 12, 1985; "Mining Development Committee," and Butte-Silver Bow Mining Development Committee Minutes, Sept. 12, 1985, Sept. 16, 1985, Fldr. See also Don Peoples to Ted Schwinden, n.d., Fldr. "Mining Development Committee 1986," Box 36, BSBCE, BSBPA; Mike R. Bowlin to Dennis R. Washington, Sept. 19, 1985, VF 1414, BSBPA; Gardiner interview; Peoples interview, by Jordan; Peoples interview, by Leech; Dennis E. Curran, "State Mining Industry No Longer 'the Pits,'" *Mont. Std.*, Dec. 7, 1986; Jim Tracy, "Mine Reopening Means $30 Million for State," *Mont. Std.*, June 11, 1986.

41. See in Fldr. "Mining Development Committee 1986," Box 36, BSBCE, BSBPA: Paddy Dennehy, Gary A. Taylor, Ron H. James, Joe C. Maynard, Wolf Heitzenroder, and Ron Singer to Montana Resources Incorporated, Jan. 29, 1986; Donald R. Peoples and Evan Barrett to Ray Tilman, July 21, 1986; Donald R. Peoples and Evan Barrett to Mining Development Committee, July 24, 1986. See also Jim Tracy, "Butte Unions Vow to Organize MRI Workers," *Mont. Std.*, June 10, 1986; State Bureau and Standard Staff, "MRI

Pamphlet 'Chills' Labor," *Mont. Std.*, July 26, 1986; Jim Tracy, "Irate Union Boss Sets Meet on Hiring," *Mont. Std.*, July 16, 1986; Peoples interview, by Leech; Ray Tilman, interview by Teresa Jordan, Jan. 9, 1987, Butte, MT, partial transcript and notes, OHC, BSBPA.

42. John McNay, "Economist's Advice to Butte Unions: Don't Knock Mine's Great 'Opportunity,'" *Mont. Std.*, July 24, 1986. On pickets, see Jim Tracy, "Miners' Political Mar. Gets Mixed Reception," *Mont. Std.*, July 18, 1986; Jim Tracy, "Hiring Methods Prompt Union Picket at MRI," *Mont. Std.*, July 22, 1986; Jim Tracy, "Unions Say Mine Pickets Premature," *Mont. Std.*, July 23, 1986; Donald R. Peoples and Evan Barrett to Mining Development Committee, Aug. 14, 1986, Fldr. "Mining Development Committee 1986," Box 36, BSBCE, BSBPA.

43. Foote interview; Dockins interview.

44. Moriarty, "Tough, But Crumbling"; Carle interview; Mary Berg, interview by Teresa Jordan, May 15, 1985, Butte, MT, partial transcript and notes, OHC, BSBPA.

45. Jonathan D. Rosenblum, *Copper Crucible: How the Arizona Miners' Strike of 1983 Recast Labor Management Relations in America*, 2nd ed. (Ithaca, NY: ILR Press, 1998). On labor's problems, starting in the 1970s, see particularly Jefferson Cowie, *Stayin' Alive: The 1970s and the Last Days of the Working Class* (New York: New Press, 2010).

46. Barbara LaBoe, "Mine Closure Revisited," *Mont. Std.*, Jan. 6, 2003; Tilman interview.

47. Jim Tracy, "Workers' Plight Shows Not All's Rosy, Murry Says," *Mont. Std.*, May 1, 1988.

48. Malone, "Collapse of Metal Mining," quotes on 463, 455. On mining's comeback elsewhere in the West, see Patricia Nelson Limerick, *Something in the Soul: Legacies and Reckonings in the New West* (New York: W. W. Norton: 2000), 220.

49. Butte-Silver Bow Urban Revitalization Agency, Urban Renewal Plan, Sept. 1980, p. 24, Fldr. "Urban Renewal," Box 55, BSBCE, BSBPA.

50. Kearney, *Butte's Pride*, 92–96; "Butte Group to Apply for Federal Funding to Build Gardens II," *Mont. Std.*, Aug. 30, 1975.

51. Andrea Ciabattari, "Aura of Columbia Gardens Refuses to Die Despite Mining Venture," *Mont. Std.*, Sept. 1, 1975. For childhood memories, see examples in Bone interview; Jack Atcheson and Mary Claire Atcheson, interview by Andrew Atcheson, Mar. 6, 2005, Butte, MT, transcript, OHC, BSBPA; Patricia Seccomb, interview by Joanne Seccomb, Mar. 2, 2002, Butte, MT, transcript, OHC, BSBPA; Quinn, "Memories of Columbia Gardens."

52. Warren interview.

53. Dave Cassidy, Mike A. Soto, and Greg Danskin, *The Columbia Gardens: A Social Impact Study of Its Closure* (Report Prepared for Urban Sociology 305, Montana State University, Bozeman, 1983), 96–112, in Grace Sicotte Personal Collection 111, BSBPA.

54. Kearney, *Butte's Pride*, 92–112; Daily interview; Dan Carter, "Gardens Group Recruits Members," *Mont. Std.*, July 12, 1985; Dan Carter, "Gardens Options Explored," *Mont. Std.*, Aug. 16, 1985; Dan Carter, "Citizens Get Say on Gardens Gear," *Mont. Std.*, Sept. 8, 1985; Dan Carter, "Gardens Equipment to Be on June Ballot," *Mont. Std.*, Apr. 4, 1986; Dan Carter, "Gardens Vote: Costly Choices," *Mont. Std.*, May 16, 1986.

55. Morin, *Legacy of American Copper Smelting*, 120–23; Shovers, "Remaking the Wide Open Town."

56. Quote is "Uptown Restoration Committee," Minutes, Dec. 17, 1977, Fldr. "The Anaconda Company 1977–78," Box 5, BSBCE, BSBPA. On the city, ARCO, and uptown, see "The CBD Problem: Background and Strategy for Redevelopment," Nov. 7, 1977, Fldr. "CBD Background and Strategy Redevelopment," Box 12, BSBCE, BSBPA; Cox to Micone, June 1, 1978, Fldr. "The Anaconda Company, 1977–1978"; Mario Micone to Ralph Cox, Jan. 11, 1978, and "Mario Micone, Press Release," Aug. 27, 1976, Fldr. "CBD Background and Strategy Redevelopment," Box 12, BSBCE, BSBPA; "Butte Group to Support Renovation," *Mont. Std.*, Aug. 3, 1976.

57. Janet A. Cornish to Butte–Silver Bow Council of Commissioners, Aug. 28, 1980, Fldr. "Urban Renewal," Box 55, BSBCE, BSBPA; Butte–Silver Bow Urban Revitalization Agency, Urban Renewal Plan, Sept. 1980, Fldr. "Urban Renewal," Box 55, BSBCE, BSBPA; Fought interview; Morin, *Legacy of American Copper Smelting*, 122–24; Richard I. Gibson, "Butte, 1950–2009: Decline, Loss, and the Rise of Historic Preservation and Cultural Tourism" (Vernacular Architecture Forum, 30th Annual Meeting, Butte, MT, 2009), www.gravmag.com/gibson-20th.pdf.

58. On headframes, see Al Darr, "Headframes Will Survive," *Mont. Std.*, Oct. 23, 1977; Jerri Pullum, "Emergency Team Will Get Mines on Film," *Mont. Std.*, Jan. 18, 1980; Edwin Dobb, "Viewpoint: Location, Occupation, Juxtaposition, Interpenetration: Notes on an Erotics of the Mining City," *Buildings and Landscapes* 17, no. 1 (Spring 2010): 1–12. On the historical society, see Morin, *Legacy of American Copper Smelting*, 124. On the World Museum of Mining, see Marcia Lee, "World Museum of Mining Attracting People, Growing," *Mont. Std.*, July 9, 1972; John McNay, "Machine Shop 'Antiques' Go on Block," *Mont. Std.*, Sept. 28, 1986.

59. Pat Kearney, *Miracle on the East Ridge* (Butte, MT: Artcraft Printers, 1990); Finn, *Tracing the Veins*, 226.

60. Barry, "Toxic Tourism," 42–51; "Friday Evening and Saturday's Schedule for Evel Knievel Days," *Mont. Std.*, July 28, 2017; An Ri Ra Montana Irish Festival, www.mtgaelic.org/an-ri-ra-montana-irish-festival; Montana Folk Festival, montanafolkfestival.com./.

61. Morin, *Legacy of American Copper Smelting*, 114–25; quote on 147. See similar claims in Mary Elizabeth Curran, "The Contested Terrain of Butte, Montana: Social Landscapes of Risk and Resiliency" (master's thesis, University of Montana, Missoula, 1996), 272.

62. Bode J. Morin, "Reflection, Refraction, and Rejection: Copper Smelting Heritage and the Execution of Environmental Policy" (PhD diss., Michigan Technological University, Houghton, 2009), 280.

63. McCarthy interview, by Leech.

64. David Schuyler, *A City Transformed: Redevelopment, Race, and Suburbanization in Lancaster, Pennsylvania* (University Park: Pennsylvania State University Press, 2002), 170. On outmigration as a loss of ethnic identity and sense of place, see Stephen Michael Kolman, "We'll Take Manhattan: The Appropriation of Immigrant Space and the Transformation of Urban Geography in New York City, 1925–1975" (PhD diss., University of Wisconsin, Madison, 2002).

65. Kathleen Neils Conzen, "Immigrants, Immigrant Neighborhoods, and Ethnic Identity: Historical Issues," *Journal of American History* 66, no. 3 (Dec. 1979), 603–15; Lee, "Ethnic Limits."

66. Stepan and Stepan interview; Ciabattari, "McQueen Spirit."

67. Maureen Crossman, interview by Pat Crossman, Apr. 3, 2002, Butte, MT, transcript, OHC, BSBPA; Fenlon interview; Leeper, "Changing Urban Landscape," 98–101.

68. Ralph interview; Tom Lester, interview by Teresa Jordan, May 10, 1985, Butte, MT, partial transcript and notes, OHC, BSBPA; Goldberg interview.

69. Downing, "Avoiding New Poverty."

70. Nicholls interview. See also Sparks interview, by Leech; Petroni interview, by Magnaghi.

71. Goldberg interview. See also Cramer interview.

72. Angelo Petroni, interview by Teresa Jordan, OHC, MHS. See similar comments in Holter interview, by Leech; deBarathy interview; Thornton, "'Little Italy'"; McCormick, "Given a Choice."

73. Leeper, "Changing Urban Landscape," 80.

74. Matteuchi interview.

75. Bone interview; Stepan and Stepan interview.

76. Kearney, *Butte Voices,* 211.

77. Franconi, Maffei, and Petroni, BHS meeting; Alan Goddard, interview by Teresa Jordan, Jan. 24, 1986, Butte, MT, partial transcript and notes; Petroni and Petroni interview; B. Lazzari interview; Holayter interview.

78. Bone interview; Fenlon interview; Petroni and Petroni interview.

79. See a similar situation in Parr, *Sensing Changes*, 79–102.

80. Holayter interview.

81. Ralph interview.

82. Angelo Petroni in Franconi, Maffei, and Petroni, BHS meeting.

83. Franconi, Maffei, and Petroni, BHS meeting; Petroni and Petroni interview; Matteuchi interview; Petroni interview, by Magnaghi.

84. Kelling, "Finntown."

85. Petroni interview, by Magnaghi.

86. Foote interview.

87. Matthew Frye Jacobsen, *Roots Too: White Ethnic Revival in Post-Civil Rights America* (Cambridge, MA: Harvard University Press, 2006).

88. "Today Is the Feast of the Sacred Heart," June 26, 1987, Box 1, Fldr. 1, Personal Collection 96 "Nick Badovinac—East Side," BSBPA; "Homily for the Annual Observance of Sacred Heart Parish Mass and Festivity," June 1, 1990, Box 1, Fldr. 8, Personal Collection 96 "Nick Badovinac—East Side," BSBPA.

89. Franconi, Maffei, and Petroni, BHS meeting; Thornton, "'Little Italy'"; Pasini and Pasini interview.

90. Barbara LaBoe, "Butte High History Students to Recreate Famed Meaderville Christmas Displays," *Mont. Std.*, Sept. 9, 2003; "Memories of Meaderville," *Mont. Std.*, Nov. 24, 2003.

91. Mary Pickett, "Butte Finns Toast St. Urho," *Billings Gazette*, Feb. 27, 2011. On the day's origins, see Eleanor Palo Stoller, "Sauna, Sisu, and Sibelius: Ethnic Identity Among Finnish Americans," *Sociological Quarterly* 37, no. 1 (Winter 1996): 145–75.

92. Filpula, *Meaderville Revisited,* 164. See also Holayter interview.

93. Bone interview; Ralph interview; Michelotti interview.

94. Pauline Mencarelli deBarathy, *Meaderville Reunion Memory Book* (Butte, MT: Sept. 9, 2006); deBarathy interview.

95. Ralph interview. See also Bone interview.

96. Thornton, "'Little Italy'"; Roberta Stauffer, "Of Places Known By Heart," *Mont. Std.*, July 28, 2008; Ralph interview.

97. Thad Kelling, "Talking About Old Times," *Mont. Std.*, Aug. 20, 2005; Kelling, "Finntown."

98. John Grant Emeigh, "East Butte Recollections," *Mont. Std.*, Feb. 2, 2009; Bone interview; Stepan and Stepan interview; Kelling, "Finntown."

99. Thornton, "'Little Italy'"; Holayter interview; Stepan and Stepan interview; Bone interview; Fisher and Corcoran interview.

100. deBarathy interview.

101. Megaprojects are discussed fully in Parr, *Sensing Changes*.

102. "Friends of Quabbin," www.foquabbin.org; *Haunting the Quabbin*, Sean Cole, Producer, 90.9WBUR, Boston Public Radio, insideout.wbur.org/documentaries/hauntingquabbin/default.asp; Ben Cosgrove, "Drowned Towns: Preserving the Lost Communities of the Swift RiverValley," *Harvard Advocate*, Winter 2009, www.theharvardadvocate.com/content/drowned-towns-preserving-lost-communities-swift-river-valley; "A Half-Century of Nostalgia for 4 Submerged Towns," *New York Times*, Dec. 27, 1987.

103. Sheila L. Steinberg and Karen Brinton, *Santa Rita, New Mexico Community Report*, Community Studies Series Report no. 4, Mar. 2003, Department of Sociology, Humboldt State University, CA.

104. Holayter interview.

105. Janie Payne, interview by James Hadley, Nov. 28, 2005, Butte, MT, transcript, OHC, BSBPA.

106. Holayter interview.

107. Foote interview; John Orth, interview by Teresa Jordan, Jan. 15, 1986, Butte, MT, partial transcript and notes, OHC, BSBPA.

108. Gary interview.

109. Kriskovich interview. See also D. C. Reynolds interview; Kovacich interview.

110. Ralph interview. See similar sentiments in Matteuchi interview; deBarathy interview; Stepan and Stepan interview.

111. Michelotti interview. See also Goldberg interview.

112. Kelling, "Finntown"; Bone interview.

113. Michelotti interview.

114. Holayter interview.

115. Doreen Massey, "Traveling Thoughts," in *Without Guarantees: In Honor of Stuart Hall*, ed. Paul Gilroy, Lawrence Grossberg, and Angela McRobbie (London and New York: Verso, 2000), 225–32; quote on 225.

116. Georgene Cachola, letter to the editor, "Lost Neighborhoods Still Bring Tears," *Mont. Std.*, July 30, 2008. Cachola was responding to Stauffer, "Of Places Known."

117. Kelling, "Finntown"; Stepan and Stepan interview; Vidrich interview.

118. deBarathy interview; Stepan and Stepan interview; Michelotti interview.

119. David M. Emmons, "Expert Report of David Emmons, Ph.D.," July 15, 1996, in U.S. District Court, District of Montana—Helena Division, *United States of America v. Atlantic Richfield Company, et. al.*, CV-89-039-BU-PGH, 294–96; Kelly Younggren, "Missoula's Copper King Legacy: Environmental Responsibility and the Milltown Dam, 1981–2004" (master's thesis, University of Wyoming, Laramie, 2010), 36–44; David Brooks, *Restoring the Shining Waters: Superfund Success at Milltown, Montana* (Norman: University of Oklahoma Press, 2015), 35–63.

120. On Love Canal and Superfund, see Elizabeth D. Blum, *Love Canal Revisited: Race, Class, and Gender in Environmental Activism* (Lawrence: University Press of Kansas. 2008); Rae Tyson, "The Intergovernmental Cleanup at Love Canal: A First Crack at the 'Sleeping Giant of the Decade,'" *Publius: The Journal of Federalism* 10, no. 1 (Winter 1980): 101–9; Carole Stern Switzer and Peter Gray, *CERCLA: Comprehensive Environmental Response, Compensation, and Liability Act (Superfund)* (Chicago: American Bar Association, 2008).

121. ARCO, "Environmental Action Plan for the Upper Clark Fork River Basin," VF 1452, BSBPA; Sara Godbout Sparks, interview by Ellen Crain, Apr. 2, 2004, Butte, MT, transcript, OHC, BSBPA; John McNay, "Twenty More Jobs Cut," *Mont. Std.*, Feb. 1, 1985.

122. For examples of the group's rhetoric about mining, see the following articles in *Clark Fork Coalition Currents* (Missoula, MT), Maureen and Mike Mansfield Library, UMM: Curran, "The Contested Terrain of Butte," 273; Peter Nielsen, "Are Headwaters Still an Industrial Ditch?," May 1987; Peter Nielsen, "Fish Kills in the Headwaters—When Will this Stop?," July 1987; "Mining in the Clark Fork Country: Industry Revival Requires Public Vigilance," Dec. 1989; Clark Fork Coalition Staff, "Prying Open Superfund: Milltown Committee Making Itself Heard," Feb. 1990. On Missoula and the Clark Fork Coalition, see Brooks, *Restoring the Shining Waters*, 10, 35–158.

123. McCarthy interview, by Leech.

124. "Clark Fork Coalition Directory," *Clark Fork Coalition Currents*, May 1988.

125. Peter Nielsen quoted in Douglas Isokait, "Natural Resource Damage in the Upper Clark Fork River Basin: A Policy Perspective" (PhD diss., University of Maryland, 2010), 112.

126. Isokait, "Natural Resource Damage," 173.

127. See examples in Curran, "The Contested Terrain of Butte," 224–26; Finn, *Mining Childhood*, 64–71.

128. Isokait, "Natural Resource Damage," 174. See also Sparks interview, by Crain.

129. McCarthy interview, by Leech; Peoples interview, by Leech.

130. Sparks interview, by Leech.

131. Curran, "Contested Terrain," 267–69.

132. McCarthy interview, by Leech; Shovers, "Remaking the Wide Open Town," 48–53; Brian Shovers, "From the Big Hole to Butte, MT: The Origins and Evolution of a Rocky Mountain Municipal Water System," *IA: The Journal of the Society for Industrial Archaeology* 37, no. 1–2 (2011): 79–96.

133. Terence E. Duaime and Nicholas J. Tucci, *History of Flooding of the Butte Underground Mines and Berkeley Pit Water Level Monitoring and Water-Quality Sampling, 2006 Consent Decree Update, Butte, Montana, 1982–2006*, Montana Bureau of Mines and Geology, 2007, Report #566, 7.

134. Patrick Fleming to Frank Gardiner, June 21, 1982, Fldr. "The Anaconda Company 1982," Box 5, BSBCE, BSBPA.

135. Rick Foote, "Arco Holding on Tight to Reports of Butte Mine Flooding," *Mont. Std.*, July 16, 1982. See also "Federal EPA Asked to Intervene in Butte Mine Flooding," *Mont. Std.*, July 22, 1982; Patrick Fleming to Frank Gardiner, July 1, 1982, in Fldr. "The Anaconda Company 1982," Box 5, BSBCE, BSBPA.

136. Rick Foote, "Water Is Rising in Butte's Mines Faster Than Expected," *Mont. Std.*, June 15, 1982; Rick Foote, "Anaconda Officials Say Flooding Mines Won't Cause Subsidence in Uptown," *Mont. Std.*, June 21, 1982; Dave Alt, "Floods or a Quake for Butte?," *Montana Magazine* 73 (Sept.-Oct. 1985): 49–50.

137. Donald R. Peoples to Butte's Legislative Delegation, July 19, 1982, Fldr. "The Anaconda Company 1982," Box 5, BSBCE, BSBPA; J. F. Gardner to Patrick Fleming, July 2, 1982, Fldr. "The Anaconda Company 1982," Box 5, BSBCE, BSBPA; Pat Williams to Steven Durham, July 21, 1982, Fldr. "The Anaconda Company 1982," Box 5, BSBCE, BSBPA.

138. Headwaters Research Institute, "Berkeley Pit Re-Industrialization and Mineral Recovery Project," Expanded Steering Committee Meeting, Jan. 20–23, Butte, MT, Fldr. "Berkeley Pit," Box 7, BSBCE, BSBPA; Fritz Daily to John Wardell, EPA, Mar. 7, 1987, Fldr. "Berkeley Pit," Box 7, BSBCE, BSBPA; Donald R. Peoples to Fritz Daily, Mar. 12, 1987, Fldr. "Berkeley Pit," Box 7, BSBCE, BSBPA.

139. Pit Watch, 2010, pitwatch.org; Fritz Daily, "Berkeley Pit Facts and Statistics," Jan. 31, 1991, Fldr. "Berkeley Pit," Box 7, BSBCE, BSBPA; Sparks interview, by Leech.

140. E. P. Myrick to Dave Fisher, Jan. 30, 1990, Fldr. "Berkeley Pit," Box 7, BSBCE, BSBPA.

141. John F. Wardell to Fritz Daily, Mar. 17, 1987, Fldr. "Berkeley Pit," Box 7, BSBCE, BSBPA; Creighton Barry to Jack Lynch, Mike Shea, Jim Johnston, Dan Denehy, and Bob McCarthy, State of Montana—Butte Silver Bow Staff Meeting Notes, Dec. 4, 1991, Fldr. "Berkeley Pit," Box 7, BSBCE, BSBPA. See also Duaime and Tucci, *History of Flooding*, 1–7.

142. British Petroleum (BP) purchased ARCO in 2000.

143. EPA, *Record of Decision: Silver Bow Creek/Butte Area oU3: Butte Mine Flooding Operable Unit* (Helena, MT: U.S. EPA Area 8, Sept. 29, 1994), 1–35, quote on 30. For similar concerns, see "Berkeley Pit," Feb. 6, 1991, Fldr. "Berkeley Pit," Box 7, BSBCE, BSBPA; Citizens Technical Environmental Committee, Open Letter, May 30, 1990, Fldr. "Berkeley Pit," Box 7, BSBCE, BSBPA; Lloyd Berg to Senator Conrad Burns, Feb. 20, 1989, Fldr. "Berkeley Pit," Box 7, BSBCE, BSBPA.

144. EPA, *Record of Decision*, 2–3; Montana Bureau of Mines and Geology, "Butte Mine Flooding Consent Decree 02-35-BU-SHE, 2012-Data Summary," Mar. 2013, www.epa.gov/sites/production/files/documents/bmf-2012datasummarywithattachments.pdf; Peoples interview, by Leech; Duaime and Tucci, *History of Flooding*, 90–113.

145. Fritz Daily, "Letter: Annual Earth Day facts about Berkeley Pit and Butte Mining," *Mont. Std.*, Apr. 22, 2012. See also Daily interview; Sparks interview, by Leech.

146. Citizens Technical Environmental Committee, Open Letter. See also "Pit Water No Danger Until 2020—Arco," *Mont. Std.*, Dec. 13, 1990. On the limits to public participation after a Consent Decree, see Sparks interview, by Leech; Daily interview.

147. Pitwatch.org; Atlantic Richfield Company, "Clark Fork River Basin Environmental Action Plan," Summer 2002, 25, VF 1452, BSBPA; Sparks interview, by Leech.

148. Alison Bolen, "Regulating the Unknown: Pit Lake Policies State by State," *Southwest Hydrology* (2002): 22–23, www.swhydro.arizona.edu/archive/V1_N3/feature6.pdf; Lisa Shevenell, Katherine A. Connors, and Christopher D. Henry, "Controls on Pit Lake Water at Sixteen Open-Pit Mines in Nevada," *Applied Geochemistry* 14, no. 5 (July 1999): 669–87.

149. Susan Dunlap, "What's in the Berkeley Pit Fog? The Answer Is, Well, Foggy," *Mont. Std.*, Mar. 5, 2016.

150. Pickett, "Butte Finns."

151. Dan Baum, "Butte, America: Poisoned, Ruined and Self-Cannibalized," *American Heritage* 48, no. 2 (Apr. 1997): 57–66.

152. *An Injury to One*, directed by Travis Wilkerson (2003; Brooklyn, NY: Icarus Films, 2004), VHS; *Butte, America*, produced by Pamela Roberts and Jane Greenberg, directed by Pamela Roberts, Rattlesnake Productions, 2008. For articles, see Edwin Dobb, "Pennies from Hell: In Montana, the Bill for America's Copper Comes Due," *Harper's Magazine* (Oct. 1996): 39–54; Jeffrey St. Clair, "Something About Butte," *Counterpunch*, Jan. 4–6, 2003, www.counterpunch.org/2003/01/04/something-about-butte/.

153. Adams, "Did Toxic Stew"; Melanie LeTourneau, "Top Job," *Current Science* 84, no. 10 (Jan. 22, 1999); "Seasonal Bird Efforts Ongoing," *Pit Watch*, 2009, www.pitwatch.org/seasonal-bird-mitigation-efforts-ongoing/.

154. Jim Robbins, "Hordes of Geese Die on a Toxic Lake in Montana," *New York Times*, Dec. 12, 2016; "Thousands of Snow Geese Die in Montana after Landing on Contaminated Water," *The Guardian*, Dec. 6, 2016; Ben Guarino, "Thousands of Montana Snow Geese Die After Landing in Toxic, Acidic Mine Pit," *Washington Post*, Dec. 7, 2016; Eve Peyser, "Thousands of Snow Geese Die After Landing on Toxic Mining Water in Montana," *Gizmodo Australia*, Dec. 8, 2016; "'Thousands' of Geese Die in Toxic Montana Mine Pit," *BBC News*, Dec. 7, 2016.

155. Ellen Crain and Janet L. Finn, "Sara Godbout Sparks: Cleanup Is Women's Work," in *Motherlode: Legacies of Women's Lives and Labors in Butte, Montana*, eds. Janet L. Finn and Ellen Crain (Livingston, MT: Clark City Press, 2005), 224–28; Sparks interview, by Crain; "Floyd Bossard to CTEC members, Oct. 21, 1991, Fldr. "Berkeley Pit," Box 7, BSBCE, BSBPA.

156. Karen Dorn Steele, "Superfund Revived Butte," *Spokesman Review* (Spokane, WA), July 28, 2002. See also Gibson, "Don Peoples," 5–43; Shovers, "Remaking the Wide Open Town," 52–53; Langewiesche, "Profits of Doom"; "Cleanup Boom Rescues Butte From Economic Pit," *National Geographic* 186, no. 6 (Dec. 1994): 172; Jim Robbins, "Butte Breaks New Ground to Mop Up a World-Class Mess," *New York Times*, July 21, 1998.

157. Brad Tyer, *Opportunity, Montana: Big Copper, Bad Water, and the Burial of an American Landscape* (Beacon Press, Boston, 2013), 197–99.

158. ARCO, "Environmental Action Plan," VF 1452, BSBPA. See also Tyer, *Opportunity, Montana*, 17, 176; LeCain, *Mass Destruction*, 225–26; Susan Dunlap, "Parrot Tailings Removal Plan Draws Praise, Criticism," *Mont. Std.*, Jan. 17, 2016.

159. Frederic L. Quivik, "Integrating the Preservation of Cultural Resources with

Remediation of Hazardous Materials: An Assessment of Superfund's Record," *Public Historian* 23, no. 2 (Spring 2001): 47–61.

160. Barry, "Coming to the Surface," 153–67, 192–205; Morin, *Legacy of American Copper Smelting*, 130–36.

161. Daily interview.

162. Gwen Florio, "The Berkeley Pit, a Former Copper Mine, Stands as a Toxic Reminder of What Unchecked Prospecting Can Unleash: Environmentalists Fear a Repeat of History in Montana," *Philadelphia Enquirer*, Sept. 1, 1998; Sarah Gilman, "Beauty and the Beast," *Denver Post*, Dec. 25, 2011; Richard Woodbury, "The Giant Cup of Poison," *Time*, Mar. 30, 1998.

163. Cherie Newman, "Hula on the Hill," *High Country News*, Aug. 2, 2010; Jamie Kelly, "Symbolic Cool Water Hula Dancing at Pit for 10 Years," *Missoulian*, Apr. 16, 2010.

164. On the dog, see Matt Vincent, "Like Butte, a Lonely Dog Hangs On," *High Country News*, Dec. 9, 2003; "Statue of Auditor, the Strip Mine Dog," *Roadside America*, Oct. 2006, www.roadsideamerica.com/story/12137; Soren Wheeler, "Stray Dog's Survival in Polluted Montana Landscape Prompts Research," *Plenty Magazine*, Aug. 2007, www.mnn.com/earth-matters/wilderness-resources/stories/stray-dogs-survival-in-polluted-montana-landscape-prompts; "Surviving Butte, MT: The Story of the Auditor," Nov. 2010, www.ohmidog.com/2010/11/02/surviving-butte-the-story-of-the-auditor/. On small organisms, "True Chemistry: a Husband-and-Wife Team Empties the Joint Checking Account to Search for Cancer Cures in Toxic Sludge," *Popular Science*, Feb. 2008, 34–35; Catherine Arnst, "A Superfund Site Helps Fight Cancer," *Businessweek*, Aug. 7, 2006, 79; Christopher Maag, "In the Battle Against Cancer, Researchers Find Hope in a Toxic Wasteland," *New York Times*, Oct. 9, 2007; "Even the Worst Laid Plans?," *Radiolab*, June 28, 2010, www.radiolab.org/2010/jun/28/even-the-worst-laid-plans/.

165. Montana Department of State Lands, *Preliminary Environmental Review for a Proposed Expansion of Anaconda Company's Berkeley Mining Complex*, Helena, MT: Sept. 20, 1976, Miscellaneous Box 2, BMCC, BSBPA.

166. On the effects of such stigma on cities, see Russo and Linkon, *Steeltown U.S.A.*

167. Marco Lucich, Butte–Silver Bow Chamber of Commerce, e-mail to the author, Aug. 19, 2010.

168. Nathaniel Miller, "The Pit," *Virginia Quarterly Review* 86, no. 4 (Fall 2010): 147; Annie Pentilla, "Rise of Selfie: Butte Chamber Capitalizes on Social Media Trend," *Mont. Std.*, June 28, 2016.

169. Jones, "Bad Pit." See also "Hey, Tourists: Check Out Our Toxic Water," *St. Petersburg Times*, Apr. 1, 2006; "An Irishman's Diary," *Irish Times*, Feb. 12, 2009; David Goldenberg, "The Toxic Tourist Attraction," *Gelf Magazine*, July 17, 2006, www.gelfmagazine.com/archives/the_toxic_tourist_attraction.php; Michael Martinez, "Mining for the Silver Lining," *Chicago Tribune*, Dec. 22, 2006; Al Lewis, "A Pit So Deep It Has Become a Tourist Attraction," *Fox Business*, Apr. 20, 2011, www.foxbusiness.com/markets/2011/04/20/al-lewis-pit-deep-tourist-attraction/.

170. Miller, "The Pit," 142–48. See also Susan Gallagher, "Ka-Ching! Chamber of Commerce Cashes in on Berkeley Pit Tourism," *Mont. Std.*, Mar. 30, 2006.

171. Barry, "Toxic Tourism," 65–76.

172. On toxic tourism at the Berkeley Pit, see Eric Smillie, "Beautiful Messes: A Travel Guide to Man-Made Disasters," *Good*, June 24, 2008, www.good.is/post/beautiful-messes-a-travel-guide-to-man-made-disasters; Brooks Barnes, "Wacky Attractions Made Summer 2000 Really Hop," *Wall Street Journal*, Sept. 6, 2000. On toxic and disaster tourism generally, see Phaedra C. Pezzullo, *Toxic Tourism: Rhetorics of Pollution, Travel, and Environmental Justice* (Tuscaloosa, AL: University of Alabama Press, 2007); Dean Irvine, "Trend Watch: Disaster Tourism," *CNN World Weekly*, Sept. 28, 2007, www.cnn.com/2007/WORLD/europe/09/27/ww.trends.disastertourism/.

173. Barry, "Toxic Tourism," 65–76.

174. Miller, "The Pit," 150.

175. All Butte festival attendance numbers in Annie Pentilla, "TBID Officials Worry About Festivals Ability to Be Self-Sustaining," *Mont. Std.*, May 3, 2016.

176. "Anaconda Closing Open Pit Copper Mine at Butte," *Mont. Std.*, Apr. 25, 1982.

177. Heritage Strategies and Walden Mills Group, *Montana's Copperway: An Action Plan for Heritage Tourism in the Butte-Anaconda Heritage Corridor*, Oct. 2, 2012, www.adlc.us/wp-content/uploads/Planning%20Dept./Planning-ComPlans_Butte-Anaconda.HeritageTourism-2012.pdf; *Preserving Our Mining Heritage*, 2006, VF 1418, BSBPA; Sparks interview, by Leech.

178. On mining ghost towns, see Dydia Delyser, "'Good, by God, We're Going to Bodie!' Ghost Towns and the American West," in *Western Paces, American Myths: How We Think About the West*, ed. Gary J. Hausladen (Reno and Las Vegas: University of Nevada Press, 2003), 273–95. On mining towns becoming ski towns, see Rogers, "'We Didn't Think'"; Rothman, *Devil's Bargains*, 202–26. On mining towns as gambling centers, see Katherine Jensen and Audie Blevins, *Last Gamble: Betting on the Future in Four Rocky Mountain Mining Towns* (Tucson: University of Arizona Press, 1998). On battles over industrial heritage tourism on the Iron Range, see Manuel, *Taconite Dreams*, 163–216.

179. Dobb, "Viewpoint," 12. See also Daily interview.

# CONCLUSION

No matter how vehemently Butte leaders argue for a new perception of their city, the Berkeley Pit remains, as does its image, as a big waste receptacle. One recent announcement makes this clear. On December 29, 2015, a decade-long battle involving the state, the county, the EPA, BP, and many Butte residents came to a head with a state decision to move the Parrot Tailings, a buried pile of waste left over from the old Parrot Smelter. More than a century ago the smelter sat at the bottom of the Butte hill in a spot behind what is now the Butte Civic Center and under the county's shops. Its 270,000 cubic yards of waste had initiated a plume of highly polluted water. According to concerned locals in the Restore Our River Coalition, the plume threatened cleanup of the still-fragile Silver Bow Creek and might, one day, infiltrate other groundwater. News articles labeled the plume even worse in quality than the Berkeley Pit's waters. BP, which had inherited ARCO's cleanup responsibilities, refused to pay for the Parrot's removal. The EPA mostly sided with BP; it long held that the Parrot Tailings were well-contained and the plume was stable, and hence did not pose an urgent threat to human or environmental health. Not disturbing such waste fit one typical EPA solution: waste in place. Because it is hard to find somewhere else to move contaminated material and expensive to do so, Butte's waste had typically been capped with new dirt. The Montana DEQ and the state's Natural Resource Damage Council, on the other hand, had joined with Butte locals, who have increasingly become environmental advocates, to formulate a new plan. That day, the state announced plans to dump most of the tailings directly into the Berkeley Pit.

The plan to dump more waste into the Berkeley Pit came mostly due to practical considerations. Montana agency officials originally looked at trucking the waste to a repository, two miles north of the pit. This plan would have meant 22,000 round trips, though, and so the state entered into an agreement with MRI, which now owns the Berkeley Pit. With the pit so close to the tailings, it seemed like a logical fit. Studies suggested the new material would neither change the pit's chemistry nor its volume to a significant degree. The pit's water would still need to be perpetually pumped and treated after the level reached the alluvial aquifer, and hence would be in danger of overflow. The tailings dump would only move the date for pumping up one month earlier, to be initiated in July 2023.[1]

Reaction to the plan was mixed. Some celebrated the decision, suggesting that it cut through the EPA's attempt to hide behind technical language and its insistence that Butte accept its condition as permanently polluted. Others worried that it meant another piecemeal step in cleaning up Butte, with no bigger plan in sight. Still others wondered what effect the decision would have on negotiations over the consent decree, a legal agreement between the various parties that would determine final cleanup liability on the Butte hill and upper Silver Bow Creek.[2] During a meeting about the plan, Butte-Silver Bow commissioner Cindy Perdue-Dolan asked whether dumping more waste in the pit simply meant "moving one problem to another problem." The Butte project manager for the program, Pat Cunneen, replied that water flowing into the pit from the mines cannot be cleaned up, partly because that aquifer lies in bedrock, but that the groundwater in the tailings site could be cleaned. Hence, the state would accomplish what it could.[3]

The exchange between Perdue-Dolan and Cunneen showcases the Berkeley Pit's continued image as a zone of sacrifice, a wasteland. Perhaps now it would also serve as a garbage can. Because hard-rock open-pits are typically impossible to reclaim, they continue to shape attitudes about mining in North America and around the world. Some pit lakes have fewer metals and hence can function like true lakes, but many leftover open pits share the Berkeley Pit's fate. Rather than serving as a testament to technological prowess, they increasingly have become targets of environmentalist ire. It is hard to love a big hole in the ground.

The public perception of mining today differs greatly when it comes to underground mining. One major Butte tourist attraction, the World Museum of Mining, opened in 1963. By the early 1970s, the museum was growing in popularity as visitors became more interested in the old days of Butte mining, before the modern open pits. The museum's centerpiece became its growing "Hellroarin' Gulch," a set of newly built, but old-looking, buildings to show visitors what early mining towns looked like. Constructed around boardwalks, the miniature town included a drug store, barber shop, blacksmith, post office, school, lawyer's office, doctor's office, Chinese laundry, mineral assay office, and saloon. Other exhibits contain an assortment of underground equipment, while the Orphan Girl's mine headframe, related outbuildings, and a very popular underground tour, typically led by a former miner, can make mining seem thrilling, if also distant and strange. Few of the TripAdvisor reviews for the museum mention the large truck outside, meant to represent the new era of surface work. Visitors instead gain an understanding of an earlier time, when the American West was full of vibrant new communities, a time when Americans did not depend on unsatisfying service jobs or boring white-collar work and instead did dangerous, exciting, real labor with their hands. The underground's hidden nature helps it stay heroic. People can either forget about environmental impact when a mine is underground (and, as in Butte's case, the smelter is far away), or they can try to remediate it—covering up or removing refuse, as they are attempting to do with the Parrot's smelter waste.

Surface mines, however, are impossible to romanticize or ignore in the same way. Often, as in the case of the Berkeley Pit, they cannot be erased, or even simply cleaned. A Google image search for "mining" and words like "environment" or "nature" tends to call up photographs of enormous pits and massive equipment. If that same search includes "Montana," the Berkeley Pit unsurprisingly dominates the results.

Butte's residents and visitors hence continue to face the legacy of industrial work begun in the 1950s. Over the course of sixty years, Butte's residents found that open-pit mining remade their lives at work, home, and play, while altering the shape of their city. Mining and its employees seemed to alternately benefit from and suffer from the open nature of the pit. The pit's scale made mechanization and efficiency more possible, giving the Anaconda Company new advantages and Butte a longer life as an important mining community. However, the size of the pit also gave miners a bigger target for their fears about the mechanized workplace. Similarly, the openness of the pit provided employees with a safety that had been impossible in the confined, unpredictable underground, but it also exposed miners' labor to greater observation, convincing workers that independence and camaraderie were increasingly unachievable. Labor's power and unity declined, partly because the pit's efficiency gave companies a new bargaining tool, but also because the organization of pit labor across wider spaces encouraged more division of duties.

Residents also became increasingly worried by the open pit's expansive nature. As it entered their living space, residents confronted industrial work in the form of new hazards. Sometimes these encounters led to protest by residents and pricey payoffs for the company, but their broad dispersal into the environment meant it became tougher to nail down liability questions. As industrial work intruded even farther, the company had to buy lots of property. These purchases resulted in the breakup of working-class neighborhoods, challenging white ethnics' traditions and bonds. Open-pit mining's continued growth then contributed to inner-city decline, influencing increasingly negative public attitudes about the mining landscape. Despite residents' attempts to repair and clean up damage wrought by its expansion, the Berkeley Pit continues to play a role for the region, intruding into residents' lives and journalists' reports. It is now either an environmental pariah or an industrial oddity, depending on whom a person asks.

Companies and communities involved in new resource extraction projects should take note. Like past mining projects, recent surface excavations can provide good and increasingly safe jobs and economic development, but they can also expose locals and their environment to industrial hazards. Sometimes they even cause what social scientists call development-induced displacement and resettlement, the phenomenon that affected Butte.[4] Despite many companies' attempts to ameliorate the worst effects, the results of open-pit mining can be painful. Consumers understandably want wiring made out of copper and cell phones containing rare earths, which means that metal mining projects will continue. Therefore, it is perhaps as much up to consumers as it is to mining companies to examine what benefits such projects might provide, and then to weigh such benefits with the costs host communities face. Unless something dramatically changes, the modern world will continue to dig big holes. We should try to do so in a way that causes the least harm, while recognizing the communities that host them as more than poisoned places.

## Notes

1. David McCumber, "State Plans to Dump Parrot Tailings into the Berkeley Pit," *Mont. Std.*, Dec. 29, 2015; Pat Munday, "Butte's Parrot Tailings (Again)," Clark Fork River Technical Assistance Committee, KUFM/Montana Public Radio Commentary, Oct. 27, 2011, cfrtac.org/102711.html; John Metesh and James Madison, "Summary of Investigation Upper Silver Bow Creek, Butte, Montana," *Montana Bureau of Mines and Geology*, Open File Report 507, Jan. 6, 2013, mbmg.mtech.edu/pdf-open-files/mbmg 507.pdf; Susan Dunlap, "Deep in the Heart of Butte, MT: A Special Report," *Mont. Std.*, July 19, 2015.

2. Dunlap, "Parrot Tailings Removal Plan"; Susan Dunlap, "Analysis: Forum Points Up Decades of Work—and Mistrust—in Superfund Cleanup," *Mont. Std.*, Jan. 20, 2016; Susan Dunlap, "Tester Says He'll Meet with EPA Boss on Butte Cleanup," *Mont. Std.*, Jan. 18, 2016.

3. McCumber, "State Plans to Dump."

4. On recent development-induced displacement and resettlement events, see Willie Neuman, "Hundreds in Peru Balk at Relocation from Site of Mine," *New York Times*, Jan. 6, 2013; Susanne Bonnell, "The Landowner Relocation Programme," in *Dilemmas of Development: The Social and Economic Impact of the Porgera Gold Mine*, ed. Colin Filer (Canberra: Australian National University E-Press, 2012), 128–59; South African Human Rights Commission, *Mining-Related Observations and Recommendations: Anglo Platinum, Affected Communities and Other Stakeholders, in and around the PPL Mine, Limpopo*, Nov. 2008, www.sahrc.org.za/home/21/files/Reports/Anglo%20 Report%20Final%202008_Chp%201%20to%203.pdf.

# BIBLIOGRAPHY

### Abbreviations

AHC: American Heritage Center, Laramie, WY
BFC: Butte Forward Collection
BMU: Butte Miners' Union
BSBPA: Butte-Silver Bow Public Archives
Fldr.: Folder
GPO: Government Publishing Office (formerly Government Printing Office)
Info. Cir.: Information Circular
IUMMSW: International Union of Mine, Mill, and Smelter Workers (also known as Mine-Mill)
MHS: Montana Historical Society Library and Archives
*MMBM: Mine-Mill Butte Miner*
*Mont. Std.: Montana Standard* (newspaper)
OHC: Oral History Collection
PSHCLA: Penn State Historical Collections and Labor Archives
Reg. Sess.: regular session
UC: University of Colorado—Boulder, Special Collections
UMM: University of Montana, K. Ross Toole Archives
USS: Utah State Archives
USWA: United Steelworkers of America
VF: vertical file

### Abbreviations for Manuscript Collections and Oral Histories

American Heritage Center, University of Wyoming, Laramie (AHC)
  Anaconda Geological Documents Collection (AGDC)
Butte-Silver Bow County Courthouse, Butte, Montana
  Butte City Council Minutes
  Butte-Silver Bow County Court, Register of Actions, vol. 33, vol. 34.
Butte-Silver Bow Public Archives, Butte, Montana (BSBPA)
  Butte Chamber of Commerce Collection (PC127)
  Butte Forward Collection (BFC)
  Butte Miners Union #1/United Steelworkers Labor History Collection (LH12)
  Butte Model Cities Collection (BMCC)
  Dave Piper Collection (MC388)
  Grace Sicotte Personal Collection (PC111)
  James Shea Personal Collection (PC104)

Butte-Silver Bow Chief Executive's Office, Micone/Peoples/Fisher Collection (BSBCE)
McQueen Fire Department Organization Collection (OC19)
Nick Badovinac—East Side Personal Collection (PC96)
Oral History Collection (OHC, BSBPA)
Montana Historical Society, Library and Archives, Helena, Montana (MHS)
43rd Montana Legislative Assembly Records (LR43)
Anaconda Company Records (MC169)
Montana Attorney General's Office Records (RS111)
Oral History Collection (OHC, MHS)
Rick Reese Collection: Montana Environmental Issues, 1970–79 (A8-2-7)
Penn State Historical Collections and Labor Archives, Eberley Family Special Collections, Paterno Library, University Park, Pennsylvania (PSHCLA).
Bernard Greenberg Papers (collection 1573)
International Union of Mine, Mill, and Smelter Workers/United Steelworkers of America Collection (USWA)
United Steelworkers of America and Labor Oral History Collection (USWAOHC)
University of California Berkeley, Regional Oral History Office
Western Mining in the Twentieth Century Oral History Series, available on Internet archive (ROHO).
University of Colorado, Boulder, CO, Norlin Library, Special Collections (UC).
Western Federation of Miners (WFA)/International Union of Mine, Mill, and Smelter Workers Collection (IUMMSW)
University of Montana Missoula, K. Ross Toole Archives and Special Collections, Mansfield Library (UMM)
Clarence C. Gordon Papers (MSS 611)
Mike Mansfield Papers (MSS65)
Oral History Collection (OHC, UMM)
Utah State Archives, Salt Lake City, Utah (USS)
Bingham City Correspondence and Subject Files, 1904–72 (Series 4590)
Bingham Canyon City Council Minutes (Series 4587)

## Oral History Interviews

Aguilar, Dan. Interview by Teresa Jordan. Butte, MT Apr. 10, 1986. OHC, BSBPA.
Antonioli, William. Interview by Chris Antonioli. Butte, MT, Mar. 27, 2002. OHC, BSBPA.
Archer, Joe. Interview by Jennifer Miller. Butte, MT, Oct. 17, 1998. OHC, BSBPA.
Atcheson, Jack, and Mary Claire Atcheson. Interview by Andrew Atcheson. Butte, MT, Mar. 6, 2005. OHC, BSBPA.
Ayres, Ernest Robert. Interview by Brenda Ayres. Butte, MT, Oct. 26, 1998. OHC, BSBPA.
Ballard, Cheryl, and Jim Ballard. Interview by Teresa Jordan, Wright, WY, June 10, 1986. OHC, BSBPA.
Barraza, Maclovio. Interview by Alice M. Hoffman. Denver, CO, Dec. 10, 1969. USWAOHC, PSHCLA.
Bartoletti, Pat. Interview by Camie Hoppe. Butte, MT, Nov. 18, 2005. OHC, BSBPA.

Bartscher, Jane. Interview by Teresa Jordan. Butte, MT, Nov. 22, 1985. OHC, BSBPA.
Beavis, Al. Interview by Teresa Jordan. Walkerville, MT, Dec. 3, 1985. OHC, BSBPA.
Beech, Ted. Interview by Ray Calkins. Butte, MT, Oct. 27, 1979. OHC, BSBPA.
Berg, Mary. Interview by Teresa Jordan. Butte, MT, May 15, 1985. OHC, BSBPA.
Bone, Ann. Interview by Brian Leech. Butte, MT, July 30, 2010. OHC, BSBPA.
Bontempo, Dominick. Interview by Judy Cusick. Butte, MT, Nov. 26, 1982. OHC, BSBPA.
Bossard, Floyd. Interview by Dawn Ingersoll. Butte, MT, Nov. 29, 1999. OHC, BSBPA.
Braach, Art. Interview by Teresa Jordan. Butte, MT, July 13, 1984. OHC, BSBPA.
Bradley, Dorothy. Interview by Bob Brown. Missoula, MT, Sept. 29, 2009. Oral History 396-62, OHC, UMM.
Breslin, Tom. Interview by Arthur S. Weinberg. July 15, 1968. USWAOHC, PSHCLA.
Brown, Candy. Interview by Teresa Jordan. Helena, MT, Jan. 31, 1986. OHC, BSBPA.
Bugni, Virginia. Interview by Brian Leech. Butte, MT, Oct. 22, 2009. OHC, BSBPA.
Burns, Ray. Interview by Brad Richins. Missoula, MT, Nov. 23, 2005. OHC, BSBPA.
Carden, Jim. Interview by Joe McCarthy. Butte, MT, Apr. 7, 2001. OHC, BSBPA.
Carle, Gary. Interview by Teresa Jordan, Butte, MT, Feb. 5, 1986. OHC, BSBPA.
Carriger, Helmie. Interview by Thomas Gordon. Butte, MT, 2001. OHC, BSBPA.
Chamberlain, Bruce, and Carol Chamberlain. Interview by Teresa Jordan. Elko, NV, June 23, 1986. OHC, BSBPA.
Connors, John. Interview by Laurie Mercier. Butte, MT, Dec. 9, 1982. OHC, BSBPA.
Corbett, Robert P. Interview by Laurie Mercier. Butte, MT, Apr. 17, 1984. Oral History 674, OHC, MHS.
Corcoran, Fran, and Trux Fisher. Interview by Brian Leech. Butte, MT, Oct. 21, 2009. OHC, BSBPA.
Cornish, Janet. Interview by Teresa Jordan. Butte, MT, Nov. 20, 1985. OHC, BSBPA.
Coupal, Joyce. Interview by John Terreo. Noxon, MT, Apr. 3, 1990. OHC, BSBPA.
Cramer, Agnes. Interview by Mary Murphy. Butte, MT, Mar. 28, 1980. OHC, BSBPA.
Crossman, Maureen. Interview by Pat Crossman. Butte, MT, Apr. 3, 2002. OHC, BSBPA.
Curtis, Vern. Interview by Greg Giebel. Denver, CO, Dec. 11, 1969. USWAOHC, PSHCLA.
Daily, Fritz. Interview by Brian Leech, Butte, MT, Aug. 4, 2010. OHC, BSBPA.
deBarathy, Pauline. Interview by Brian Leech. Butte, MT, July 20, 2010. OHC, BSBPA.
Dick, Jim. Interview by Teresa Jordan. Butte, MT, May 28, 1985. OHC, BSBPA.
Dockins, Ruth. Interview by Teresa Jordan. Butte, MT, Feb. 7, 1986. OHC, BSBPA.
Dow, Frances. Interview by Ray F. Calkins. Butte, MT, Feb. 1980. OHC, BSBPA.
Dudding, William. Interview by Logan Dudding. Butte, MT, Nov. 20, 2005. OHC, BSBPA.
Fenlon, Father Tom. Interview by Ellen Crain. Butte, MT, Oct. 23, 2006. OHC, BSBPA.
Finch, Tom. Interview by Darby Stacey. Butte, MT, Mar. 14, 2002. OHC, BSBPA.
Foote, Rick. Interview by Teresa Jordan. Butte, MT, Jan. 7, 1987. OHC, BSBPA.
Forsell, Louis. Interview by Teresa Jordan. Clancy, MT, Jan. 2, 1986. OHC, BSBPA.
Fought, Jana. Interview by Teresa Jordan. Butte, MT, Nov. 23, 1985. OHC, BSBPA.
Franconi, Esther, Maurice Maffei, and Angelo Petroni. Butte Historical Society Meeting, Topic: Meaderville Italians. Butte, MT, Nov. 19, 1992. OHC, BSBPA.
Gardiner, John Frank. Interview by Brian Leech. Wise River, MT, Aug. 3, 2010. OHC, BSBPA.
Gary, Frank. Interview by Teresa Jordan. Butte, MT, Mar. 30, 1986. OHC, BSBPA.

Godard, Charles C. Interview by Ray F. Calkins. Butte, MT, Mar. 21, 1979. OHC, BSBPA.
Goldberg, Aili Maria. Interview by Mary Murphy. Butte, MT, Feb. 29, 1980. OHC, BSBPA.
Hart, Al, and Tim Little. Interview by Teresa Jordan. Butte, MT, Apr. 17, 1985. OHC, BSBPA.
Hart, Laura. Interview by Teresa Jordan. Butte, MT, Dec. 9, 1985. OHC, BSBPA.
Hendrickson, Leonard. Interview by Ronald D. Hendrickson. Butte, MT, Apr. 17, 2005. OHC, BSBPA.
Henningsen, Rex J. Interview by Teresa Jordan. Butte, MT, Apr. 12, 1988. OHC, BSBPA.
Hoffman, Larry. Interview by Teresa Jordan. Butte, MT, Apr. 12, 1988. OHC, BSBPA.
Holayter, Marlene. Interview by Brian Leech. Anaconda, MT, July 29, 2010. OHC, BSBPA.
Holleran, Tom. Interview by Taylor Flotre. Butte, MT, Nov. 12, 2005. OHC, BSBPA.
Holter, Tom. Interview by Brian Leech. Butte, MT, Oct. 22, 2009. OHC, BSBPA.
———. Interview by Mary Ferriter. Butte, MT, Apr. 12, 2005. OHC, BSBPA.
———. Interview by Teresa Hayward. Butte, MT, Nov. 12, 1998. OHC, BSBPA.
Hooper, Al. Interview by Marc J. Lee. Butte, MT, Dec. 1, 1999. OHC, BSBPA.
Hovan, Judy, and Paul Hovan. Interview by Teresa Jordan. Elko, NV, June 22, 1986. OHC, BSBPA.
Hughes, John. Interview by Teresa Jordan. Butte, MT, Feb. 7, 1985. OHC, BSBPA.
Humphrey, William A. "Mining Operations and Engineering Executive for Anaconda, Newmont, Homestake, 1950 to 1995." Interview by Eleanor Swent. San Francisco, CA, 1994–95, ROHO.
Jursnick, Frank. Interview by Ray F. Calkins. Butte, MT, June 6, 1979. OHC, BSBPA.
Kenck, Hugo. Interview by Laurie Mercier. Butte, MT, Dec. 8, 1982. OHC, MHS.
Killoy, Jim. Interview by Teresa Jordan. Butte, MT, Apr. 10, 1986. OHC, BSBPA.
Kovacich, Bob. Interview by Teresa Jordan. Butte, MT, Mar. 19, 1986. OHC, BSBPA.
Krause, Chuck. Interview by Brian Leech. Butte, MT, Oct. 22, 2009. OHC, BSBPA.
Kriskovich, John. Interview by Teresa Jordan. Butte, MT, June 10, 1986. OHC, BSBPA.
Lapham, Erin. Interview by Teresa Jordan. Butte, MT, Mar. 7, 1986. OHC, BSBPA.
Lazzari, Bernie. Interview by Matt Moore. Butte, MT, Nov. 27, 2005. OHC, BSBPA.
Lazzari, Lorraine. Interview by Callie Boyle. Butte, MT, Nov. 27, 2005. OHC, BSBPA.
Lemm, Margaret R. Interview by Mary Murphy. Butte, MT, May 9, 1980. OHC, BSBPA.
Lester, Tom. Interview by Teresa Jordan. Butte, MT, May 10, 1985. OHC, BSBPA.
Maffei, Camille, and Lucy Maffei. Interview by Russ Magnaghi. Butte, MT, May 4, 1983. Oral History 583, OHC, MHS.
Malloy, Tom. Interview by Teresa Jordan. Butte, MT, May 21, 1985. OHC, BSBPA.
Matteuchi, Jim. Interview by Brian Leech. Butte, MT, Oct. 19, 2009. OHC, BSBPA.
McCarthy, Bob. Interview by Brian Leech. Butte, MT, Oct. 26, 2009. OHC, BSBPA.
McCarthy, Bob J. Interview by Teresa Jordan. Butte, MT, Mar. 29, 1986. OHC, BSBPA.
McCloskey, Shirley. Interview by Sarah Nyland. Butte, MT, Apr. 9, 2002. OHC, BSBPA.
McCullogh, Robin. Interview by Adam Lawrence. Butte, MT, Nov. 26, 2002. OHC, BSBPA.
Melvin, Robert. Interview by Mary Murphy. Butte, MT, Mar. 17, 1980. OHC, BSBPA.
Merzlak, June. Interview by Ray F. Calkins. Butte, MT, July 16, 1980. OHC, BSBPA.

Michelotti, Jim. Interview by Brian Leech. Butte, MT, July 28, 2010. OHC, BSBPA.
Micone, Mike. Interview by Teresa Jordan. Helena, MT, Jan. 5, 1987. OHC, BSBPA.
Miller, Dana. Interview by Teresa Jordan. Butte, MT, June 8, 1986. OHC, BSBPA.
Mitchell, Gwen. Interview by Mary Murphy. Butte, MT, May 29, 1987. OHC, BSBPA.
Murphy, Dennis J. Interview by Teresa Jordan. Butte, MT, May 30, 1986. OHC, BSBPA.
Murphy, Manila. Interview by Ray F. Calkins. Butte, MT, Jan. 1980. OHC, BSBPA.
Murry, James W. (Jim). Interview by Paul E. Verdon. Butte, MT, Apr. 22, 2004. OHC, MHS.
———. Interview by Brian Leech. Montana City, MT, Aug. 2, 2010. OHC, BSBPA.
Nicholls, Helen. Interview by Teresa Jordan. Butte, MT, Apr. 23, 1985. OHC, BSBPA.
Olsen, Arnold. Interview by Teresa Jordan. Butte, MT, Dec. 16, 1986. OHC, BSBPA.
Onkalo, John. Interview by Ray F. Calkins. Butte, MT, Feb. 1980. OHC, BSBPA.
Orizotti, John. Interview by Cody Maldonado. Butte, MT, Apr. 22, 2005. OHC, BSBPA.
Orth, John. Interview by Teresa Jordan. Butte, MT, Jan. 15, 1986. OHC, BSBPA.
Osier, Robert. Interview by Ed Brown. Butte, MT, Mar. 30, 2002. OHC, BSBPA.
Ossello, Guy J. Interview by Brad Spear. Butte, MT, Nov. 9, 2001. OHC, BSBPA.
O'Sullivan, Father Sarsfield. Interview by Teresa Jordan. Butte, MT, Feb. 10, 1985. OHC, BSBPA.
Panisco, Frank. Interview by Ray F. Calkins. Butte, MT, May 6, 1980. OHC, BSBPA.
Pasini Luigi, and Norma Pasini. Interview by Russ Managhi. Butte, MT, May 4, 1983. Oral History 582, OHC, MHS.
Payne, Janie. Interview by James Hadley. Butte, MT, Nov. 28, 2005. OHC, BSBPA.
Pentilla, Ann. Interview by Caroline Smithson and Ray F. Calkins. Butte, MT, Apr. 27, 1979. OHC, BSBPA.
Peoples, Don. Interview by Teresa Jordan. Butte, MT, Jan. 14, 1987. OHC, BSBPA.
———. Interview by Brian Leech. Butte, MT, Oct. 26, 2009. OHC, BSBPA.
Perdita, Duncan. Interview by Mary Murphy. Butte, MT, Mar. 18, 1980. OHC, BSBPA.
Perry, Vincent D. "A Half Century as Mining and Exploration Geologist with the Anaconda Company." Interview by Eleanor Swent. Carmel, CA, 1990. ROHO.
Peterson, Debbie. Interview by Jori Liva. Butte, MT, Apr. 19, 2005. OHC, BSBPA.
Petritz, Rudolph. Interview by Keri Petritz. Butte, MT, Nov. 15, 1998. OHC, BSBPA.
Petroni, Angelo. Interview by Russ Magnaghi. Butte, MT, May 4, 1983. Oral History 581, OHC, MHS.
Petroni, Angelo, and Virginia Petroni. Interview by Brian Leech. Butte, MT, Oct. 21, 2009. OHC, BSBPA.
Piazzola, Minnie. Interview by Laurie Mercier. Butte, MT, Mar. 11, 1982. Oral History 454, OHC, MHS.
Platisha, Eli. Interview by Lisa Speegle. Butte, MT, Nov. 23, 2001. OHC, BSBPA.
Reynolds, Ann Marie Quilici. Interview by Roxanna Lynn Reynolds. Butte, MT, Oct. 30, 2005. OHC, BSBPA.
Reynolds, Darrell C. Interview by Cindy Wymer. Butte, MT, Nov. 13, 2005. OHC, BSBPA.
Rae, Nancy, and Ronald Rae. Interview by Teresa Jordan. Butte, MT, June 21, 1986. OHC, BSBPA.
Ralph, Rose. Interview by Brian Leech. Butte, MT, Aug. 2, 2010. OHC, BSBPA.

Ramirez, Daniel R., and Wilma J. Ramirez. Interview by Teresa Jordan. Grand Junction, CO, June 20, 1986. OHC, BSBPA.
Rodin, Robert. Interview by Teresa Jordan. Butte, MT, Jan. 16, 1986. OHC, BSBPA.
Rowe, Allen L. Interview by Sheila Franey. Butte, MT, Mar. 5, 2002. OHC, BSBPA.
Scofienza, John. Interview by Ray F. Calkins. Butte, MT, June 27, 1979. OHC, BSBPA.
Seccomb, Patricia. Interview by Joanne Seccomb. Butte, MT, Mar. 2, 2002. OHC, BSBPA.
Shone, Frank. Interview by John Terreo. Helena, MT, Apr. 3, 1989. Oral History 1229, OHC, MHS.
Shutey, Kathy, and Mike Shutey. Interview by Teresa Jordan. Butte, MT, Apr. 12, 1988. OHC, BSBPA.
Skates, Dick. Interview by Teresa Jordan. Butte, MT, Feb. 5, 1986. OHC, BSBPA.
Skinner, Al. Interview by Helmut Golatz. Tucson, AZ, Mar. 3, 1975. USWAOHC, PSHCLA.
——. Interview by Alice Hoffman. Silverthorn, CO, June 19, 1973. USWAOHC, PSHCLA.
Skinner, Bert, and Donnetta Skinner. Interview by Teresa Jordan. Butte, MT, June 22, 1986. OHC, BSBPA.
Sparks, Sara Godbout. Interview by Ellen Crain. Butte, MT, Apr. 2, 2004. OHC, BSBPA.
——. Interview by Brian Leech. Butte, MT, July 29, 2009. OHC, BSBPA.
Stepan, Arlene, and Ivan Stepan. Interview by Brian Leech. Butte, MT, Aug. 5, 2010. OHC, BSBPA.
Street, Jim. Interview by Craig M. Chatriand. Butte, MT, Nov. 25, 1998. OHC, BSBPA.
Sullivan, Dolores, and Jim Sullivan. Interview by Teresa Jordan. Wright, WY, June 9, 1986. OHC, BSBPA.
Sullivan Sr., Michael J. Interview by Michael J. Sullivan Jr. Butte, MT, Oct. 18, 1999. OHC, BSBPA.
Tarrant, George. Interview by Adam Lawrence. Butte, MT, Apr. 9, 2002. OHC, BSBPA.
Thatcher, John. Interview by Lucas Thatcher. Butte, MT, Apr. 5, 2005. OHC, BSBPA.
Tierney, Theresa. Interview by Teresa Jordan. Butte, MT, Mar. 12, 1986. OHC, BSBPA.
Tilman, Julie, and Ray Tilman. Interview by Teresa Jordan. Butte, MT, Jan. 21, 1986. OHC, BSBPA.
Tilman, Ray. Interview by Teresa Jordan. Butte, MT, Jan. 9, 1987. OHC, BSBPA.
Ulrich, Don. Interview by Teresa Jordan. Butte, MT, Jan. 7, 1987. OHC, BSBPA.
Vidrich, George. Interview by Brian Leech. Butte, MT, Aug. 2, 2010. OHC, BSBPA.
Warren, Reverend Joe. Interview by Teresa Jordan. Butte, MT, Apr. 7, 1986. OHC, BSBPA.
Watters, Perle. Interview by Laurie Mercier. Butte, MT, Mar. 25, 1983. Oral History 484, OHC, MHS.
Wendel, Herbert A. Interview by Ray F. Calkins. Butte, MT, July 1979. OHC, BSBPA.
Wold, George. Interview by Alexis Wold. Butte, MT, Apr. 7, 2002. OHC, BSBPA.
Wood, Wilbur. Interview by Laurie Mercier. Roundup, MT, Mar. 28, 1986. Oral History 900, OHC, MHS.
Zanchi, Louise, and Victor Giuseppe Segna. Interview by Russ Magnaghi. Butte, MT, May 3, 1983. Oral History 579, OHC, MHS.
Zanchi, Peter. Interview by Brian Leech. Butte, MT, July 30, 2010. OHC, BSBPA.

### Newspapers and Periodicals

*American Forests*
*American Heritage*
*American Mercury*
*American Metal Market*
*Anaconda Company Trailsman* (in BSBPA and MHS)
*The Anacondan* (Butte, in UMM)
*Anode* (Butte, in UMM)
*ArcoSpark* (in VF 1376, BSBPA)
*Atlantic Monthly*
*Billings Gazette* (Billings, MT)
*Bingham Bulletin* (Bingham, UT)
*Borrowed Times* (Missoula, MT, in UMM)
*Bozeman Daily Chronicle* (Bozeman, MT)
*Bulletin of the Atomic Scientists*
*Businessweek*
*Butte Daly Post*
*Butte Miner*
*Chicago Tribune*
*Chinorama* (at University of Western New Mexico Library, Silver City, New Mexico)
*Christian Science Monitor*
*City: Magazine of Urban Life and Environment*
*Clark Fork Coalition Currents* (Missoula, MT)
*Craftsman*
*Colorado Gazette Telegraph* (Colorado Springs, CO)
*Counterpunch*
*Current Science*
*Daily Inter Lake* (Kalispell, MT)
*Daily Inter-Mountain* (Butte, MT)
*Denver Post*
*Diggers' Truth* (in USWA, PSHCLA)
*Engineering and Mining Journal*
*Forbes*
*Fortune*
*Garden and Home Builders Magazine*
*Good Housekeeping*
*Great Falls Tribune* (Great Falls, MT)
*Greely Daily Tribune* (Greely, CO)
*Harper's Magazine*
*Harvard Advocate* (Cambridge, MA)
*The Headframe* (in MHS)
*Helena Independent Record* (Helena, MT)
*High Country News*
*The Irish Times*

*International Union of Mine, Mill, and Smelter Workers Officer's and Stewards Bulletin* (in USWA, PSHCLA)
*Journal of Commerce*
*Kennescope* (Salt Lake City, in J. Willard Marriott Library Special Collections, University of Utah (UU))
*Life* magazine
*Loveland Colorado Reporter*
*Metal Mining and Processing*
*Mine-Mill Butte Miner* (in IUMMSW, UC)
*Mine-Mill News Bulletin* (in USWA, PSHCLA)
*Mine-Mill Research Report* (in USWA, PSHCLA)
*Mining Congress Journal*
*Mining Engineering*
*Mining and Scientific Press*
*Mining Technology*
*Mining World*
*Missoulian*
*Montana Magazine*
*Montana Post*
*Montana Standard* (Butte, MT)
*National Geographic*
*Newsweek*
*New Times* (New York, NY)
*The New Yorker*
*The New York Times*
*Nonferrous Report* (in USWA, PSHCLA)
*Nonferrous Strike Bulletin* (in USWA, PSHCLA)
*Northern Lights*
*Northern Plains Resource Council Newsletter* (Helena, MT)
*The People's Voice* (Helena, MT)
*The Philadelphia Enquirer*
*Popular Science*
*Professional Montanan*
*Reader's Digest*
*Rocky Mountain News* (Denver, CO)
*San Diego Union*
*Salt Lake Tribune*
*Saturday Evening Post*
*Scribner's Magazine*
*Seattle Times*
*Spokesman Review* (Spokane, WA)
*Star Citizen* (Tucson, AZ)
*St. Petersburg Times*
*Survey Graphic*
*Time*

*Transactions of the American Institute of Mining and Metallurgical Engineers*
*Tucson Daily Citizen*
*U.S. News and World Report*
*Virginia Quarterly Review*
*The Wall Street Journal*
*The Washington Post*
*Wilson Quarterly*
*World Dredging and Marine Construction*

## Government Documents, Reports, Cases, and Rulings

### *Federal*

*Automation and Technological Change*. Hearings Before the Subcommittee on Economic Stabilization of the Joint Committee on the Economic Report. 84th Cong., 1st sess. Washington, DC: GPO, 1955.

Boyd, Eugene. "Economic Development Administration: A Review of Elements of Its Statutory History." *Congressional Research Service Report for Congress*. June 3, 2011. 7-5700. R41241. Accessed June 3, 2012. (Online article no longer available.)

Cirkel, Fritz. *Asbestos: Its Occurrence, Exploitation, and Uses*. Ottawa: Mines Branch, Canadian Department of the Interior, Toronto, 1905.

*Closure of Underground Copper Mines: Hearing to Review the Anaconda Company's Plans to Close Its Underground Copper Mines in Butte, Montana*. Hearing before the U.S. Senate Subcommittee on Minerals, Materials, and Fuels of the Committee on Interior and Insular Affairs. 93rd Cong., 2nd sess. Nov. 25, 1974. Washington, DC: GPO, 1974.

*Cue for Survival: Operation Cue, A.E.C. Nevada Test Site, May 3, 1955; a Report*. Washington, DC: GPO, 1956.

Emmons, David M. "Expert Report of David Emmons, Ph.D." July 15, 1996. In U.S. District Court, District of Montana—Helena Division. *United States of America v. Atlantic Richfield Company, et al.*, CV-89-039-BU-PGH.

"Transcript of Proceedings of the Special Panel Appointed to Assist in Settlement of the Copper Dispute." Feb. 1, 1968. Washington, DC: CSA Reporting Co., 1968. Copy in USWA, PSHCLA.

U.S. Department of the Interior. Bureau of Land Management. "Do You Really Own the Minerals Under Your Land." BLM/MT/GI-06/012. 2012. www.blm.gov/pgdata/etc/medialib/blm/mt/blm_programs/mining.Par.72349.File.dat/minerals.pdf

U.S. Department of the Interior. Bureau of Mines. "Blast Vibrations and Other Potential Causes of Damage in Homes Near a Large Surface Coal Mine in Indiana." David E. Siskind, Steven V. Crum, and Matthew N. Pils. Report of Investigations 9455. Washington, DC: GPO, 1993.

U.S. Department of the Interior. Bureau of Reclamation. *North Central Power Study, Report of Phase I*, vol. 1, Study of Mine-Mouth Powerplants with Extra-High Voltage Transmission for Delivery of Power to Load Centers. Coordinating Committee of North Central Power Study, Oct. 1971.

———. *Block-Caving at Kelley Mine, The Anaconda Company, Butte, Mont.* Constantine C. Popoff. Info. Cir. 7758. Washington, DC: GPO, 1956.

———. "Health and Safety Inspection: Berkeley Pit, The Anaconda Company, Butte, Silver Bow County, Montana." T. J. Shepich. Metal and Nonmetal Mine Safety: Rocky Mountain District. Washington, DC: GPO, 1970. Copy in USWA, PSHCLA.

———. "Miners' Consumption in the Mines of Butte, Montana." Daniel Harrington and A. J. Lanza. Technical Paper 260. Washington, DC: GPO, 1921.

———. *Mining Methods and Costs at the Anaconda Company Berkeley Pit, Butte, Mont.* John R. McWilliams. Info. Cir. 7888. Washington, DC: GPO, 1959.

———. *Open-Cut Metal Mining*. Eugene Delos Gardner and McHenry Mosier. Bulletin 433. Washington, DC: GPO, 1941.

———. "Structure Response and Damage Produced by Ground Vibration from Surface Mine Blasting." David E. Siskind, Mark S. Stagg, John W. Kopp, and Charles H. Dowding. Report of Investigations 8057. Washington, DC: GPO, 1980.

———. "Surface Response and Damage Produced by Airblast from Surface Mining." David E. Siskind, Virgil J. Stachura, Mark S. Stagg, and John W. Kopp. Report of Investigations 8485. Washington, DC: GPO, 1980.

———. *Undercut Block-Caving Method of Mining in Western Copper Mines.* Eugene Delos Gardner. Info. Cir. 6350. Washington, DC: GPO, 1930.

U.S. Department of the Interior. Environmental Protection Agency, Area 8. *Record of Decision: Silver Bow Creek/Butte Area oU3: Butte Mine Flooding Operable Unit*. Helena, MT, Sept. 29, 1994.

U.S. Department of the Interior. United States Geological Survey. "Historical Statistics for Minerals and Material Commodities in the United States." Data Series 140. U.S. Department of the Interior, Washington, DC. minerals.usgs.gov/ds/2005/140/#data.

U.S. Department of Labor. Bureau of Labor Statistics. *Impact of Automation: A Collection of Twenty Articles About Technological Change, From the Monthly Labor Review*. Bulletin 1287. Washington, DC: GPO, 1960.

———. *Wage Chronology: The Anaconda Co. (Montana Mining Division) and the Steelworkers, 1941–77.* Bulletin 1953. Washington, DC: GPO, 1977.

### State

Arizona Bureau of Mines. *A Brief History of the United Verde Open Pit, Jerome, Arizona*. E. M. J. Alenius. Bulletin 178. Tucson: University of Arizona, 1968.

*House Journal for the State of Montana*, 37th Legislative Assembly, Reg. Sess., 1961.

*Kadillak v. The Anaconda Co.*, 184 M 127, 602 P2d 147 (1979).

*Laws, Resolutions, and Memorials of the State of Montana*, 37th Legislative Assembly, Reg. Sess., 1961, Chapter 216

Montana Bureau of Mines and Geology, "Butte Mine Flooding Consent Decree 02-35-BU-SHE, 2012-Data Summary." Mar. 2013. www.epa.gov/sites/production/files/documents/bmf-2012datasummarywithattachments.pdf.

———. *History of Flooding of the Butte Underground Mines and Berkeley Pit Water Level Monitoring and Water-Quality Sampling, 2006 Consent Decree Update, Butte, Montana, 1982-2006*. Terence E. Duaime and Nicholas J. Tucci. Report 566 (2007).

———. "Summary of Investigation Upper Silver Bow Creek, Butte, Montana." John Metesh and James Madison. Open File Report 507. mbmg.mtech.edu/pdf-open-files/mbmg507.pdf.

*Montana Constitutional Convention*. Transcript of Proceedings, vol 6. Mar. 2, 1972–Mar. 5, 1972. Helena, MT: Montana Legislature & Constitutional Convention Editing and Publishing Committee, 1972. Copy in Montana State University Library, Internet archive.

*Montana Constitutional Convention Enabling Act, Convention Rules, Delegate Information, Delegate Proposals, Committee Proposals*, vol. 1. Helena, MT: Montana Legislature & Constitutional Convention Editing and Publishing Committee, 1979.

Montana Department of State Lands, *Preliminary Environmental Review for a Proposed Expansion of Anaconda Company's Berkeley Mining Complex*. Helena, MT: State Publishing Co., 1976. Copy in BMCC, BSBPA.

*Schara v. Anaconda Co.*, 187 Mont. 377, 610 P.2d 132 (1980).

*Senate Journal for the State of Montana*, 37th Legislative Assembly, Reg. Sess., 1961.

### Pamphlets and Unpublished Sources

*Anaconda Copper Mining Company: Its Properties and Operations in Montana*, n.d. Copy at BSBPA

Atlantic Richfield Company. "Clark Fork River Basin Environmental Action Plan." Summer 2002. Copy in VF 1452, BSBPA.

"Beautiful Columbia Gardens: 'Senator Clark's Masterpiece.'" 1899. Copy in VF 1045, BSBPA.

"Belmont Subsidence Evaluation and Redevelopment Project." Prepared for the City and County of Butte-Silver Bow by Portage Environmental. Jan. 2002. Copy at Butte Urban Redevelopment Assistance Office, Butte, Montana.

Butte Chamber of Commerce, *Seven Talks about Mines*. n.d. (1939–40). Copy at BSBPA.

*Butte City Polk Directory*. 1955. Copy at BSBPA.

Butte Community Union. "Preservation of a Neighborhood: A Neighborhood Preservation Plan for the Central Butte Neighborhood." Butte, Montana: Nov. 1985. Copy at Renewable Technologies, Inc. office, Butte, Montana.

"Butte Street Maps." VF 1361, BSBPA.

Byrne, John. "Pieces from the Past: The Central Butte Neighborhood of Butte, Montana." 1982. Copy in VF 1609, BSBPA.

Cassidy, Dave, Mike A. Soto, and Greg Danskin. *The Columbia Gardens: A Social Impact Study of Its Closure*. Report Prepared for Urban Sociology 305, Montana State University, 1983. Copy in PC111, BSBPA.

"Columbia Gardens." Copy in VF 1040, BSBPA.

Heilbronner, Adolph H. *Sights and Scenes and a Brief History of Columbia Gardens: Butte's Only Pleasure Resort*. Butte: Butte Miner Co., 1902. Copy in VF 1041, BSBPA.

Lindberg, Paul. Personal communication through e-mail with the author. July 30, 2012.

Lucich, Marco. Personal communication through e-mail with the author. Aug. 19, 2010.

Madison, Cheryl Lynn. "Butte's 'East Side': Memories and History of the 1960s." Oct. 19, 1999. Copy in VF 1614, BSBPA.

*Meaderville Reunion Book*. Butte, 2006. Copy in Montana Tech Library, Butte, Montana.

"Meaderville Volunteer Fire Department." Copy in VF 1117, BSBPA.

"Meaderville Volunteer Fire Department: Fifty Years of Community Service." 1960. Copy in VF 1117, BSBPA.

Murphy, Mary, and Brian Shovers. "The Decline of a Neighborhood: Butte's Eastside." Butte Historical Society. 1986. Copy in VF 1615, BSBPA.

Nyland, Waino S. "A Famous Mining Camp in Literature, Written in 1926 While a Student at the University of Colorado." Nyland File Folder, BSBPA.

*Preserving Our Mining Heritage*. 2006. Copy in VF 1418, BSBPA.

Quinn, Frank. "Memories of Columbia Gardens." Butte, MT: Quinn and Associates, 1973.

Shovers, Brian. "Walkerville: Butte's Sister Silver Town." *Report for the Butte Historic Landmark District Project*. June 1986. Copy in VF 1637, BSBPA.

Steinberg, Sheila L., and Karen Brinton. *Santa Rita, New Mexico Community Report*. Community Studies Series Report no. 4. Mar. 2003. Department of Sociology, Humboldt State University, Humboldt, CA.

"Tardy Arson Report." Jan. 20, 1981. Copy in VF 1111, BSBPA.

## Websites

An Ri Ra Montana Irish Festival. www.mtgaelic.org/an-ri-ra-montana-irish-festival/.

Brewer, Leo, George H. Brimhall, Charles M. Gilbert, Richard L. Hay, and Joseph A. Pask. "Charles Meyer, 1915–1987." Earth and Planetary Science, University of California Berkeley. eps.berkeley.edu/content/charles-meyer.

Charter, Anne Goddard. 2002 Alumnae Award Acceptance Address. Wells College. Jan. 22, 2002. www.wells.edu/advancement/wells-college-alumnae-alumni-award-recipients.

Clark Fork River Technical Assistance Committee. cfrtac.org/102711.html.

"The Daniel C. Jackling Award." Society for Mining, Metallurgy, and Exploration. test.smenet.org/membership/awards/division-awards/mining-exploration-(m-e)/daniel-c-jackling-award.

"Friends of Quabbin." www.foquabbin.org.

Goldenberg, David. "The Toxic Tourist Attraction." *Gelf Magazine*. July 17, 2006. www.gelfmagazine.com/archives/the_toxic_tourist_attraction.php.

Irvine, Dean. "Trend Watch: Disaster Tourism." *CNN World Weekly*. Sept. 28, 2007. www.cnn.com/2007/WORLD/europe/09/27/ww.trends.disastertourism/.

Lewis, Al. "A Pit So Deep It Has Become a Tourist Attraction." *Fox Business*. Apr. 20, 2011. www.foxbusiness.com/markets/2011/04/20/al-lewis-pit-deep-tourist-attraction/.

*Montana Bureau of Mines and Geology*. www.mbmg.mtech.edu/env/env-berkeley.html

*Montana's Copperway: An Action Plan for Heritage Tourism in the Butte-Anaconda Heritage Corridor*. Heritage Strategies and Walden Mills Group. Oct. 2, 2012. www.adlc.us/wp-content/uploads/Planning%20Dept./Planning-ComPlans_Butte-Anaconda.HeritageTourism-2012.pdf.

Montana Folk Festival. montanafolkfestival.com/.

Montana Tech. "History." mtech.edu/about/history.htm.

Pitwatch. pitwatch.org.

Smillie, Eric. "Beautiful Messes: A Travel Guide to Man-Made Disasters." *Good*. June 24, 2008. www.good.is/post/beautiful-messes-a-travel-guide-to-man-made-disasters.

"Statue of Auditor, the Strip Mine Dog." *Roadside America*. Oct. 2006. www.roadsideamerica.com/story/12137.
"Stray Dog's Survival in Polluted Montana Landscape Prompts Research." *Plenty Magazine*. Aug. 2007. www.mnn.com/earth-matters/wilderness-resources/stories/stray-dogs-survival-in-polluted-montana-landscape-prompts.
"Surviving Butte: The Story of the Auditor." Ohmidog.com, Nov. 2010, www.ohmidog.com/2010/11/02/surviving-butte-the-story-of-the-auditor/.

## Audio-Visual Sources

*An Injury to One*. Directed by Travis Wilkerson. Brooklyn, NY: Icarus Films, 2004. VHS.
"Bad Pit." *The Daily Show with Jon Stewart*. Episode 11079. June 22, 2006. www.cc.com/video-clips/iunb4u/the-daily-show-with-jon-stewart-jones---bad-pit.
*Butte, America*. Produced by Pamela Roberts and Jane Greenberg. Directed by Pamela Roberts. Rattlesnake Productions. 2008. DVD.
"Even the Worst Laid Plans?" *Radiolab*. June 28, 2010. www.radiolab.org/2010/jun/28/even-the-worst-laid-plans/.
*Haunting the Quabbin*. Produced by Sean Cole. 90.9WBUR, Boston Public Radio. archives.wbur.org/insideout/documentaries/hauntingquabbin/index.php.html.

## Dissertations and Theses

Barry, Bridget R. "Toxic Tourism: Promoting the Berkeley Pit and Industrial Heritage in Butte, Montana." Master's thesis, University of Nebraska, Lincoln, 2012.
Barry, Stacie Lynn. "Coming to the Surface: The Environment, Health, and Culture in Butte, Montana, 1950–2010." PhD diss., University of Montana, Missoula, 2012.
Bozorgebrahimi, Enyatollah. "The Evaluation of Haulage Truck Size Effects on Open Pit Mining." PhD diss., University of British Columbia, Vancouver, 2004.
Carl, Matthew Ronald. "The Columbia Gardens Amusement Park: Company Sponsored Community in Butte, Montana." Master's thesis, University of Montana, Missoula, 2011.
Curran, Mary Elizabeth. "The Contested Terrain of Butte, Montana: Social Landscapes of Risk and Resiliency." Master's thesis, University of Montana, Missoula, 1996.
Curtis, Kent. "An Ecology of Industry: Mining and Nature in Western Montana." PhD diss., University of Kansas, Lawrence, 2001.
Foor, Forrest L. "The Senatorial Aspirations of William A. Clark, 1898–1901: A Study in Montana Politics." PhD diss., University of California, Berkeley, 1941.
Isokait, Douglas. "Natural Resource Damage in the Upper Clark Fork River Basin: A Policy Perspective." PhD diss., University of Maryland, College Park, 2010.
Kirk, Kelly. "State of Change: Gender and the 1972 Montana Constitutional Convention." Master's thesis, Montana State University, Bozeman, 2011.
Kolman, Stephen Michael. "We'll Take Manhattan: The Appropriation of Immigrant Space and the Transformation of Urban Geography in New York City, 1925–1975." PhD diss., University of Wisconsin, Madison, 2002.
Lee, Margaret Brennan. "Ethnic Limits: Negotiating Ethnicity, Race, and Community Boundaries in Suburban Chicago." PhD diss., University of Wisconsin, Madison, 2010.

Leech, Brian. "Corporate Tyranny and Foreign Labor: Populism, Nativism, and Montana's 1894 State Capital Election." Master's thesis, University of Wisconsin, Madison, 2007.

Leeper, Joseph Samuel. "The Changing Urban Landscape of Butte, Montana." PhD diss., University of Oregon, Eugene, 1974.

Luft, Stephen James. "Virtual Reconstruction: A Primer to the Digital Reconstruction of McQueen." Master's thesis, Montana Tech, Butte, 2004.

Mihelich, John. "The Richest Hill on Earth: An Ethnographic Account of Industrial Capitalism, Religion, and Community." PhD diss., Washington State University, Pullman, 1999.

Morin, Bode J. "Reflection, Refraction, and Rejection: Copper Smelting Heritage and the Execution of Environmental Policy." PhD diss., Michigan Technological University, Houghton, 2009.

Ore, Janet. "Labor and the New Deal in Butte, Montana: The International Union of Mine, Mill and Smelter Workers' Strike of 1934." Master's thesis, Washington State University, Pullman, 1987.

Poe, Tracy Nicole. "Food, Culture, and Entrepreneurship Among African-Americans, Italians, and Swedes in Chicago." PhD diss., Harvard University, Cambridge, MA, 1999.

Quivik, Fredric L. "Smoke and Tailings: An Environmental History of Copper Smelting Technologies in Montana, 1880–1930." PhD diss., University of Pennsylvania, Philadelphia, 1998.

Savage, Carletta H. "Women Coal Miners: Another Chapter in Central Appalachia's Struggle Against Hegemony." Master's thesis, West Virginia University, Morgantown, 1999.

Shovers, Brian Lee. "Miners, Managers, and Machines: Industrial Accidents and Occupational Disease in the Butte Underground, 1880–1920." Master's thesis, Montana State University, Bozeman, 1987.

Snow, Bradley Dean. "From the Sixth Floor to the Copper Dome: 'The Company's' Political Influence in Montana, 1920–1959." Master's thesis, Montana State University, Bozeman, 2003.

Thompson, Judith Diane. "Goodbye to the Gardens: A Look at the Controversial Closure of Butte's Columbia Gardens and How the *Montana Standard* Covered the Event." Bachelor's thesis, University of Montana, Missoula, 1987.

Toole, K. Ross. "A History of the Anaconda Copper Mining Company: A Study in the Relationships between a State and Its People and a Corporation, 1880–1950." PhD diss., University of California, Los Angeles, 1954.

———. "Marcus Daly: A Study of Business in Politics." Master's thesis, University of Montana, Missoula, 1948.

Younggren, Kelly. "Missoula's Copper King Legacy: Environmental Responsibility and the Milltown Dam, 1981–2004." Master's thesis, University of Wyoming, Laramie, 2010.

Yount, Kristen R. "Women and Men Coal Miners: Coping with Gender Integration Underground." PhD diss., University of Colorado, Boulder, 1986.

## Published Articles and Books

Abrams, Lynn. *Oral History Theory*. New York: Routledge, 2010.

Aiken, Katherine G. *Idaho's Bunker Hill: The Rise and Fall of a Great Mining Company, 1885–1981*. Norman: University of Oklahoma Press, 2005.

Allen, James B. *The Company Town in the American West*. Norman: University of Oklahoma Press, 1966.

Ammon, Fracesca Russello. "Unearthing Benny the Bulldozer: The Culture of Clearance in Postwar Children's Books." *Technology and Culture* 53, no. 2 (Apr. 2012): 306–36.

*Anaconda Company Annual Report, 1971*. Copy in VF 1432, BSBPA.

Andrews, Thomas G. *Killing for Coal: America's Deadliest Labor War*. Cambridge, MA: Harvard University Press, 2008.

Archdeacon, Thomas J. *Becoming American: An Ethnic History*. New York: The Free Press, 1983.

Ard, John V. "Economic Crunch Shakes Financial Footing on New Town Developments." *Industrial Development* (Jan./Feb. 1975): 11–14.

Arrington, Leonard J., and Gary B. Hansen. *"The Richest Hole on Earth": A History of the Bingham Copper Mine*. Logan: Utah State University Press, 1963.

Avila, Eric. *Popular Culture in the Age of White Flight: Fear and Fantasy in Suburban Los Angeles*. Berkeley and Los Angeles: University of California Press, 2004.

Barber, William J., James L. Cochrane, Neil de Marchi, and Joseph A. Yager. "Energy: 1945–1980: From John F. Kennedy to Jimmy Carter." *Wilson Quarterly* 5, no. 2 (Apr. 1981): 70–90.

Bartos, Paul J. "A Light in the Darkness: Modern U.S. Mine Lamps—Carbide and Electric Cap Lamps." *Mining History Journal* 17 (2010): 90–102.

Bartos, Paul J. "A Light in the Darkness: U.S. Mine Lamps, the Early Years—Candlesticks, Oil Lamps, and Safety Lamps." *Mining History Journal* 16 (2009): 45–63.

Basso, Matthew L. *Meet Joe Copper: Masculinity & Race on Montana's World War II Home Front*. Chicago, IL: University of Chicago Press, 2013.

Basso Matthew L., and William Allen Burke. "Another Look at Burke's Butte: The Great Depression and William Allen Burke's 'Greenhorn Miner.'" *Montana: The Magazine of Western History* 56, no. 4 (Winter 2006): 18–31, 97–98.

Baumler, Ellen. "Devil's Perch: Prostitution from Suite to Cellar in Butte, Montana." *Montana: The Magazine of Western History* 28, no. 3 (Autumn 1998): 4–21.

Beck, Ulrich. *Risk Society: Towards a New Modernity*. Translated by Mark Ritter. Newbury Park, CA: Sage, 1992.

Bederman, Gail. *Manliness and Civilization: A Cultural History of Gender and Race in the United States, 1880–1917*. Chicago, IL: University of Chicago Press, 1996.

Berkman, Herman G. "The New Town and Urban Change Form." *Land Economics* 48, no. 2 (May 1972): 93–103.

Biles, Roger. *The Fate of Cities: Urban America and the Federal Government, 1945–2000*. Lawrence: University Press of Kansas, 2011.

———. "New Towns for the Great Society: A Case Study in Politics and Planning." *Planning Perspectives* 13 (1998): 113–32.

Bijsterveld, Karin. *Mechanical Sound: Technology, Culture, and Public Problems of Noise in the Twentieth Century*. Boston: Massachusetts Institute of Technology, 2008.

Bisht, Tulsi Charan. "Development-Induced Displacement and Women: The Case of the Tehri Dam, India." *Asia Pacific Journal of Anthropology* 10, no. 4 (2009): 301–17.

Bix, Amy Sue. *Inventing Ourselves Out of Jobs? Americas's Debate over Technological Unemployment, 1929–1981*. Baltimore, MD: Johns Hopkins University Press, 2000.

Blum, Elizabeth D. *Love Canal Revisited: Race, Class, and Gender in Environmental Activism*. Lawrence: University Press of Kansas. 2008.

Boese, Donald L. *John C. Greenway and the Opening of the Western Mesabi*. Bovey-Coleraine Joint Bi-Centennial Commission, 1975.

Bolen, Alison. "Regulating the Unknown: Pit Lake Policies State by State." *Southwest Hydrology* (2002): 22–23. www.swhydro.arizona.edu/archive/V1_N3/feature6.pdf.

Bonnell, Susanne. "The Landowner Relocation Programme." In *Dilemmas of Development: The Social and Economic Impact of the Porgera Gold Mine*, edited by Colin Filer, 128–59. Canberra: Australian National University E-Press 2012.

Braverman, Harry. *Labor and Monopoly Capital: The Degradation of Work in the Twentieth Century*. New York: Monthly Review Press, 1974.

Bridge, Gavin. "Contested Terrain: Mining and the Environment." *Annual Review of Environment and Resources* 29 (2004): 205–29.

Bright, James R. *Automation and Management*. Boston: Harvard University Press, 1958.

Brooks, David. *Restoring the Shining Waters: Superfund Success at Milltown, Montana*. Norman: University of Oklahoma Press, 2015.

Brothers, Beverly J. *Sketches of Walkerville*. Butte, MT: Ashton Printing, 1973.

Brüggemeir, Franz-Josef. "The Ruhr Basin 1850–1980: A Case Study of Large-Scale Environmental Pollution." In *Silent Countdown: Essays in European Environmental History*, edited by Christian Pfister and Peter Brimblecombe, 210–27. Berlin, Germany: Springer-Verlag, 1990.

Burkhart, Lynne C. *Old Values in a New Town: The Politics of Race and Class in Columbia, Maryland*. New York: Praeger, 1981.

Butte–Silver Bow Public Archives, The Montana Standard, and The World Museum of Mining, *Remembering Butte, Montana's Richest City*. Helena, MT: Far Country Press, 2001.

Calvert, Jerry W. *The Gibraltar: Socialism and Labor in Butte, Montana, 1895–1920*. Helena: Montana Historical Society Press, 1988.

Campbell, Carlos C. *New Towns: Another Way to Live*. Reston, VA: Reston Publishing, 1976.

Capek, Stella M. "The Environmental Justice Frame: A Conceptual Discussion and an Application." *Social Problems* 40, no. 1 (1993): 5–24.

Chandler, Alfred D. *The Visible Hand: The Managerial Revolution in American Business*. Cambridge, MA: Harvard University Press, 1977.

Christensen, Bonnie. *Red Lodge and the Mythic West: Coal Miners to Cowboys*. Lawrence: University Press of Kansas, 2002.

Churchill, David S. "Making Broad Shoulders: Body-Building and Physical Culture in Chicago 1890–1920." *History of Education Quarterly* 48, no. 3 (Fall 2008): 341–70.

Clement, Wallace. *Hardrock Mining: Industrial Relations and Technological Change at INCO*. Toronto, Ont.: McClelland and Stewart, 1981.

Clements, Eric L. *After the Boom in Tombstone and Jerome, Arizona: Decline in Western Resource Towns*. Reno: University of Nevada Press, 2003.

Coates, Peter A. "The Strange Stillness of the Past: Toward an Environmental History of Sound and Noise." *Environmental History* 10, no. 4 (Oct. 2005): 636–65.

Colquhoun, James. *The Story of the Birth of the Porphyry Coppers*. London: William Clowes and Sons, 1933.

Conzen, Kathleen Neils. "Immigrants, Immigrant Neighborhoods, and Ethnic Identity: Historical Issues." *Journal of American History* 66, no. 3 (Dec. 1979), 603–15.

Couch, Stephen R. "Environmental Contamination, Community Transformation, and the Centralia Mine Fire." In *The Long Road to Recovery: Community Responses to Industrial Disaster*, edited by James K. Mitchell, 60–85. New York: United Nations University Press, 1996.

Cowie, Jefferson. *Stayin' Alive: The 1970s and the Last Days of the Working Class*. New York: New Press, 2010.

Cox, Robert M. "Rock Blasting in the Urban Environment—Legal and Technical Problems." *Proceedings of the 18th U.S. Symposium on Rock Mechanics*. June 22–24, 1977. Colorado School of Mines, Golden, CO.

Cross, Gary S., and John K. Walton. *The Playful Crowd: Pleasure Places in the Twentieth Century*. New York: Columbia University Press, 2005.

Cross, C. Louise. "The Battle for the Environmental Provisions in Montana's 1972 Constitution." *Montana Law Review* 51 (1990): 449–57.

Curtis, Kent A. *Gambling on Ore: The Nature of Metal Mining in the United States, 1860–1910*. Boulder: University Press of Colorado, 2013.

———. "Producing a Gold Rush: National Ambitions and the Northern Rocky Mountains, 1853–1863." *Western Historical Quarterly* 40, no. 3 (Autumn 2009): 275–97.

Delyser, Dydia. "'Good, by God, We're Going to Bodie!' Ghost Towns and the American West." In *Western Paces, American Myths: How We Think About the West*, edited by Gary J. Hausladen, 273–95. Reno and Las Vegas: University of Nevada Press, 2003.

Dewey, Scott. "Working for the Environment: Organized Labor and the Origins of Environmentalism in the United States." *Environmental History* 3 (Jan. 1998): 45–63.

Diner, Hasia R. *Hungering for America: Italian, Irish, and Jewish Foodways in the Age of Migration*. Cambridge, MA: Harvard University Press, 2001.

Dinius, Oliver J., and Angela Vergara, eds. *Company Towns in the Americas: Landscape, Power, and Working-Class Communities*. Athens: University of Georgia Press, 2011.

Dix, Keith. *What's a Coal Miner to Do? The Mechanization of Coal Mining*. Pittsburgh, PA: University of Pittsburgh Press, 1988.

Dobb, Edwin. "Viewpoint: Location, Occupation, Juxtaposition, Interpenetration: Notes on an Erotics of the Mining City." *Buildings and Landscapes* 17, no. 1 (Spring 2010): 1–12.

Doig, Ivan. *English Creek*. New York: Penguin, 1984.

Downing, Theodore E. *Avoiding New Poverty: Mining-Induced Displacement and Resettlement*. London: International Institute for Environment and Development, 2002.

Dublin, Thomas. *When the Mines Closed: Stories of Struggles in Hard Times*. New York: Cornell University Press, 1998.

Dubofsky, Melvyn. "Technological Change and American Worker Movements." In *Technology, the Economy, and Society*, edited by Joel Colton and Stuart Bruchey, 162–85. New York: Columbia University Press, 1987.

Dunaway, Finis. *Seeing Green: The Use and Abuse of American Environmental Images* Chicago, IL: University of Chicago Press, 2015.

Edwards, Richard. *Contested Terrain: The Transformation of The Workplace in the Twentieth Century*. New York: Basic, 1979.

Einzig, Paul. *The Economic Consequences of Automation*. New York: W. W. Norton, 1956.

Emmons, David M. *The Butte Irish: Class and Ethnicity in an American Mining Town, 1875–1925*. Urbana and Chicago: University of Illinois Press, 1989.

———. "The Orange and the Green in Montana: A Reconsideration of the Origins of the Clark-Daly Feud." *Arizona and the West* 28 (1986): 225–45.

———. "The Price of Freedom: Montana in the Late and Post-Anaconda Era." *Montana: The Magazine of Western History* 44, no. 4 (Autumn 1994): 66–73.

Enyeart, John P. *The Quest for 'Just and Pure Law': Rocky Mountain Workers and American Social Democracy, 1870–1924*. Stanford, CA: Stanford University Press, 2009.

Ferguson, Cody. *This Is Our Land: Grassroots Environmentalism in the Late Twentieth Century*. New Brunswick, NJ: Rutgers University Press, 2015.

Ferril, William Columbus. *Sketches of Colorado*, vol. I. Denver, CO: Western Press Bureau Company, 1911.

Figes, Orlando. "Private Life in Stalin's Russia: Narratives, Memory, and Oral History." *History Workshop Journal* 65 (2008): 117–37.

Filpula, Joan M. *Meaderville Revisited*. Butte, MT: Pit Printers, 2008.

Finn, Janet L. *Mining Childhood: Growing Up in Butte, Montana, 1900–1960*. Helena: Montana Historical Society Press, 2012.

———. "A Penny for Your Thoughts: Stories of Women, Copper, and Community." *Frontiers: A Journal of Women Studies* 19 no. 2 (1998): 231–49.

———. *Tracing the Veins: Of Copper, Culture, and Community from Butte to Chuquicamata*. Berkeley: University of California Press, 1998.

Finn, Janet L., and Ellen Crain, eds. *Motherlode: Legacies of Women's Lives and Labors in Butte, Montana*. Livingston, MT: Clark City Press, 2005.

Flippen, J. Brooks. *Nixon and the Environment*. Albuquerque: University of New Mexico Press, 2000.

Flynn, Ted P. "From Bulls to Bulldozers." In *Forest History*, edited by Joseph A. Miller (Fall 1963): 14–17.

Foucault, Michel. *Discipline and Punish: The Birth of the Prison*. Translated by Alan Sheridan. New York: Pantheon Books, 1977.

Francaviglia, Richard. *Hard Places: Reading the Landscape of America's Historic Mining Districts*. Iowa City: University of Iowa, 1991.

Franklin, Jonathan S. "Working-Class Themes in the New Environmental History." *Labor: Studies in the Working-Class History of the Americas* 9, no. 1 (2012): 119–31.

Fritz, Harry W. "The 1972 Montana Constitution in a Contemporary Context." *Montana Law Review* 51 (1990): 270–74.

———. "The Origins of Twenty-First-Century Montana." *Montana: The Magazine of Western History* 42, no. 1 (Winter, 1992): 77–81.

Gamm, Gerald. *Urban Exodus: Why the Jews Left Boston and the Catholics Stayed*. Cambridge, MA: Harvard University Press, 1999.

Gans, Herbert J. *The Urban Villagers: Group and Class in the Life of Italian-Americans*. New York: Macmillan, 1962.

Gibson, Richard I. "Butte, 1950–2009: Decline, Loss, and the Rise of Historic Preservation and Cultural Tourism." Vernacular Architecture Forum, 30th Annual Meeting, Butte, MT, 2009. www.gravmag.com/gibson-20th.pdf.

Gier, Jaclyn J., and Laurie Mercier, eds. *Mining Women: Gender in the Development of a Global Industry, 1670–2005*. New York: Palgrave Macmillan, 2006.

Gini, Albert R. "A 'Butte' of a Hole in Montana." *Journal of Business Ethics* 5 (1986): 79–83.

Glasscock, C. B. *The War of the Copper Kings*. New York: Grosset and Dunlap, 1935.

Goin, Peter, and Elizabeth Raymond. "Living in Anthracite: Mining Landscape and Sense of Place in Wyoming Valley, Pennsylvania." *The Public Historian* 23, 2 (2001): 29–45.

Graham, Amanda. "Copper Commando and the Anaconda Company's Wartime Production." *Montana: The Magazine of Western History* 59, no. 4 (Winter 2009): 67–72.

Green, Hardy. *The Company Town: The Industrial Edens and Satanic Mills That Shaped the American Economy*. New York: Basic, 2010.

Greenberg, Amy S. *Cause for Alarm: The Volunteer Fire Department in the Nineteenth-Century City*. Princeton, NJ: Princeton University Press, 1998.

Gressley, Gene M. "Colonialism: The Perpetual Pendulum?" *Montana: The Magazine of Western History* 38, no. 4 (Autumn, 1988): 70–73.

Gunther, John. *Inside USA*. New York: Harper & Brothers, 1947.

Haber, Samuel. *Efficiency and Uplift: Scientific Management in the Progressive Era 1890–1920*. Chicago, IL: University of Chicago Press, 1964.

Hammett, Dashiell. *Red Harvest*. New York: Alfred A. Knopf, 1929.

Hand, Wayland. "The Folklore, Customs, and Traditions of the Butte Miner." *California Folklore Quarterly* 5 (Jan. 1946): 19–20.

Hawes, William. "A Brief History of Block-Caving." *Mining History News* (July 2008): 5.

Haycraft, William. *Yellow Steel: The Story of the Earthmoving Equipment Industry*. Urbana and Chicago: University of Illinois Press, 2000.

Hays, Samuel P. *Conservation and the Gospel of Efficiency: The Progressive Conservation Movement, 1890–1920*. Cambridge, MA: Harvard University Press, 1959.

Herberg, Will. *Protestant-Catholic-Jew*. New York: Doubleday, 1955.

High, Steven. "Deindustrializing Youngstown: Memories of Resistance and Loss Following 'Black Monday' 1977–1997." *History Workshop Journal* (Great Britain) 54 (2002): 100–21.

Highsmith, Andrew R. "Demolition Means Progress: Urban Renewal, Local Politics, and State-Sanctioned Ghetto Formation in Flint, Michigan." *Journal of Urban History* 35, no. 3 (Mar. 2009): 348–68.

Hounshell, David A. *From the American System to Mass Production: The Development of Manufacturing Technology in the United States*. Baltimore, MD: Johns Hopkins University Press, 1984.

Hovis, Logan, and Jeremy Mouat. "Miners, Engineers, and the Transformation of Work in the Western Mining Industry, 1880–1930." *Technology and Culture* 37, no. 3 (1996): 429–56.

Howard, Joseph Kinsey. *Montana: High, Wide, and Handsome*. New Haven, CT: Yale University Press, 1943.

Huggard, Christopher J. "Squeezing Out the Profits: Mining and the Environment in the U.S. West, 1945-2000." In *The American West in 2000: Essays in Honor of Gerald D. Nash*, edited by Richard W. Etulain and Ferenc Morton Szasz, 105–26. Albuquerque: University of New Mexico Press, 2003.

Huggard, Christopher J., and Terrence M. Humble. *Santa Rita Del Cobre: A Copper Mining Community in New Mexico*. Boulder: University Press of Colorado, 2012.

Hurley, Andrew. *Environmental Inequalities: Class, Race, and Institutional Pollution in Gary, Indiana, 1945–1980*. Chapel Hill: University of North Carolina Press, 1995.

Hyde, Charles K. *Copper for America: The United States Copper Industry from Colonial Times to the 1990s*. Tucson: University of Arizona Press, 1998.

Isenberg, Andrew C. *Mining California: An Ecological History*. New York: Hill and Wang, 2005.

Jacobs, Jane. *The Death and Life of Great American Cities*. New York: Random House, 1961.

Jacobsen, Matthew Frye. *Roots Too: White Ethnic Revival in Post-Civil Rights America*. Cambridge: Harvard University Press, 2006.

Jacoby, Sanford M. *Modern Manors: Welfare Capitalism Since the New Deal*. Princeton, NJ: Princeton University Press, 1998.

James, Daniel. *Doña María's Story: Life History, Memory, and Political Identity*. Durham, NC: Duke University Press, 2001.

James, Ronald M. *The Roar and the Silence: A History of Virginia City and the Comstock Lode*. Reno: University of Nevada Press, 1998.

Jensen, Katherine, and Audie Blevins. *Last Gamble: Betting on the Future in Four Rocky Mountain Mining Towns*. Tucson: University of Arizona Press, 1998.

Johnson, Aili K. "Lore of the Finnish-American Sauna." *Midwest Folklore* 1, no. 1 (Apr. 1951): 33–39.

Johnson, Benjamin Herber. "Conservation, Subsistence, and Class at the Birth of Superior National Forest." *Environmental History* 4, no. 1 (Jan. 1999): 80–99.

Johnson, Susan Lee. *Roaring Camp: The Social World of the California Gold Rush*. New York: W. W. Norton, 2000.

Joramelon, Ira B. *Copper: The Encompassing Story of Mankind's First Metal*. Berkeley, CA: Howell-North Books, 1973.

Jursnich, Edward. *Butte's East Side: Gone, But Not Forgotten*. Denver: Outskirts Press, 2009.

Kearney, Pat. *Butte's Pride—The Columbia Gardens*. Butte, MT: Skyhigh Communications, 1994.

———. *Butte Voices: Mining, Neighborhoods, People*. Butte, MT: Skyhigh Communications, 1998.

Keeling, Arn, and John Sandlos. "Environmental Justice Goes Underground? Historical Notes from Canada's Northern Mining Frontier." *Environmental Justice* 2, no. 3 (2009): 117–25.

———. "Introduction: The Complex Legacy of Mining in Northern Canada." In *Mining and Communities in Northern Canada: History, Politics, and Memory*, edited by Arn Keeling and John Sandlos, 1–32. Calgary: University of Calgary Press, 2015.

Keyes, Sarah. "'Like a Roaring Lion': The Overland Trail as a Sonic Conquest." *Journal of American History* 96, no. 1 (June 2009): 19–43.

Kilpinen, Jon T. "Finnish Cultural Landscapes in the Pacific Northwest." *Pacific Northwest Quarterly* 86, no. 1 (Dec. 1994): 25–34.

Kirk, Kelly. "State of Change: Women and the 1972 Montana Constitutional Convention." *Montana: The Magazine of Western History* 66, no. 2 (Summer 2016): 3–24.

Klass, Alexandra. "The Frontier of Eminent Domain." *University of Colorado Law Review* 79, no. 651 (2008): 654–61.

Kroll-Smith, Stephen, and Stephen R. Couch. "What Is a Disaster? An Ecological-Symbolic Approach to Resolving the Definitional Debate." *International Journal of Mass Emergencies and Disasters* 9 (1991): 355–66.

Krueger, Rob. "Relocating Regulation in Montana's Gold Mining Industry." *Environment and Planning A* 34 (2002): 867–81.

Lahiri-Dutt, Kuntala, and Martha Macintyre, eds. *Women Miners in Developing Countries: Pit Women and Others*. Burlington, VT: Ashgate, 2006.

Langford, Letitia C., and Gwen Bell. "Federally Sponsored New Towns of the Seventies." *Growth and Change* (Oct. 1975): 24–31.

Lankton, Larry. *Hollowed Ground: Copper Mining and Community Building on Lake Superior, 1840s–1990s*. Detroit, MI: Wayne State University Press, 2010.

LeCain, Timothy J. *Mass Destruction: The Men and Giant Mines that Wired America and Scarred the Planet*. New Brunswick, NJ: Rutgers University Press, 2009.

———. "'See America the Bountiful': Butte's Berkeley Pit and the American Culture of Consumption." *Montana: The Magazine of Western History* 56, no. 4 (Winter 2006): 5–17, 97.

Levenstein, Harvey. "The American Response to Italian Food, 1880–1930." In *Food in the USA*, edited by Carole M. Counihan, 75–90. New York: Routledge, 2002.

———. *Communism, Anticommunism, and the CIO*. Westport, CT: Greenwood, 1981.

Libecap, Gary D. "The Assignment of Property Rights on the Western Frontier: Lessons for Contemporary Environmental and Resource Policy." *Journal of Economic History* 67, no. 2 (June 2007): 266–71.

Limerick, Patricia Nelson. *Something in the Soul: Legacies and Reckonings in the New West*. New York: W. W. Norton: 2000.

Lindley, Curtis H. *A Treatise on the American Law Relating to Mines and Mineral Land within the Public Land States and Territories and Governing the Acquisition and Enjoyment of Mining Rights in Lands of Public Domain*. San Francisco: Bancroft-Whitney Company, 1914.

Lingenfelter, Richard E. *Bonanzas and Borrascas: Copper Kings and Stock Frenzies, 1885–1918*. Norman: Arthur H. Clark Company—University of Oklahoma Press, 2012.

Lockwood, Yvonne R. "The Sauna: An Expression of Finnish-American Identity." *Western Folklore* 36, no. 1 (Jan. 1977): 71–84.

Long, Bill. *Pit Driver: Reminiscences of a Berkeley Pit Haul-Truck Driver*. La Habra, CA: Aljon Graphics, 2007.

MacLane, Mary. *The Story of Mary MacLane by Herself*. Chicago, IL: Herbert S. Stone, 1902.

MacMillan, Donald. *Smoke Wars: Anaconda Copper, Montana Air Pollution, and the Courts, 1890–1920*. Helena: Montana Historical Society Press, 2000.

Malone, Michael P. *The Battle for Butte: Mining and Politics on the Northern Frontier, 1846–1906*. Seattle: University of Washington Press, 1981.

———. "The Close of the Copper Century." *Montana: The Magazine of Western History* 35 (Spring 1985): 69–72.

———. "The Collapse of Metal Mining: An Historical Epitaph." *Pacific Historical Review* 55, no. 3 (Aug. 1986): 455–64.

———. "Montana as a Corporate Bailiwick: An Image in History." In *Montana: Past and Present*, edited by Peter J. Powell and Michael P. Malone, 55–76. Los Angeles: William Andrews Clark Memorial Library, University of California at Los Angeles, 1976.

Malone, Michael P., and Dianne G. Dougherty. "Montana's Political Culture: A Century of Evolution." *Montana: The Magazine of Western History* 31, no. 1 (Winter 1981): 44–58.

Malone, Michael P., and Richard W. Etulain. *The American West: A Twentieth-Century History*. Lincoln: University of Nebraska Press, 1989.

Malone, Michael P., Richard B. Roeder, and William L. Lang. *Montana: A History of Two Centuries*. Seattle: University of Washington, 1991.

Manuel, Jeffrey T. "Mr. Taconite: Edward W. Davis and the Promotion of Low-Grade Iron Ore, 1913–1955." *Technology and Culture* 54, no. 2 (Apr. 2013): 317–45.

March, Richard. "Polka and Tamburitza: Ethnic Music and Dance Traditions in the Upper Midwest." In *Ethnic and Border Music*, edited by Norm Cohen, 139–70. Westport, CT: Greenwood Press, 2007.

Marcosson, Isaac F. *Anaconda*. New York: Dodd, Mead, & Co., 1957.

Marsh, Ben. "Continuity and Decline in the Anthracite Towns of Pennsylvania." *Annals of the Association of American Geographers* 77, no. 3 (1987): 337–52.

Martin, Susan R. *Wonderful Power: The Story of Ancient Copper Workings in the Lake Superior Basin*. Detroit, MI: Wayne State University Press, 1999.

Massey, Doreen. "Traveling Thoughts." In *Without Guarantees: In Honor of Stuart Hall*, edited by Paul Gilroy, Lawrence Grossberg, and Angela McRobbie, 225–32. London and New York: Verso, 2000.

Mather, Cotton, and Matti Kaups. "The Finnish Sauna: A Cultural Index to Settlement." *Annals of the Association of American Geographers* 53, no. 4 (Dec. 1963): 494–504.

McCarthy, Bob J. "Re-Claiming Butte: The Doctrine of Subjacent Support." *Montana Law Review* 49, no. 2 (Summer 1988): 267–83.

Mead, Margaret. "New Towns to Set New Lifestyles." In *New Towns—Why and For Whom?*, edited by H. S. Perloff and N. C. Sandberg, 117–29. New York: Praeger, 1973.

Mehlhorn, Dmitri. "A Requiem for Blockbusting: Law, Economics, and Race-Based Real Estate Speculation." *Fordham Law Review* 67 (Dec. 1998): 1145–61.

Mercier, Laurie. *Anaconda: Labor, Community, and Culture in Montana's Smelter City*. Urbana: University of Illinois, 2001.

———. "Bordering on Equality: Women Miners in North America." In *Gendering the Field: Towards Sustainable Livelihoods for Mining Communities*, Asia-Pacific

Monograph 6, edited by Kuntala Lahiri-Dutt, 33–48. Canberra: Australian National University E-Press, 2011.

Meyer, Charles, Edward P. Shea, Charles C. Goddard Jr., and Staff. "Ore Deposits of Butte, Montana." In *Ore Deposits of the United States, 1933–1967*, Graton-Sales, Vol. II. New York: AIME, 1968, 1363–1416.

Mihelich, John. "Baseball Was Our Life: Amateur Baseball in Butte, Montana, 1920–1960." *Montana: The Magazine of Western History* 59, no. 2 (Summer 2009): 62–73.

Montrie, Chad. Review of *Killing for Coal: America's Deadliest Labor War*, by Thomas G. Andrews. *Journal of American History* 114, no. 3 (June 2009): 772–73.

———. *To Save the Land and People: A History of Opposition to Surface Coal Mining in Appalachia*. Chapel Hill: University of North Carolina Press, 2003.

Morin, Bode J. *The Legacy of American Copper Smelting: Industrial Heritage versus Environmental Policy*. Knoxville: University of Tennessee Press, 2013.

Morrissey, Katherine G. "Rich Crevices of Inquiry: Mining and Environmental History." In *A Companion to American Environmental History*, edited by Douglas Cazaux Sackman, 394–409. Malden, MA: Wiley-Blackwell, 2010.

Muggah, Robert. "A Tale of Two Solitudes: Comparing Conflict and Development-induced Internal Displacement and Involuntary Resettlement." *International Migration* 41, no. 5 (2003): 5–31.

Murphy, Mary. *Hope in Hard Times: New Deal Photographs of Montana, 1936–1942*. Helena: Montana Historical Society Press, 2003.

———. *Mining Cultures: Men, Women, and Leisure in Butte, 1914–41*. Urbana: University of Illinois Press, 1997.

Murphy, Michelle. *Sick Building Syndrome and the Problem of Uncertainty: Environmental Politics, Technoscience, and Women Workers*. Durham: Duke University Press, 2006.

Myers, A. J. "The Block Caving System at Pewabic Mine." *Proceedings of the Lake Superior Mining Institute* 21, 25–28. Ishpeming, MI: Lake Superior Mining Institute, 1917.

Navin, Thomas R. *Copper Mining and Management*. Tucson: University of Arizona Press, 1978.

Nelson, Daniel. "Mass Production and the U.S. Tire Industry." *Journal of Economic History* 47, no. 2 (June 1987): 329–39.

*New Towns and the Suburban Dream: Ideology and Utopia in Planning and Development*, edited by Irving Lewis Allen, 3–17. Port Washington, NY: Kennikat Press, 1977.

Noble, David F. *Forces of Production: A Social History of Industrial Automation*. New York: Alfred A. Knopf, 1984.

Nuwer, Michael. "From Batch to Flow: Production Technology and Work-Force Skills in the Steel Industry, 1880–1920." *Technology and Culture* 29, no. 4 (Oct. 1988): 808–38.

Nye, David E. *The American Technological Sublime*. Boston: MIT, 1996.

Nystrom, Eric C. *Seeing Underground: Maps, Models, and Mining Engineering in America*. Reno: University of Nevada Press, 2014.

———. "Underground Mine Maps and the Development of the Butte System at the Turn of the Twentieth Century." *IA: The Journal of the Society for Industrial Archaeology* 37, no. 1–2 (2011): 97–113.

O'Brien, F. S. "The 'Communist-Dominated' Unions in the United States Since 1950." *Labor History* 9 (Spring 1968): 184–209.

Orenic, Liesl Miller. *On the Ground: Labor Struggle in the American Airline Industry*. Urbana: University of Illinois Press, 2009.

Park, Roberta J. "'Soldiers May Fall but Athletes Never!': Sport as an Antidote to Nervous Diseases and National Decline in America, 1865–1905." *International Journal of the History of Sport* 29, no. 6 (Apr. 2012): 792–812.

Parr, Joy. "Local Water Diversely Known: Walkerton, Ontario 2000 and After." *Environment and Planning D: Society and Space* 22 (2004): 347–48.

———. *Sensing Changes: Technologies, Environments, and the Everyday, 1953–2003*. Vancouver: University of British Columbia Press, 2010.

Parsons, A. B. *The Porphyry Coppers*. New York: The American Institute of Mining and Metallurgical Engineers, 1933.

Paul, Rodman Wilson, and Elliott West. *Mining Frontiers of the Far West, 1848–1880*. Rev. and expanded ed. Albuquerque, NM: University of New Mexico Press, 2001, original 1963.

Peck, Gunther. "The Nature of Labor: Fault Lines and Common Ground in Environmental and Labor History." *Environmental History* 11 (Apr. 2006): 212–38.

———. "Padrones and Protest: 'Old' Radicals and 'New' Immigrants in Bingham, Utah, 1905–1912." *Western Historical Quarterly* 24, no. 2 (1993): 157–78.

Pentland, Brenda. "Letters from World's End." *Montana: The Magazine of Western History* 53, no. 4 (Dec. 2003): 37–49.

Petersen, William. "The Ideological Origins of Britain's New Towns." *Journal of the American Institute of Planners* 34, no. 3 (May 1968): 160–70.

Pezzullo, Phaedra C. *Toxic Tourism: Rhetorics of Pollution, Travel, and Environmental Justice*. Tuscaloosa, AL: University of Alabama Press, 2007.

Piper, Liza. *The Industrial Transformation of Subarctic Canada*. Vancouver and Toronto: University of British Columbia Press, 2009.

———. "Subterranean Bodies: Mining the Large Lakes of North-west Canada, 1921–1960." *Environment and History* 13 (2007): 155–86.

Poe, Tracy Nicole. "The Labour and Leisure of Food Production as a Mode of Ethnic Identity Building Among Italians in Chicago, 1890–1940." *Rethinking History* 5, no. 1 (2001): 131–48.

Portelli, Alessandro. "On the Peculiarities of Oral History." *History Workshop Journal* 12 (Spring 1981): 96–107.

Porteous, J. Douglas, and Sandra E. Smith. *Domicide: The Global Destruction of Home*. Toronto: McGill Queens, 2001.

Pratt, William C. "Rural Radicalism on the Northern Plains, 1912–1950." *Montana: The Magazine of Western History* 42, no. 1 (Winter 1992): 42–55.

Pritchard, Sara B. "An Envirotechnical Disaster: Nature, Technology, and Politics at Fukushima." *Environmental History* 17, no. 2 (Apr. 2012): 219–43.

Punke, Michael. *Fire and Brimstone: The North Butte Mining Disaster of 1917*. New York: Hyperion, 2006.

Pyke, Magnus. *Automation: Its Purpose and Future*. New York: Philosophical Library, 1957.

Quivik, Frederic L. "Integrating the Preservation of Cultural Resources with Remediation of Hazardous Materials: An Assessment of Superfund's Record." *Public Historian* 23, no. 2 (Spring 2001): 47–61.

Reid, Donald. *The Miners of Decazeville: A Geneaology of Deindustrialization*. Cambridge, MA: Harvard University Press, 1985.

Reinhardt, Bob H. "Drowned Towns in the Cold War West: Small Communities and Federal Water Projects." *Western Historical Quarterly* (Summer 2011): 149–72.

Reuss, Martin, and Stephen H. Cutcliffe, eds. *The Illusory Boundary: Environment and Technology in History*. Charlottesville: University of Virginia Press, 2010.

Richter, F. Ernest. "The Amalgamated Copper Company: A Closed Chapter in Corporation Finance." *Quarterly Journal of Economics* 30, no. 2 (Feb. 1916): 387–407.

Rickard, T. A. *Interviews with Mining Engineers*. San Francisco, CA: Mining and Scientific Press, 1922.

Ridge, Martin, ed. *Frederick Jackson Turner: Wisconsin's Historian of the Frontier*. Madison, WI: State Historical Society of Wisconsin, 1986.

Robbins, William G. "The 'Plundered Province' Thesis and the Recent Historiography of the American West." *Pacific Historical Review* 55, no. 4 (Nov. 1986): 577–97.

Robbins, William G., and Katrine Barber. *Nature's Northwest: The North Pacific Slope in the Twentieth Century*. Tucson: University of Arizona Press, 2011.

Robertson, David. *Hard as the Rock Itself: Place and Identity in the American Mining Town*. Boulder: University Press of Colorado, 2006.

Roediger, David R. *Working Toward Whiteness: How America's Immigrants Became White: The Strange Journey from Ellis Island to the Suburbs*. New York: Basic Books, 2005.

Rogers, Kristen Smart. "'We Didn't Think He Was Gonna Build It': Skiing Hits a Mining Town." *Utah Historical Quarterly* 69, no. 4 (2001): 310–25.

Rolston, Jessica Smith. *Mining Coal and Undermining Gender: Rhythms of Work and Family in the American West*. New Brunswick, NJ: Rutgers University Press, 2014.

Rosenblum, Jonathan D. *Copper Crucible: How the Arizona Miners' Strike of 1983 Recast Labor Management Relations in America*, 2nd ed. Ithaca, NY: ILR Press, 1998.

Rosner, David, and Gerald Markowitz. *Deadly Dust: Silicosis and the On-Going Struggle to Protect Workers' Health*, rev. ed. Ann Arbor: University of Michigan Press, 2006.

Rothman, Hal K. *Devil's Bargains: Tourism in the Twentieth-Century American West*. Lawrence: University Press of Kansas, 1998.

———. *The Greening of a Nation?: Environmentalism in the United States Since 1945*. Fort Worth, TX: Harcourt/Brace, 1998.

Rotter, Andrew J. "Empire of the Senses: How Seeing, Hearing, Smelling, Tasting, and Touching Shaped Imperial Encounters." *Diplomatic History* 35, no. 1 (Jan. 2011): 3–19.

Russo, John, and Sherry Lee Linkon. *Steeltown U.S.A.* Lawrence: University Press of Kansas, 2002.

Sanders, Jeffrey Craig. *Seattle and the Roots of Urban Sustainability: Inventing Ecotopia*. Pittsburgh: University of Pittsburgh Press, 2010.

Sandlos, John, and Arn Keeling. "Claiming the New North: Development and Colonialism at the Pine Point Mine, Northwest Territories, Canada." *Environment and History* 18 (2012): 5–34.

———. "Zombie Mines and the (Over)burden of History." *Solutions* 4, no. 3 (June 2013): 80–83.

Satter, Beryl. *Family Properties: How the Struggle Over Race and Real Estate Transformed Chicago and Urban America*. New York: Henry Holt and Co., 2009.

Schmidt, Deborah Beaumont, and Robert J. Thompson. "The Montana Constitution and the Right to a Clean and Healthful Environment." *Montana Law Review* 51 (1990): 414–22.

Schmitz, Christopher. "The Rise of Big Business in the World Copper Industry 1870–1930." *Economic History Review* 39, no. 3 (Aug. 1986): 392–410.

Schuyler, David. *A City Transformed: Redevelopment, Race, and Suburbanization in Lancaster, Pennsylvania*. University Park: Pennsylvania State University Press, 2002.

Seefeldt, Douglas. "Creating Kearney: Forging a Historical Identity for a Central Arizona Mining Community." *Journal of Arizona History* 46, no. 1 (2005): 1–32.

Self, Robert O. *American Babylon: Race and the Struggle for Postwar Oakland*. Princeton, NJ: Princeton University Press, 2003.

Seligman, Amanda I. *Block by Block: Neighborhoods and Public Policy on Chicago's West Side*. Chicago, IL: University of Chicago Press, 2005.

Seligman, Ben B. *Most Notorious Victory: Man in an Age of Automation*. New York: Macmillan, 1966.

Shevenell, Lisa, Katherine A. Connors, and Christopher D. Henry. "Controls on Pit Lake Water at Sixteen Open-Pit Mines in Nevada." *Applied Geochemistry* 14, no. 5 (July 1999): 669–87.

Shopes, Linda. "Oral History and the Study of Communities: Problems, Paradoxes, and Possibilities." *Journal of American History* 89, no. 2 (Sept. 2002): 588–98.

Shovers, Brian. "From the Big Hole to Butte: The Origins and Evolution of a Rocky Mountain Municipal Water System." *IA: The Journal of the Society For Industrial Archaeology* 37, no. 1–2 (2011): 79–96.

———. "The Perils of Working in the Butte Underground: Industrial Fatalities in the Copper Mines, 1880–1920." *Montana: The Magazine of Western History* (Spring 1987): 26–39.

———. "Remaking the Wide Open Town: Butte at the End of the Twentieth Century." *Montana: The Magazine of Western History* 48, no. 3 (Autumn 1998): 40–53.

Shriver, Thomas E., and Dennis K. Kennedy. "Contested Environmental Hazards and Community Conflict Over Relocation." *Rural Sociology* 70, no. 4 (2005): 491–513.

Simpson, Charles R. "A Fraternity of Danger: Volunteer Fire Companies and the Contradictions of Modernization." *American Journal of Economics and Sociology* 55, no. 1 (Jan. 1996): 17–34.

Smilor, Raymond Wesley. "Personal Boundaries in the Urban Environment: The Legal Attack on Noise: 1865–1930." *Environmental Review* 3 (1979): 24–36.

———. "Toward an Environmental Perspective: The Anti-Noise Campaign, 1893–1932." In *Pollution and Reform in American Cities, 1870–1930*, edited by Martin V. Melosi, 135–51. Austin: University of Texas Press, 1980.

Smith, Duane A. *Mining America: The Industry and the Environment, 1800–1980*. Lawrence: University of Kansas, 1987.

———. *Rocky Mountain Heartland: Colorado, Montana, and Wyoming in the Twentieth Century*. Tucson: University of Arizona Press, 2008.

Soneryd, Linda. "Hearing as a Way of Dwelling: The Active Sense-making of Environmental Risk and Nuisance." *Environment and Planning D: Society and Space* 22 (2004): 737–53.

Sorensen, Mads P., and Allan Christiansen. *Ulrich Beck: An Introduction to the Theory of Second Modernity and the Risk Society*. New York: Routledge, 2013.

South African Human Rights Commission. *Mining-Related Observations and Recommendations: Anglo Platinum, Affected Communities and Other Stakeholders, in and around the PPL Mine, Limpopo*. Nov. 2008. www.sahrc.org.za/home/21/files/Reports /Anglo%20Report%20Final%202008_Chp%201%20to%203.pdf.

Spence, Clark C. *Territorial Politics and Government in Montana, 1864–89*. Urbana: University of Illinois Press, 1975.

Steigerwald, David. "Walter Reuther, the UAW, and the Dilemmas of Automation." *Labor History* 51, no. 3 (Aug. 2010): 429–53.

Stine, Jeffrey K., and Joel A. Tarr. "At the Intersection of Histories: Technology and the Environment, Updated with a Postscript by the Authors." *H-Environment Global Historiography Series*. Dec. 2001. The article was originally published in *Technology and Culture* 39, no. 4 (1998): 601–40.

Stoller, Eleanor Palo. "Sauna, Sisu, and Sibelius: Ethnic Identity Among Finnish Americans." *Sociological Quarterly* 37, no. 1 (Winter 1996): 145–75.

Stradling, David, and Richard Stradling. "Perceptions of the Burning River: Deindustrialization and Cleveland's Cuyahoga River." *Environmental History* 13 (July 2008): 515–35.

Sugrue, Thomas J. *The Origins of the Urban Crisis: Race and Inequality in Postwar Detroit*. Princeton, NJ: Princeton University Press, 2005.

Summerby-Murray, Robert. "Interpreting Industrial Heritage in the Mining Towns of Cumberland County, Nova Scotia: Landscape Examples from Springhill and River Herber." *Urban History Review* 35, no. 2 (Spring 2007): 51–59.

Swett, Ira L. *Montana's Trolley's 2: Butte Anaconda BAP, Interurbans Magazine*, Special 50, vol. 26, no. 4, Winter 1969.

Swibold, Dennis L. *Copper Chorus: Mining, Politics, and the Montana Press, 1889–1959*. Helena: Montana Historical Society Press, 2006.

Switzer, Carole Stern, and Peter Gray. *CERCLA: Comprehensive Environmental Response, Compensation, and Liability Act (Superfund)*. Chicago, IL: American Bar Association, 2008.

Taylor, Alan. "Unnatural Inequalities: Social and Environmental Histories." *Environmental History* 1 (Oct. 1996): 6–19.

Terborgh, George. *The Automation Hysteria*. New York: Norton, 1966.

Thompson Jr., Barton H. "Constitutionalizing the Environment: The History and Future of Montana's Environmental Provisions." *Montana Law Review* 64 (2003): 157–98.

Thomson, Alistair. "Anzac Memories: Putting Popular Memory Theory into Practice in Australia." *Oral History* 18, no. 2 (1990): 25–31.

Tone, Andrea. *The Business of Benevolence: Industrial Paternalism in Progressive America*. Ithaca, NY: Cornell University Press, 1997.

Toole, K. Ross. *Montana: A State of Extremes*. Norman: University of Oklahoma Press, 1972.

———. *Montana: An Uncommon Land*. Norman: University of Oklahoma Press, 1959.

———. *The Rape of the Great Plains: Northwest America, Cattle and Coal*. Boston: Little, Brown, & Co., 1976.

Tucker, Eric, ed. *Working Disasters: The Politics of Recognition and Response*. Amityville, NY: Baywood Publishing, 2006.

Tulloch, John, and Deborah Lupton. *Risk and Everyday Life*. London and Thousand Oaks, CA: SAGE, 2003, 34.

Twitty, Eric. *Blown to Bits in the Mine: A History of Mining and Explosives in the United States*. Ouray, CO: Western Reflections, 2001.

Tyer, Brad. *Opportunity, Montana: Big Copper, Bad Water, and the Burial of an American Landscape*. Beacon Press, Boston, 2013.

Tyson, Rae. "The Intergovernmental Cleanup at Love Canal: A First Crack at the 'Sleeping Giant of the Decade.'" *Publius: The Journal of Federalism* 10, no. 1 (Winter 1980): 101–9.

Van Horssen, Jessica. *A Town Called Asbestos: Environmental Contamination, Health, and Resilience in a Resource Community*. Vancouver: University of British Columbia Press, 2016.

Van West, Carroll. "The Pit and the Stack: Decoding the Southwestern Montana Landscape." *Drumlummon Views* 3, no. 1 (Spring 2009): 21–26.

Vonnegut, Kurt. *Player Piano*. New York: Dell, 1952.

Voyles, Traci Brynne. *Wastelanding: Legacies of Uranium Mining in Navajo Country*. Minneapolis: University of Minnesota Press, 2015.

Waldron, Ellis, and Paul B. Wilson. *Atlas of Montana Elections, 1889–1976*. Missoula, MT: University of Montana Publications in History, 1978.

Walker, David A. *Iron Frontier: The Discovery and Early Development of Minnesota's Three Ranges*. Minneapolis: Minnesota Historical Society Press, 1979.

Ward, Ashley. "Reclaiming Place through Remembrance: Using Oral Histories in Geographic Research." *Historical Geography* 40 (2012): 133–45.

Weber, Bret A. "Denver's Model Cities Program." PhD diss., University of Utah, 2007.

Weber, Bret A., and Amanda Wallace. "Revealing the Empowerment Revolution: A Literature Review of the Model Cities Program." *Journal of Urban History* 38, no. 1 (2012): 173–92.

Wendel, Clarence Adami. "Mining in Butte 40 Years Ago." *Montana: The Magazine of Western History* 31, no. 3 (Summer 1981): 60–66.

White, Richard. "'Are You an Environmentalist or Do You Work for a Living?': Work and Nature." In *Uncommon Ground: Rethinking the Human Place in Nature*, edited by William Cronon, 171–85. New York: W. W. Norton, 1996.

Whitfield, Stephen J. *The Culture of the Cold War*. Baltimore, MD: The Johns Hopkins University Press, 1996.

Wiener, Norbert. *Cybernetics: Or Control and Communication in the Animal and the Machine*. New York: John Wiley & Sons, 1948.

———. *The Human Use of Human Beings: Cybernetics and Society*. Boston: Houghton Mifflin, 1950.

Work, Clemens P. *Darkest Before Dawn: Sedition and Free Speech in the American West*. Albuquerque: University of New Mexico Press, 2005.

Work Projects Administration. *Copper Camp: Stories of the World's Greatest Mining Town, Butte, Montana*. New York: Hastings House, 1943.

Wyckoff, William. "Postindustrial Butte." *Geographical Review* 85, no. 4 (Oct. 1995): 478–96.

Young, Otis E. *Western Mining: An Informal Account of Precious-Metals Prospecting, Placering, Lode Mining, and Milling of the American Frontier from Spanish Times to 1893*. Norman: University of Oklahoma Press, 1970.

Zimmerman, Jonathan. "Simplified Spelling and the Cult of Efficiency in the 'Progressive' Era." *Journal of the Gilded Age and Progressive Era* 9, no. 3 (July 2010): 365–94.

# ABOUT THE AUTHOR

Brian Leech is a Montana native and assistant professor of history at Augustana College in Rock Island, Illinois. He completed his doctorate in American history at the University of Wisconsin, Madison, and currently serves as secretary of the Mining History Association. He won the 2012 Vivian Paladin Award for the best article published in the journal *Montana: The Magazine of Western History.* Although they all enjoy the Midwest, Leech, his wife, and their two children wish they made it back to Montana's mountains more often.

# INDEX